# 10th International Seminar on Medical Physics and 1st AMDI International Oncology Symposium (ISMP & IOS 2016)



Journal of Physics: Conference Series Volume 851

Penang, Malaysia
26 - 28 August 2016

**Editor:**

**Hafiz M. Zin**

ISBN: 978-1-5108-4191-8
ISSN: 1742-6588

**Printed from e-media with permission by:**

Curran Associates, Inc.
57 Morehouse Lane
Red Hook, NY 12571

CURRAN ASSOCIATES INC. proceedings.com

**Some format issues inherent in the e-media version may also appear in this print version.**



Printed by Curran Associates, Inc. (2017)



**Additional copies of this publication are available from:**

Curran Associates, Inc.
57 Morehouse Lane
Red Hook, NY 12571 USA
Phone: 845-758-0400
Fax: 845-758-2633
Email: curran@proceedings.com
Web: www.proceedings.com

# Table of contents

## Volume 851

**10th International Seminar on Medical Physics & 1st AMDI International Oncology Symposium**

**26–28 August 2016, Penang, Malaysia**

**Accepted papers received: 19 May 2017**
**Published online: 31 May 2017**

### Preface



### Imaging Physics

**Radiotherapy Physics**

## Oncology

IOP Conf. Series: Journal of Physics: Conf. Series **851** (2017) 011001 doi:10.1088/1742-6596/851/1/011001

# Preface

The International Seminar on Medical Physics (ISMP) 2016 is a national medical physics conference organised every two years by the Malaysian Association of Medical Physics (MAMP). Previously the biennial conference is known as the National Seminar on Medical Physics (NSMP). The conference was held on 27 & 28 August 2016 in Bayview Batu Ferringhi Hotel, Penang, Malaysia in conjunction with an Oncology Symposium (IOS) organised by Advanced Medical & Dental Institute of Universiti Sains Malaysia.

The conference was renamed the International Seminar on Medical Physics (ISMP) on the 10th anniversary to recognise the diversity of the nationality of the conference speakers and attendees in the 8th (2012) and 9th (2014) conference. The change is also to encourage more international participants especially from the neighbouring countries of Malaysia and beyond to come together in "Fostering Medical Physics Research and Clinical Practice for Better Healthcare" in this region.

About 300 participants from universities and hospitals participated in the conference. 28 oral contributions and 22 posters were presented at the conference. We also had a few invited lectures by international experts. The conference was divided into two main themes, imaging physics and radiotherapy physics. The topics presented were radiation dosimetry, image quality, magnetic resonance imaging, nuclear medicine, quality assurance, particle therapy, image guided radiotherapy, brachytherapy and Monte Carlo simulation & modelling. The IOS held in parallel covered topics in clinical oncology, cancer imaging and cancer biology.

41 papers are included in the volume organised into three main themes, i.e., imaging physics (ISMP), radiotherapy physics (ISMP) and oncology (IOS). All papers published in this volume of Journal of Physics: Conference Series have been peer reviewed through processes administered by the proceedings Editor. Each paper was peer-reviewed by two reviewers from the scientific committee. Reviews were conducted by expert referees to the professional and scientific standards expected of a proceedings journal published by IOP Publishing.

**Scientific Committee and Reviewers**

***International***

Naruhiro Matsufuji, National Institute of Radiological Sciences, Japan
Tom Richards, University College London NHS, United Kingdom
Supriyanto Ardjo Pawiro, University of Indonesia
Mahbubunnabi Tamal, University of Dammam, Saudi Arabia

***Malaysia***

Ahmad Nazlim Yusoff, Universiti Kebangsaan Malaysia
Ahmad Taufek Abdul Rahman, Universiti Teknologi MARA
Ashrani Aizzuddin Abd Rahni, Universiti Kebangsaan Malaysia
Hafiz M Zin, Universiti Sains Malaysia
Jeannie Hsiu Ding Wong, University of Malaya
M Iqbal Saripan, Universiti Putra Malaysia
Mohd Zahri Abd Aziz, Universiti Sains Malaysia
Ng Kwan Hoong, University of Malaya

Noramaliza Mohd Noor, Universiti Putra Malaysia
Rafidah Zainon, Universiti Sains Malaysia
Shazril Imran Shaukat, Universiti Sains Malaysia
Suhairul Hashim, Universiti Teknologi Malaysia
Ung Ngie Min, University of Malaya
Wan Nordiana Wan Abd Rahman, Universiti Sains Malaysia
Yeong Chai Hong, Universiti Malaya

Many thanks to all invited speakers, oral presenters and poster presenters for their participation and to the Organising Committee members for all their hard work in making the conference happen. Thanks to all authors who submitted the extended write-up of the work presented at the conference to be included in the volume. The Scientific Committee members and reviewers are also thanked for reviewing the submitted manuscripts and improve the scientific quality of this proceedings. Finally, thanks to all who attended the conference and the sponsors for their financial support.

**Editor, Proceedings of ISMP & IOS 2016**
Hafiz M Zin
Universiti Sains Malaysia

**Sponsorship Acknowledgements**

The 2016 International Seminar on Medical Physics & Oncology Symposium is sponsored by:

Diamond

Gold

Silver

**Host Organisation**

Advanced Medical & Dental Institute (AMDI) of Universiti Sains Malaysia or IPPT (Malay acronym of the Institute) was established with the vision of developing state-of-the-art advanced research, conduct innovative postgraduate programmes and deliver tertiary healthcare services towards sustainable mankind.

One of the core aims of AMDI is to be forefront in the field of oncology and making oncology as the main thrust of clinical service, research and education. AMDI currently has state of the art linear accelerator, the only one in public institution in the northern region of Malaysia and is able to carry out new radiotherapy techniques such as Intensity modulated radiotherapy (IMRT). All oncology services are carried out under one roof at the Clinical Trials Complex, which itself is an approved centre for clinical trials by Clinical Research Malaysia. The organisation of this conference institute is inline with the plan to increase awareness of the scientific and medical community to AMDI at the same time enhance networking for collaborations. AMDI is based in Bertam, a thriving new township in the mainland part of Penang, in the northern part of Peninsular Malaysia.

# Peer review statement

All papers published in this volume of *Journal of Physics: Conference Series* have been peer reviewed through processes administered by the proceedings Editors. Reviews were conducted by expert referees to the professional and scientific standards expected of a proceedings journal published by IOP Publishing.

IOP Conf. Series: Journal of Physics: Conf. Series **851** (2017) 012001 doi:10.1088/1742-6596/851/1/012001

# Response of optically stimulated luminescence dosimeters subjected to X-rays in diagnostic energy range

**Y Musa[1,2], S Hashim[1], M K A Karim[1,3], K A Bakar[4], W C Ang[1], N Salehhon[1]**

[1]Department of Physics, Universiti Teknologi Malaysia, Johor, Bahru Johor, 81300, Malaysia
[2]Centre for Energy Research and Training, Ahmadu Bello University, Zaria, Kaduna State, P.M.B. 1014, Nigeria
[3]Department of Diagnostic Imaging, National Cancer Institute, Precint 7 62250 Putrajaya, Malaysia
[4] Department of Radiology, Hospital Sultanah Aminah, Johor, Bahru Johor, 80100, Malaysia

E-mail: yahaya_ms@yahoo.com

**Abstract**. The use of optically stimulated luminescence (OSL) for dosimetry applications has recently increased considerably due to availability of commercial OSL dosimeters (nanoDots) for clinical use. The OSL dosimeter has a great potential to be used in clinical dosimetry because of its prevailing advantages in both handling and application. However, utilising nanoDot OSLDs for dose measurement in diagnostic radiology can only be guaranteed when the performance and characteristics of the dosimeters are apposite. In the present work, we examined the response of commercially available nanoDot OSLD ($Al_2O_3$:C) subjected to X-rays in general radiography. The nanoDots response with respect to reproducibility, dose linearity and signal depletion were analysed using microStar reader (Landauer, Inc., Glenwood, IL). Irradiations were performed free-in-air using 70, 80 and 120 kV tube voltages and tube currents ranging from 10 – 100 mAs. The results showed that the nanoDots exhibit good linearity and reproducibility when subjected to diagnostic X-rays, with coefficient of variations (CV) ranging between 2.3% to 3.5% representing a good reproducibility. The results also indicated average of 1% signal reduction per readout. Hence, the nanoDots showed a promising potential for dose measurement in general X-ray procedure.

## 1. Introduction

Optically stimulated luminescence (OSL) is one of the modern optical techniques for detection and measurement of ionising radiation that is applicable to various areas including medical, environmental and personal dosimetry. Due to its vast applications, the use of OSL technique is on the increase due to dramatic growth in the use of artificial sources of ionising radiation especially in medicine. Measurement of organ doses in diagnostic radiology is important in order to assess the risk-to-benefit associated with the radiation. Thermoluminescence dosimeters (TLDs) were the most commonly used dosimeters for organ and tissue dose estimation, but OSL dosimeters have recently been recommended for dosimetry applications due to its promising dosimetric properties which include stable sensitivity, high precision and accuracy, high speed of readout, possibility of multiple re-analysis and dose accumulation among others [1]. Both optically stimulated luminescence (OSL) and thermoluminescence (TL) begin by irradiation with primary source of ionising radiation in which energy may be deposited in the material in the form of trapped charge carriers (i.e. electrons and holes). But the released trapped charge can then be stimulated by absorption of optical photons of appropriate wavelength, causing OSL, or by absorption of heat, causing TL. The $Al_2O_3$:C based OSL dosimeter can be stimulated with a wide spectrum of light ranging from 400 to 700 nm and emission in a broad band of wavelengths with a peak centred at 410 – 420 nm [2]. The basis of detection of ionising radiation using OSL is illustrated in figure 1, which involve three stages. In stage (a), the dosimeter is exposed to ionising radiation in which the energy deposited may give rise to excitations and ionizations. Stage (b) is the period characterized by a metastable concentration of trapped electrons and holes and it corresponds to the period in which the dosimeters are transported back to the laboratory and stored before measurement. The information stored in the dosimeter can be read by light stimulation in which light of certain wavelength $\lambda_{stim}$ stimulates the electrons to the conduction band which leads to recombination of electrons and holes. Subsequently, defect is created in the excited state and relaxes to the ground state by emission of photon with wavelength $\lambda_{OSL}$ [3].

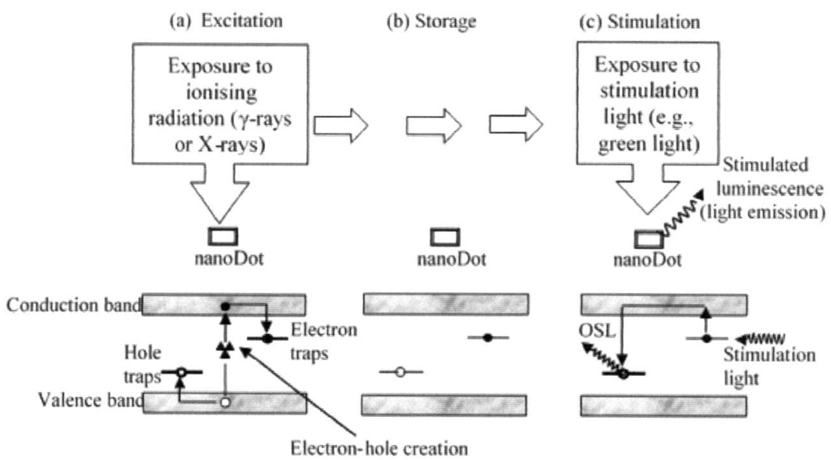


**Figure 1**. Schematic diagram illustrating the stages of OSL process

X-rays remain the most commonly used artificial source of radiation for radiological examinations in medicine which contributed enormously to the diagnosis of countless number of diseases, having the largest contribution (> 90%) to the combined doses from all artificial sources of radiation [4], [5]. But the amount of radiation used in diagnostic radiology requires quantification to ensure that the correct ration of radiation is used which allows an assessment of the risk involved. Radiation dosimetry is however required to estimate the dose delivered to patient and optically stimulated luminescence is found to be one of the most successful techniques for radiation dosimetry in recent years [6]. The dosimetric properties of OSL dosimeters have recently been characterised and demonstrated a good performance within the diagnostic energy range, 40 – 140 kV [7]–[9]. Based on literature survey, there are few studies on the performance of OSL dosimeters in diagnostic general X-ray compared to studies associated to radiation therapy [9]. The purpose of this work is to evaluate the reproducibility, linearity and signal depletion of the OSL dosimeters using X-rays in diagnostic energy range to ascertain the possibility of utilising these dosimeters for dose measurements.

## 2. Materials and method

The OSL dosimeters used in the present work were InLight nanoDots ($Al_2O_3$:C) by Landauer Inc (Glenwood, IL), which are among the recent advances in OSL dosimeters useful for dose measurements

IOP Conf. Series: Journal of Physics: Conf. Series **851** (2017) 012001 doi:10.1088/1742-6596/851/1/012001

in diagnostic radiology [7], [9]. The nanoDots were irradiated using a general X-ray facility (Siemens Multix-top Model) installed at the Diagnostic Imaging Department, Hospital Permai, Johor. Each nanoDot contain one OSL element in a single slide, basically designed for storing the detector element in a black light-tight case which protects it from visible light. The size of the nanoDot is illustrated in figure 2, with the detector element pushed out of the case alongside its adapter.

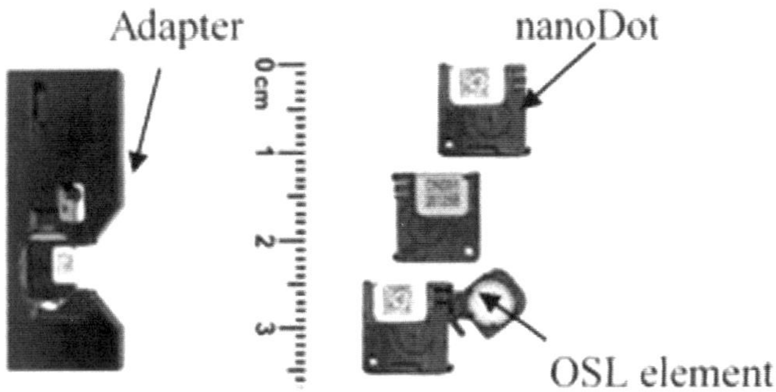


**Figure 2**. InLight nanoDot dosimeter

Measurements were carried out using microStar reader, which is a portable InLight reader available to measure clinical radiation doses in both patients and employees. The reader consists of an array of green light emitting diodes (LEDs) as a high intensity stimulating source and operates in continuous-wave optically stimulated luminescence (CW-OSL) mode to measure the OSL signal [10]. The reader was calibrated using the linear (standard) calibration method with a set of preirradiated nanoDots provided by the manufacturer and exposed free-in-air to known doses ranging from 0 – 1000 mGy using Cs-137 source having effective energy of 662 keV. The absorbed dose is calculated using equation 1.

$$Dose = \frac{PMT\ counts}{Calibration\ factor * Sensitivity} \tag{1}$$

### 2.1. *Reproducibility test*

The reproducibility test was performed using two set of dosimeters, one set was pre-irradiated with known doses of 10 and 1 Gy from Cs-137 source. The second set of dosimeters consist of 10 unexposed nanoDots having the same sensitivity, irradiated free-in-air with 70 kVp X-ray and 100 cm from the tube focus while varying the tube current from 10 to 100 mAs (see table 1). Each nanoDots was then read up to 10 times from which the mean and standard deviations were obtained.

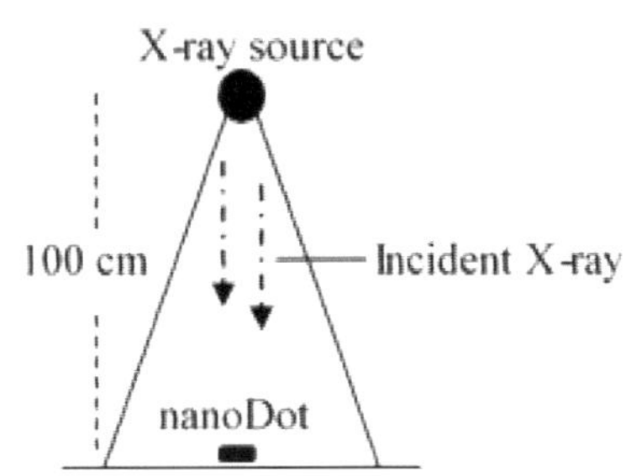


**Figure 3**. Schematic diagram of the nanoDot's irradiation set-up

### 2.2. *Dose linearity*

The linearity response was tested using two different tube voltages in general X-ray: 80 kVp and 120 kVp. For each kVp, ten nanoDots were irradiated by utilising one dosimeter per mAs setting while increasing the tube currents from 10 – 100 mAs as shown in table 1. Each dosimeter was read 3 times and average of the 3 readings was utilised.

**Table 1**. Irradiation regime of the nanoDots using general X-ray

| Test | Beam Potential (kVp) | Tube current (mAs) |
|---|---|---|
| Reproducibility | 70 | 10 – 100 |
| Linearity | 80<br>120 | |
| Signal depletion | 120 | 80 |

*2.3. Signal depletion*

The depletion test was performed by irradiating three nanoDots at 100 cm from the tube focus using 120 kVp/80 mAs setting. After 24 hours of post irradiation, the nanoDots were read up to 60 times sequentially per OSLD. The background signal of the OSLDs were measured following optical annealing before the irradiation.

## 3. Results and Discussion

*3.1. Reproducibilty test*

The reproducibility of each nanoDot was measured as the coefficients of variation (CV) of ten repeated readouts. For the two dosimeters exposed to known dose of 10 and 1Gy, the coefficients of variations were found to be 2.4% and 0.26% respectively, indicating a good reproducibility. While nanoDots irradiated with general X-ray unit also showed good reproducibility ranging between 2.3 % and 3.5% which describe the reliability of the nanoDots and microStar reader to reproduce their output with good precision. As shown in figure 4, at minimum tube current the OSLD seemed to have the highest coefficient of variation which may be attributed to the lowest dose received during the exposure, indicating better reproducibility and low standard deviation for higher mAs setting. Previous studies have also revealed nanoDot's reproducibility between 2.9% and 3.6% for X-rays in diagnostic energy range and a maximum of 6.6% for doses below 1 mGy [7], [9].

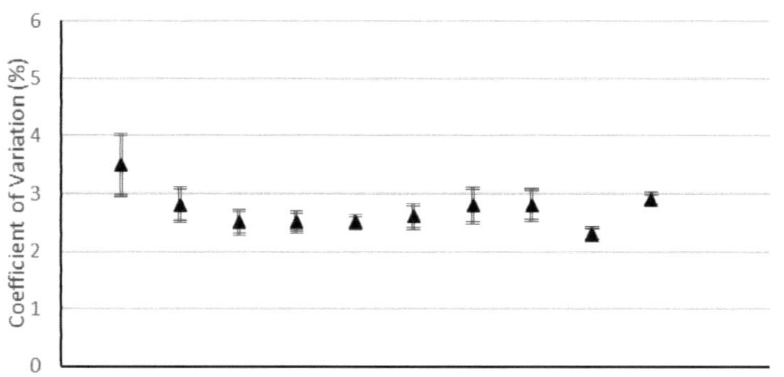


**Figure 4**. Coefficient of variation at different tube currents using 70 kVp X-ray. The error bars represent the standard deviation of the mean.

*3.2. Dose linearity*

As shown in figure 5, good linearity between the OSLD dose response and current-time product was obtained for both 80 and 120 kVp radiation qualities with values of $R^2$ found to be 0.9996 and 0.9992 respectively. This indicated that the nanoDots exhibit a linear response across the entire dose range (measured) and selected tube currents for the two radiation qualities. The OSLD dose response has earlier been reported to be linear for X-rays in diagnostic energy range [2], [3], [7], [9], [11]. Good linearity was also obtained between the PMT counts and measured dose for the two radiation qualities as shown in figure 6. The PMT counts were normalised to the minimum value and regression lines showed a nearly perfect fit with values of $R^2$ well above 0.99. The error bars in figures 5 and 6 represent the standard errors of the measurements

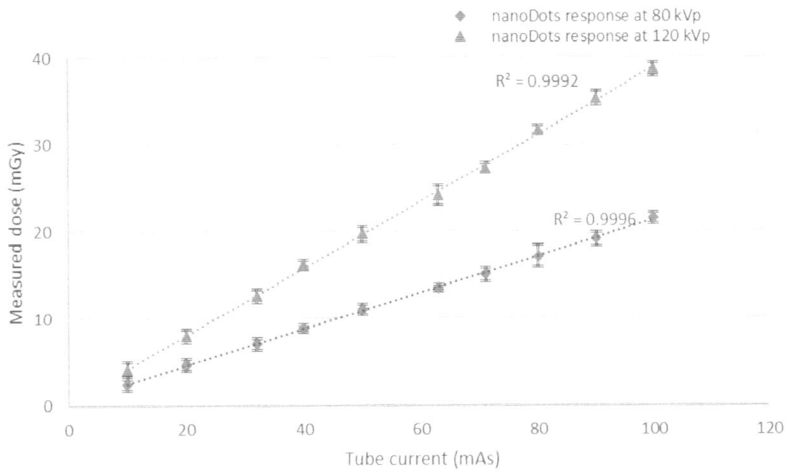


**Figure 5**. OSLD response when subjected to increasing current-time products.

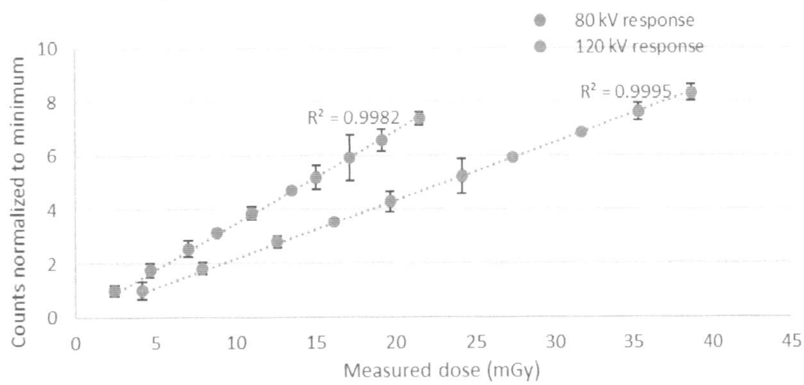


**Figure 6**. Counts versus measured dose showing the linear response of the nanoDots

### *3.3. Signal depletion*

After 24 hours of post-irradiation, each nanoDot was read sixty times and the average of three sequential readouts. As shown in figure 7, each time the dosimeter was read, certain amount of its original OSL signal is reduced which may be attributed to the partial discharge of the trapped charges as earlier demonstrated [12]. Each data point on the horizontal axes represent the mean value obtained from the 3 nanoDots and the error bars represent their standard deviation. It can be seen from figure 7 that the both count and dose deplete in the similar trend with average of 1% signal loss per readout and 2.5% signal loss per 3 repeated readouts.

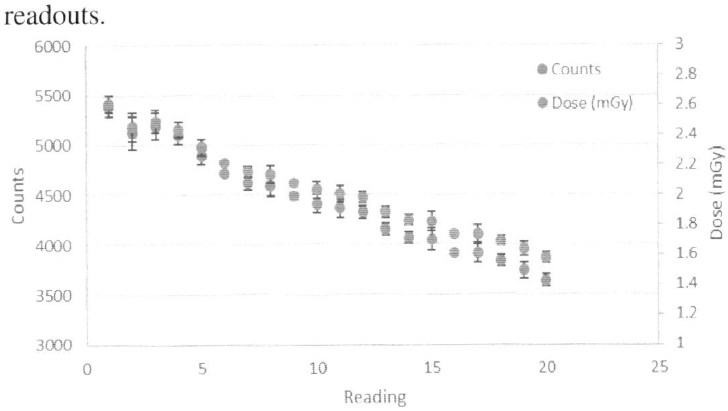


**Figure 7**. Depletion of OSL signal with respect to counts and measured dose.

## 4. Conclusion

We successfully examined the response of OSLDs (nanoDots) when subjected to X-rays in diagnostic energy range. The nanoDots are found to exhibit good linearity and reproducibility in the selected radiation qualities, with an average of less than 2.5% signal loss per 3 repeated readouts. This is an indication that the nanoDots could be utilised for radiation dosimetry in general radiography.

## 5. References

[1] McKeever SWS and Moscovitch M 2003 On the advantages and disadvantages of optically stimulated luminescence dosimetry and thermoluminescence dosimetry. *Radiat. Prot Dosimetry.* **104**(3):263–270.

[2] Jursinic PA 2010 Changes in optically stimulated luminescent dosimeter (OSLD) dosimetric characteristics with accumulated dose. *Med. Phys.* **37**(1):132–40.

[3] Yukihara EG and McKeever SWS 2011 Optically stimulated luminescence: Fundamentals and Applications. *John Wiley & Sons*, USA. ISBN:978-0-470-97706-4.

[4] IAEA 2007 Dosimetry in diagnostic radiology: an international code of practice TRS 457.

[5] Karim MKA, Suhairul H, Sabarudin A, Bradley DA, and Bahruddin NA 2016 Evaluating organ dose and radiation risk of routine CT examinations in Johor, Malaysia. *Sains Malaysiana*, **45**(4):567–73.

[6] Yukihara EG and McKeever SWS 2008 Optically stimulated luminescence (OSL) dosimetry in medicine. *Phys. Med. Biol.* **53**:R351–R79.

[7] Al-Senan RM and Hatab MR 2011 Characteristics of an OSLD in the diagnostic energy range. *Med. Phys.* **38**(7):4396–405.

[8] Endo A, Katoh T, Kobayashi I, Joshi R, Sur J and Okano T 2012 Characterization of optically stimulated luminescence dosemeters to measure organ doses in diagnostic radiology," *Dentomaxillofacial Radiol.* **41**(3):211–16.

[9] Lim CS, Lee S B, and Jin GH 2011 Performance of optically stimulated luminescence Al2O3 dosimeter for low doses of diagnostic energy X-rays," *Appl. Radiat. Isot.* **69**(10):1486–489.

[10] Bos AJJ and Wallinga J 2009 Optically stimulated luminescence signals under various stimulation modes assuming first-order kinetics. *Phys. Rev. B - Condens. Matter Material. Phys.* **79**(19):1-12.

[11] Akselrod MS, Botter-Jensen L and McKeever SWS 2006 Optically stimulated luminescence and its use in medical dosimetry. *Radiat. Meas.* **41**(1):S78–99.

[12] Jursinic PA 2007 Characterization of optically stimulated luminescent dosimeters, OSLDs, for clinical dosimetric measurements. *Med. Phys.* **34**(12):4594–604.

## Acknowledgement

The authors wish to acknowledge the support from Research University Grant Scheme with project No. QJ130000.2517.13H39 and Ahmadu Bello University, Zaria, through the Needs Assessment Intervention Fund.

IOP Conf. Series: Journal of Physics: Conf. Series **851** (2017) 012002 doi:10.1088/1742-6596/851/1/012002

# 128 slice computed tomography dose profile measurement using thermoluminescent dosimeter

**N Salehhon[1], S Hashim[1,2], M K A Karim[1,3], W C Ang[1], Y Musa[1], N A Bahruddin[1]**

[1] Department of Physics, Faculty of Science, Universiti Teknologi Malaysia, 81310 Skudai, Johor, Malaysia
[2]Centre for Sustainable Nanomaterials (CSNano), Ibnu Sina Institute for Scientific and Industrial Research (ISI-SIR), Universiti Teknologi Malaysia, 81310 UTM Skudai, Johor, Malaysia
[3]Department of Diagnostic Imaging, National Cancer Institute, Precint 7, 62250 Putrajaya, Malaysia

E-mail: nasuhasalehhon@yahoo.com

**Abstract**. The increasing use of computed tomography (CT) in clinical practice marks the needs to understand the dose descriptor and dose profile. The purposes of the current study were to determine the CT dose index free-in-air ($CTDI_{air}$) in 128 slice CT scanner and to evaluate the single scan dose profile (SSDP). Thermoluminescent dosimeters (TLD-100) were used to measure the dose profile of the scanner. There were three sets of CT protocols where the tube potential (kV) setting was manipulated for each protocol while the rest of parameters were kept constant. These protocols were based from routine CT abdominal examinations for male adult abdomen. It was found that the increase of kV settings made the values of $CTDI_{air}$ increased as well. When the kV setting was changed from 80 kV to 120 kV and from 120 kV to 140 kV, the $CTDI_{air}$ values were increased as much as 147.9% and 53.9% respectively. The highest kV setting (140 kV) led to the highest $CTDI_{air}$ value (13.585 mGy). The *p*-value of less than 0.05 indicated that the results were statistically different. The SSDP showed that when the kV settings were varied, the peak sharpness and height of Gaussian function profiles were affected. The full width at half maximum (FWHM) of dose profiles for all protocols were coincided with the nominal beam width set for the measurements. The findings of the study revealed much information on the characterization and performance of 128 slice CT scanner.

## 1. Introduction

The discovery of X-ray by Wilhelm Conrad Roentgen in 1895 marked the true start of imaging [1]. The introduction of X-ray computed tomography (CT) to medicine in the early 1970s was done by Godfrey Newbold Hounsfield [2]. The introduction of computed tomography dose index (CTDI) as a metric to quantify the radiation output from a CT examination was made by Shope et al. in 1981 [3]. The word “index” was particularly incorporated in CTDI’s name to distinguish the quantity from the radiation dose absorbed by patients [3]. The concept of CTDI was introduced as a simple way to evaluate the CT dose descriptor.

Since the advent of CT, there were a lot of reports on the radiation exposure of CT systems [4]. Those reports described various measurement methods to characterize the radiation delivered by CT and many of them were based on single scans measurements. A long (100 mm) pencil ionization chamber is commonly used to make CTDI measurements which integrates the longitudinal single scan dose profile (SSDP) using a single axial scan [5]. By far the most common method for measuring SSDP is using thermoluminescent dosimeter (TLDs) due to its abundance of advantages, though several other methods have been described as well [6].

One of the dosimetric quantities to characterize the exposure from CT scanners concerns the CTDI and can easily be determined free-in-air for a single scan on the axis of rotation of the scanner ($CTDI_{air}$) [7]. $CTDI_{air}$ is a course indicator of patient exposure for an examination and thus, it is an important element in the implementation of patient dosimetry. The SSDP is referring to the CT output characteristics of air-kerma distribution along the z-axis of CT scanner using a single axial scan [8]. The concept of dose descriptor (CTDI) and dose profile (SSDP) provide a lot of information on the characterization and performance of CT scanners. Besides, the effect of different tube potential (kV) in measuring CT dose is worthy of special attention. Therefore, the aim of this study is to determine the CT dose index free-in-air ($CTDI_{air}$) in 128 slice CT scanner and to evaluate the single scan dose profile (SSDP) using thermoluminescent dosimeter (TLD-100).

## 2. Materials and Methods

*2.1 Multi-slice CT scanner*

Free-in-air measurements were carried out in 128 slice CT scanner Siemens SOMATOM Definition AS+ in Hospital Sultanah Aminah, Johor Bahru. The protocols were based from routine CT abdominal examinations for male adult abdomen. The protocols used to measure the dose profiles consisted of single axial scan of 5.0 mm slice thickness, 32 × 1.2 mm nominal beam width and scan time of 0.5 s. The tube current was fixed at 100 mAs, while the tube potential was modified to 80 kV for Protocol I, 120 kV for Protocol II and 140 kV for Protocol III.

*2.2 Thermoluminescent dosimeter*

The type of thermoluminescent dosimeter (TLD) used in this study was lithium fluoride doped with magnesium and titanium (LiF: Mg, Ti) chips or better known as TLD-100, manufactured by Harshaw Chemicals. TLD-100 was preferred because it is well-known with its high precision and sensitivity and has long-term stability. The TLDs used were in the form of chips with dimensions of 3.2 mm × 3.2 mm and 0.89 mm thickness [9].

The TLD-100 chips were calibrated at the Secondary Standard Dosimetry Laboratory (SSDL) in Malaysia Nuclear Agency (Nuclear Malaysia), which they were irradiated at 8 mGy to determine the TL response of the dosimeter to a measured exposure or absorbed dose of radiation of clearly defined energy [10]. The TLDs were read-out using Harshaw Thermoluminescence Dosimetry (TLD) Reader Model 3500. The readings obtained from this process were used to get the calibration factor. The annealing of TLDs was performed in TLD annealing oven for 1 hour at 400°C, 2 hours at 100°C and cooled down slowly to ambient temperature [11].

*2.3 Dose profile measurements*

Three sets of TLD-100 cases custom built made of perspex were prepared (figure 1). The length of case was 150 mm and it was grooved with 5 mm width and 3.5 mm depth where sets of three TLDs can be placed. The distance between the centre of each groove was 7 mm and the central part of the case was designed in such a way that it gives the possibility to place 11 TLDs adjacent to each other. A total of 213 TLDs (three batches of 71 TLDs) were used in the measurements. The groove of the central part was stacked with 11 TLDs (high sampling area) to ensure the accuracy of measurement at the central part as the X-ray beam is concentrated at the centre of target. The quantity of chips was odd because the dose profiles were plotted according to the position along the z-axis and there was a position of z=0 or the midpoint of dose profile. To arrange the TLD chips into the cases, the vacuum tweezers (Dymax 5 - Charles Austen Pumps Ltd) was used to avoid any scratches on the chips. The cases were covered with black covers which were also made of perspex for storage purpose.

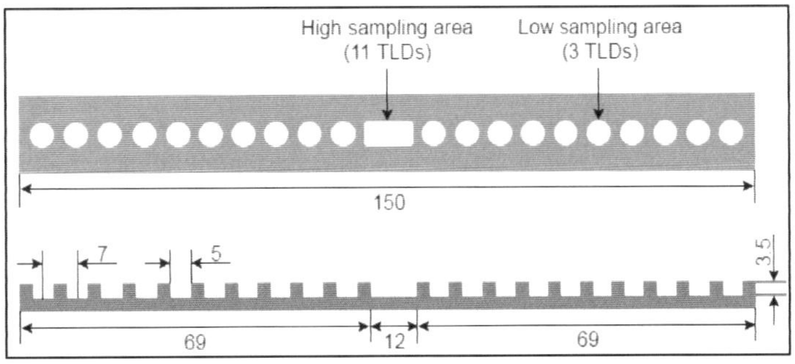


**Figure 1.** The schematic diagram of the custom-built TLD-100 case.

The case was placed at the isocenter of CT scanner as shown in figure 2 and the chips were irradiated according to the protocols that have been mentioned in subsection 2.1. For all protocols, all parameters were set constant except the tube potential (kV). The irradiated TLD chips were read out using Harshaw Thermoluminescent Dosimetry (TLD) Reader Model 3500. A list of readings from the chips was referring to the total charge collected in unit nC. The TLD reader's background and the background reading of TLD chips were also read out to get the final values of $CTDI_{air}$ for each protocol. From the air kerma values obtained from the TLDs, the SSDP was plotted in accordance to the position (mm) along the z-axis using OriginPro 9.0 software.

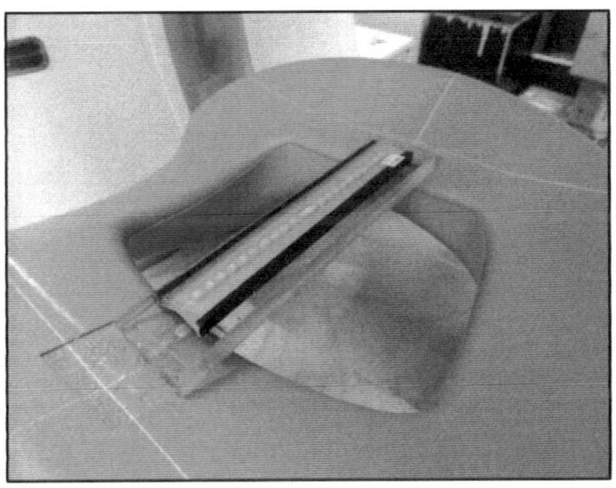

**Figure 2.** The TLD-100 case at the isocenter of CT scanner.

## 3. Results

The parameters for each protocol and the values of $CTDI_{air}$ obtained from the free-in-air measurements are shown in table 1. There were three different sets of protocols with different kV settings. While the tube current was fixed at 100 mAs for all protocols, the kV was changed to 80 kV for Protocol I, 120 kV for Protocol II and 140 kV for Protocol III. The lowest value of $CTDI_{air}$ was contributed by Protocol I which was 3.560 mGy when the kV setting was 80 kV and the highest $CTDI_{air}$ value was 13.585 mGy with 140 kV from Protocol III. It was found that as the kV values were increased, the $CTDI_{air}$ values were increased as well. When the kV setting was changed from 80 kV to 120 kV and from 120 kV to 140 kV, the $CTDI_{air}$ values were increased as much as 147.9% and 53.9% respectively. For the statistical analysis, the findings were tested with the Kruskal-Wallis test to show how much the data is significantly different from one protocol to another. According to the test, the $p$-value was less than 0.05. This test was chosen because it is a non-parametric method to compare two or more independent samples, where in this study, there were three protocols to be compared.

**Table 1.** The parameters and $CTDI_{air}$ values for each protocol.

| Protocol | Slice Thickness (mm) | Nominal beam width (number of detector × slice width) | Scan time (s) | Tube current (mAs) | Tube potential (kV) | $CTDI_{air}$ (mGy) | *p*-value |
|---|---|---|---|---|---|---|---|
| I | 5.0 | 32 × 1.2 mm | 0.5 | 100 | 80 | 3.560 | |
| II | 5.0 | 32 × 1.2 mm | 0.5 | 100 | 120 | 8.827 | < 0.05 |
| III | 5.0 | 32 × 1.2 mm | 0.5 | 100 | 140 | 13.585 | |

Figure 3 (a-c) shows the single scan dose profiles (SSDP) for the three protocols. The dose distribution profiles were plotted based on the kerma readings in unit mGy against the position in unit mm along the z-axis. The doses were distributed like a bell shape and therefore, they were fitted using Gaussian function in Origin Pro 9.0 software. It was observed that the peak of the dose profiles were getting higher and sharper when the kV increased. The full width at half maximum (FWHM) which gives the information on the nominal beam width of CT scanner can be determined from the SSDP using the same software [3]. The FWHM were found to be 36.2, 36.4 and 38.6 for Protocol I, Protocol II and Protocol III respectively.

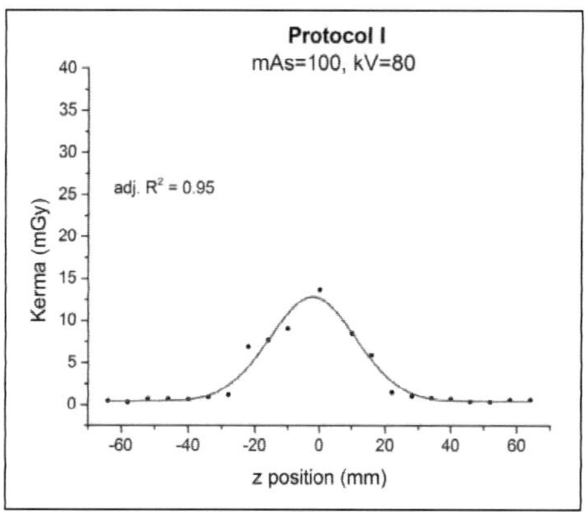


(a)

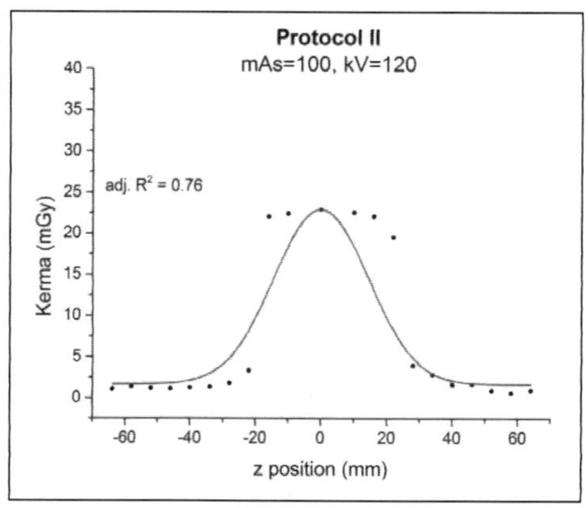


(b)

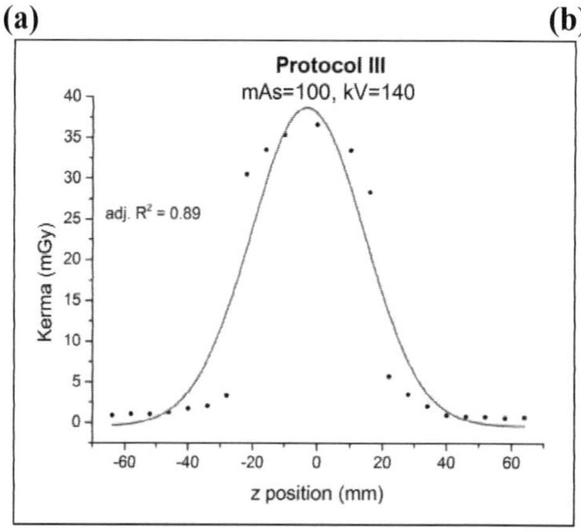


(c)

**Figure 3.** The single scan dose profile (SSDP) for (a) Protocol I, (b) Protocol II and (c) Protocol III.

## 4. Discussion

According to table 1, the increase in $CTDI_{air}$ when the kV settings were varied increasingly showed that the change of kV affected the patient exposure from CT scanner. Although $CTDI_{air}$ did not show the patient dose because CT dose index is not the same as patient dose, $CTDI_{air}$ is an important element in the implementation of patient dosimetry [7]. CTDI gives information such as on how a CT machine operates, the patient size and scanned anatomy to estimate the patient dose [3]. This is where the use of CTDI is important for the implementation of patient dosimetry. The tube potential which is measured in kilovolt (kV) shows the amount of energy supplied to the machine to radiate the X-ray and controls the quantity of the X-ray beam. For CT scan, there are a few factors that influence the value of CTDI such as kV, mAs, tube rotation time, pitch and beam collimation. In this study, the effect of kV setting was observed. The higher kV results in the higher CTDI values. According to the results obtained from the study, it was proven true that when the kV was higher, the $CTDI_{air}$ value was higher as well. The $p$-value of less than 0.05 acquired from the Kruskal-Wallis test indicated that these results were reliable and significantly different. This is important to determine the $CTDI_{air}$ and evaluate the dose profile for each protocol.

The SSDP is usually measured in 140 to 150 mm long phantoms [6]. Therefore, to imitate the measurement using phantom, the same integration length was applied for the measurement free-in-air, which was also 150 mm. The cases of TLD-100 were grooved such that there were high sampling area (11 TLDs) at the central part and low sampling areas (three TLDs) at the adjacent sides. This was done so because the CT beam was concentrated at the centre part of target. Therefore, to get more accurate results at the concentrated part, 11 TLD chips were placed adjacent to each other. Referring to figure 3 (a-c), there were apparent difference between the measured kerma values in the central part and the adjacent sides. For high sampling area which was the central part or the peak near slice centre, each measured point on the profile was derived from a single TLD value. While for the positions adjacent to the high sampling area or the long dose tail, the dose values were obtained by averaging over three TLD values.

The SSDP showed some characteristics of CT scanner's exposure. The sharp peak near the slice centre where $z=0$ resulted primarily from contribution of the primary X-ray beam and the centre part has the highest contribution of doses [12]. The maximum kerma was found to be at the midpoint of the axis, which was exactly at $z=0$. This was true since the central part of target is the position where the X-ray beam was the most concentrated. The long dose tails were caused by the scattered radiation [3]. The nominal beam width of CT scanner was fixed to be 38.4 mm (32 × 1.2 mm) for all protocols. It was found that the FWHM for the protocols were 36.2 (Protocol I), 36.4 (Protocol II) and 38.6 (Protocol III). There were an underestimation of 5.7% and 5.2% for Protocol I and II respectively, while Protocol III had an overestimation of 0.5%. For more accurate results, the measurement free-in-air for the beam width of more than 32 mm should be performed with an integration length of more than 150 mm [13]. Besides that, although TLD-100 offers many advantages as stated earlier in the Material and Methods section, this TL dosimeter also have several disadvantages. It has high uncertainty (~18.3%) and non-continuous reading, which may affect the evaluation of SSDP [14]. Overall, the FWHM were coincided with the nominal beam width used during measurements since the percentage of the under and overestimations were less than 10% (the acceptable value). The CT dose profiles reflected the efficiency of the scanner to irradiate the X-ray beam in accordance to the nominal beam width. The analysis made from the measurements of $CTDI_{air}$ and evaluation of SSDP pointed out that the performance of CT scanner were in optimum state.

## 5. Conclusion

In the current study, the computed tomography dose index free-in-air ($CTDI_{air}$) was determined and the single scan dose profile (SSDP) was evaluated in 128 slice CT scanner using thermoluminescent dosimeter (TLD-100). The changes of tube potential (kV) have influenced the $CTDI_{air}$ values and CT dose profiles. The increase in kV values led to the increasing of $CTDI_{air}$ values by as much as 147.9%

when the kV setting was changed from 80 kV to 120 kV. The statistical analysis using Kruskal-Wallis test which showed a *p*-value of less than 0.05 validated that the data were significantly different from one protocol to another. The SSDP indicated the behaviour of the exposure of CT scanner. The FWHM obtained from SSDP implied the CT scanner's nominal beam width. The FWHM were coincided with the nominal beam width used during measurements and both of them are in good agreement.

There were several limitations to this study. First, this study only involved the measurements of CTDI free-in-air using TLD-100. Future study with other dose measuring apparatus like optically stimulated luminescence dosimeter (OSLD) must be performed to confirm our findings. Second, our result was confined to one type of CT scanner which was 128 slice CT scanner Siemens SOMATOM Definition AS+. The study could be extended to different protocols with different parameter settings in various CT scanner. This work may contribute to the establishment of the national diagnostic reference level (DRLs) in term of SSDP. In conclusion, the $CTDI_{air}$ and SSDP are useful methods for characterizing and assessing the performance of the CT scanner. By applying both methods, we can conclude that the 128 slice CT scanner's performance was in optimum condition.

## 6. References

[1] Mikla V I and Mikla V V 2014 *Medical Imaging Technology* (Massachusetts, USA: Elsevier Inc)
[2] Kalender W 2011 *Computed Tomography: Fundamentals, System Technology, Image Quality, Applications* (Erlangen: Publicis Corporate Pub)
[3] McCollough C H, Leng S, Yu L, Cody D D, Boone J M and McNitt-Gray M F 2011 *Radiol.* **259** 311
[4] Shope T B, Gagne R M and Johnson G C 1981 *J. Med. Phys.* **8** 488
[5] Jucius R A and Kambic G X 1977 *Proc. Soc. Photo-Opt. Instrum. Eng.* **127** 286
[6] Nakonechny K D, Fallone B G and Rathee S 2005 *J. Med. Phys.* **32** 98
[7] Office for Official Publications of the European Communities 1999 *European guidelines on quality criteria for computed tomography* (Luxembourg: Office for Official Publications of the European Communities)
[8] ICRU 2012 ICRU Report No. 87 *J. ICRU* **12** 87
[9] McKinlay A F 1981 *Thermoluminescence Dosimetry* (Bristol: Adam Hilger in collaboration with the Hospital Physicists' Association)
[10] Kramer R, Khoury H J, Vieira J W 2008 *Phys. Med. Biol.* **53** 6437
[11] Ramli A T 1988 *Dosimetri Luminesens Terma: Pengenalan dan Penggunaannya* (Kuala Lumpur: Dewan Bahasa dan Pustaka)
[12] Tsai H Y, Tung C J, Huang M H and Wan Y L 2003 *J. Med. Phys.* **30** 1982
[13] Gancheva M, Dyakov I, Vassileva J, Avramova-Cholakova S and Taseva D 2015 *Radiat. Prot. Dosim.* **165** 190
[14] Oliveira B B, Mourão A P and Da Silva T A 2011 *World Academy of Science, Engineering and Technology* **80** 88

**Acknowledgement**
The authors would like to express their appreciation for the support of the Research University Grant Scheme with Project No Q.J130000.2517.13H39 and MyMaster MyBrain15 Scholarship.

IOP Conf. Series: Journal of Physics: Conf. Series **851** (2017) 012003 doi:10.1088/1742-6596/851/1/012003

# A single institution study of radiation dose received from CT imaging: A comparison to Malaysian NDRL

**N D Osman[1], S B M Shamsuri[1,2], Y W Tan[3], M A S M Razali[1] and S M Isa[1]**

[1] Advanced Medical and Dental Institute, Universiti Sains Malaysia, Kepala Batas, Penang, 13200, Malaysia
[2] Faculty of Science, Technology, and Human Development, Universiti Tun Hussein Onn, Batu Pahat, Johor, 86400, Malaysia
[3] School of Physics, Universiti Sains Malaysia, Minden, Penang, 11800, Malaysia

E-mail: noordiyana@usm.my

**Abstract.** Advancement of CT technology has led to an increase in CT scanning as it improves the diagnosis. However, it is important to assess health risk of patients associated with ionising radiation received from CT. This study evaluated current dose distributions at Advanced Medical and Dental Institute (AMDI), Malaysia and was used to establish Local Diagnostic Reference Level (LDRL). Dose indicators such as CT Dose Index ($CTDI_{vol}$ and $CTDI_w$) and Dose-Length Product (DLP) were gathered for all routine CT examinations performed at the Imaging Unit, AMDI from January 2015 to June 2016. The first and third quartile values for each dose indicator were determined. A total of 364 CT studies were performed during that period with the highest number of cases being Thorax-Abdomen-Pelvis (TAP) study (57% of total study). The $CTDI_w$ ranged between 2.0 mGy to 23.4 mGy per procedure. DLP values were ranged between 94 mGy.cm to 1687 mGy.cm. The local dose data was compared with the national DRL to monitor the current CT practice at AMDI and LDRL will be established from the calculated third quartile values of dose distribution. From the results, some of the local dose values exceeded the Malaysian and further evaluation is important to ensure the dose optimisation for patients.

## 1. Introduction

With the technology advancement in CT imaging, it has been performed more frequently as patients can benefit from a quicker procedure and more accurate diagnosis. However, critical attention must be given to assess the health risk associated with the ionising radiation received by patients during CT examination. According to the International Commission on Radiological Protection (ICRP), the radiation exposure received from the diagnostic radiology should be justified and optimized [1–2]. Therefore, in this study, the local radiation dose data for all CT examinations performed in AMDI, Malaysia were reviewed and compared with the standard value from the National Diagnostic Reference Level (NDRLs) provided by Ministry of Health (MOH) Malaysia [3]. By comparing the Local Diagnostic Reference Levels LDRLs with standard NDRLs, this will enable optimisation of current local CT practices on exposure settings and

IOP Conf. Series: Journal of Physics: Conf. Series **851** (2017) 012003 doi:10.1088/1742-6596/851/1/012003

protocol selection in AMDI. The local dose distribution can also improve by reducing the frequency of unjustified exposure and recommending corrective actions for dose reduction.

The first clinical CT scan on a patient in AMDI took place on 1st January 2015. In this centre, CT delivers large numbers of the collective effective dose received for diagnostic imaging procedures. This study also intends to establish a LDRL standard for each CT protocol performed in AMDI. The DRLs are recommended as an optimisation tool to help manage the radiation dose to patients so that the dose received commensurate with clinical purposes [4–9]. The establishment of LDRL can serves as a dose guideline that for diagnostic and therapeutic procedures. With the documentation of the local dose data, it will be useful for the verification of current radiation protection program in this centre.

## 2. Methodology

The dose survey was performed on patients' data from Siemens SOMATOM AS+ CT scanner (Siemens Healthcare, Germany) at Imaging Unit, Clinical Trial Complex, AMDI, in Malaysia. The dose data for all CT protocols were collected from January 2015 until June 2016. Institutional committee review for ethical approval on clinical data study was obtained (USM/JEPeM/16040164) and there were no exclusion criteria for the selection of patients.

The exposure setting and patient information were obtained from CT consoles: tube potential (kV), effective tube current (mAs), shift per rotation ratio (pitch), slice thickness, scan range, patient's gender, and age for all the routine examination. The displayed dose output such as $CTDI_{vol}$, and DLP were collected from CT scanner. However, the national DRLs recommended for CT are expressed in $CTDI_w$ (mGy) and DLP (mGy.cm) [3]. The weighted computed tomography dose index ($CTDI_w$) was calculated using the following formula:

$$CTDI_w = CTDI_{vol} \times pitch \quad (1)$$

Minimum, mean, maximum, the first and third quartile values for each $CTDI_w$, $CTDI_{vol}$ and DLP of all CT protocol were calculated respectively. The dose indicator values were then analysed and compared with the national guidelines of DRLs. The local DRLs for this centre will be established from the calculated values of the third quartile of mean of patient dose distribution for each type of examination. In accordance with the standard regulations, the LDRL should be established after a three year period of data collection or minimum of 10 cases [6–8]. All the collected dose data were then documented for annual clinical evaluation of current local practices in CT examination. If the radiation doses exceed the NDRLs, an appropriate local review should be performed and recommendation for the corrective actions to reduce radiation dose should be proposed to the Imaging Department, AMDI.

## 3. Results

From the dose survey done at Imaging Department, AMDI, a total of 364 CT examinations' data were observed for. The total number of CT study and acquisition settings for different CT protocols obtained in the survey is presented in table 1. The highest number of CT procedure collected was TAP region for 57% of the total cases; while the least number was spine, cardiac, and head–neck study (1% respectively). From table 1, it is noted that the acquisition parameters used in this survey show a wide variation, except for slice thickness selection that are typically at constant value of 0.6. The highest value for effective mAs was noted in CT TAP study with a value of 498 mAs whereas the lowest value of 17 mAs was found in the abdomen-pelvis study (Table 1).

**Table 1.** Range of CT acquisition parameters collected in this study.

| CT protocols | n[a] | Acquisition parameters | | | |
|---|---|---|---|---|---|
| | | **Tube voltage (kVp)** | **Eff. tube current (mAs)** | **Pitch** | **Slice thickness** |
| Abdomen | 58 | 80–140 | 33–421 | 0.6–1.2 | 0.6 |
| Abd-Pelvis | 10 | 80–140 | 17–334 | 0.6–1.4 | 0.6 |
| Brain / Head | 34 | 100–120 | 56–496 | 0.55–1.2 | 0.3–0.6 |
| Cardiac | 2 | 120 | 22–127 | 0.2 | 0.6 |
| Head-Neck | 2 | 100 | 125–276 | 0.8 | 0.6 |
| Neck | 6 | 80–120 | 93–330 | 0.8–1.2 | 0.6 |
| Neck-Thorax | 3 | 100–120 | 85–279 | 0.6 | 0.6 |
| NTAP[b] | 13 | 100–140 | 94–333 | 0.6 | 0.6 |
| Pelvis | 2 | 100–120 | 145–206 | 0.6–0.8 | 0.6 |
| Spine | 3 | 100–120 | 231–406 | 0.6–1.2 | 0.6 |
| TAP[c] | 209 | 100–140 | 84–498 | 0.6 | 0.6 |
| Thorax-Abd | 7 | 100–120 | 112–327 | 0.6–0.8 | 0.6 |
| Thorax | 15 | 100–140 | 101–397 | 0.6–1.3 | 0.6 |

[a] number of cases.

[b] Neck-Thorax-Abdomen-Pelvis region.

[c] Thorax-Abdomen-Pelvis region.

Figure 1 shows the distribution of $CTDI_w$ values for various CT protocols collected which includes abdomen region (Abd), abdomen-pelvis (AP), cardiac, head, neck, neck-thorax (NT), neck-thorax-abdomen-pelvis (NTAP), spine, pelvis, thorax-abdomen-pelvis (TAP), thorax, and thorax-abdomen region (TA). The $CTDI_w$ ranged between 0.90 mGy to 98.7 mGy per procedure. The highest value of cumulative $CTDI_w$ was noted in the CT neck–thorax–abdomen–pelvis (NTAP) examinations, with a maximum value of 98.7 mGy, whereas the lowest value of 0.90 mGy was recorded for the CT cardiac study.

Figure 2 shows the distribution of collected DLP values for each CT protocol of different scanning regions. As depicted in the figure, CT head study has the highest value of cumulative DLP with a maximum value of 4401 mGy.cm, whereas the lowest DLP value of 64 mGy.cm was recorded for the CT cardiac study. The CT cardiac has lowest values of both $CTDI_w$ and DLP. A great variation of DLP distribution can be observed in CT head study (Figure 2). These variations may be due to the selection of the imaging protocols markedly in greater range of the selected mAs, and the multiple exposures for contrast enhancement study during the monitoring scanning.

Table 2 shows the comparison of local dose data (the third quartile value of the dose distribution) in this study with the standard NDRL set by MOH Malaysia. As depicted in the table, some of the local dose values in specific body regions exceeded the NDRL values such as abdomen, pelvis, thorax, and spine CT study, and thorough evaluation is required to assess higher radiation dose possibly received by patients and to propose the correction procedure strategy.

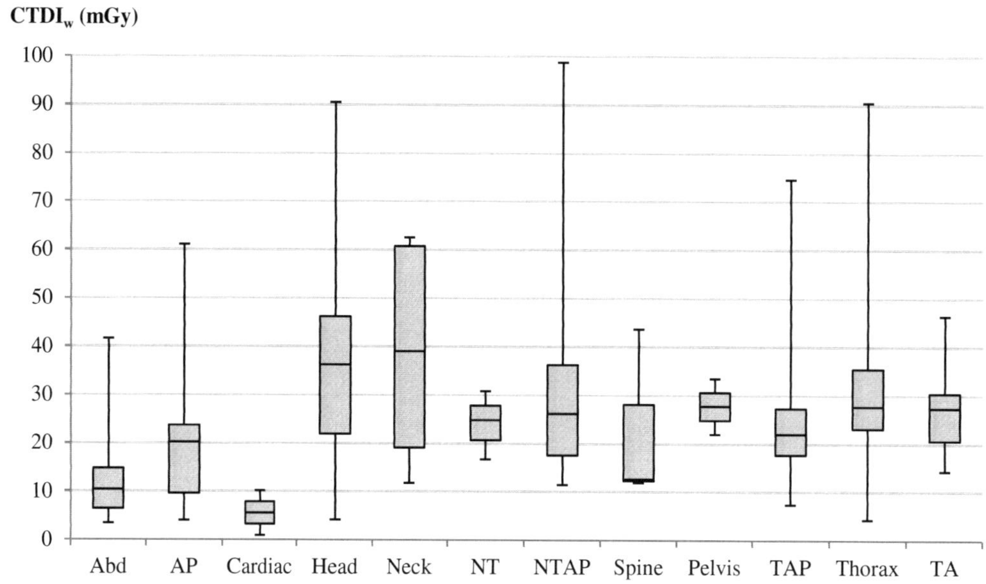


**Figure 1.** $CTDI_w$ (mGy) distribution for each CT protocol.

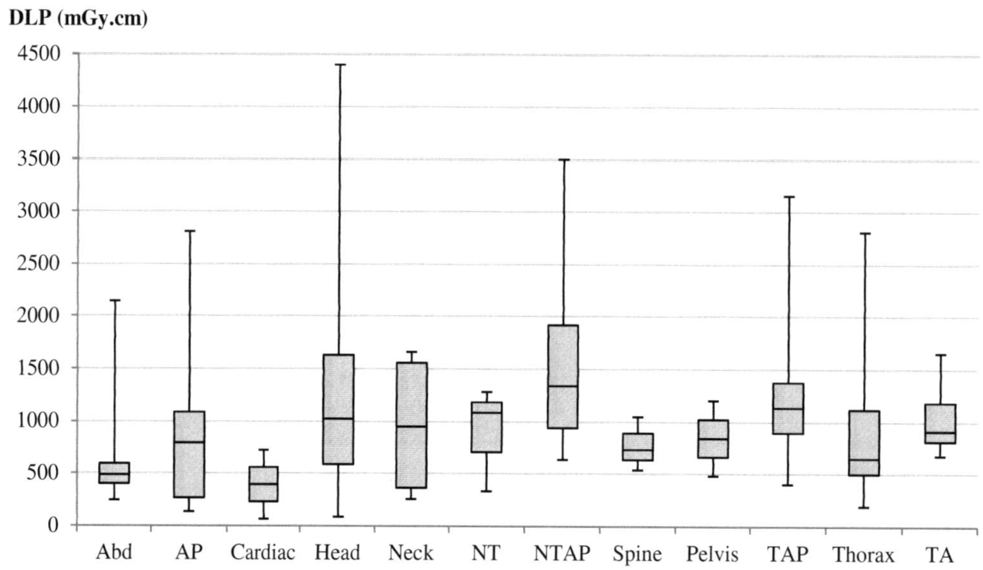


**Figure 2.** Dose-length product (DLP) distribution for each CT procedure.

**Table 2.** Comparison of local CT dose data in AMDI with the reference values of NDRL.

| Examination types | $CTDI_w$ (mGy) | | | DLP (mGy.cm) | | |
|---|---|---|---|---|---|---|
| | This study[a] | NDRL[b] | % difference | This study[a] | NDRL[b] | % difference |
| Abdomen | 12.6 | 12.8 | 1.56 | 466 | 450 | 3.56 |
| Brain / Head | 49.5 | 46.8 | 5.77 | 995 | 1050 | 5.23 |
| Cardiac | 21.0 | 11.8 | 77.9 | 295 | 870 | 66.1 |
| Pelvis | 29.4 | 39.1 | 24.8 | 632 | 730 | 13.4 |
| Spine | 21.0 | 16.3 | 28.8 | 578 | 390 | 48.2 |
| Thorax | 33.8 | 21.3 | 58.7 | 642 | 420 | 52.9 |

[a] Local dose data in this study were set at the level of third quartile value for $CTDI_w$ and DLP.

[b] NDRL values are based on reference standard established by MOH Malaysia [3].

## 4. Discussion

The radiation dose received from CT imaging mostly depends on the CT acquisition parameters, including tube potential (kVp), tube current (mAs), and shift per rotation ratio (pitch). From our observation, a great variation of dose ($CTDI_w$ and DLP) distribution was observed in certain study protocols, which were CT NTAP and CT head study, respectively (Figure 2 and 3). These variations may be due to the acquisition parameters selection because of the higher selected mAs values, and the higher dose values may be due to multiple exposures from pre- and monitoring contrast protocols. In this study, the dose survey done can help promote the improvement in patient protection at AMDI.

The third quartile values of the collected local doses were compared with the national DRL of Malaysian regulation [3]. From table 2, some of the local dose values in specific body regions exceeded the Malaysian NDRLs for $CTDI_w$ and DLP by factors ranged from 0.03 to 0.7. As reported by previous finding, the current Malaysian DRLs are based on dose surveys taken from 2007 to 2009, which comprises of dose data obtained mainly from single slice (SSCT) scanners [11]. Theoretically, the doses received from multi-slice (MSCT) scanner were slightly higher than SSCT scanner. This is also one of the major factors that contribute to higher local dose at AMDI as compared to the NDRLs dose.

As the first step in optimisation strategies at AMDI, a local DRL will be established based on the standard regulations [3,6]. The radiation dose surveys have been suggested to be performed constantly in order to monitor CT dose trends received by our local patients. Besides, this regular review can help identify the influencing features that affect patient doses in CT and to keep all doses within acceptable ranges for optimisation of radiation protection. A local dose database should also be established as a sustainable resource for monitoring dose trends in CT examinations.

However, there were several limitations in this study. This study is limited by the small sample size within certain CT protocols that constricted the establishment of local DRLs. The comparative study on the dose data is done only for selected CT protocols that are available for comparison based on NRLs, as depicted in table 2. Thus, examination-specific DRLs for the unavailable CT protocols in the current NDRLs should be established at the institutional level for local reference to portray our current practice. The Malaysian DRLs also did not specify the detailed procedures for each reference value.

## 5. Conclusion

From the local dose investigation, it indicates that some of the dose values exceeded the standard requirement set by MOH. Further evaluation is critical to ensure the dose optimisation for patients and to reduce the occurrence of excessive radiation exposure. Appropriate selection of exposure parameters and scanning techniques should be selected for dose justification. A corrective strategy for dose reduction should be proposed to the department as a part of local radiation protection program such as establishment of standard CT acquisition protocols based on individual patient's size and body-mass index (BMI).

## 6. References

[1] ICRP1990 Recommendations of the International Commission on Radiological Protection. ICRP Report 60 *Ann. ICRP* **21(1-3)** 1–201
[2] ICRP 1996 Radiological protection and safety in medicine. ICRP Report 73 *Ann. ICRP* **26(2)** 1–47
[3] Malaysian Ministry of Health (MOH) 2013 Malaysian Diagnostic Reference Levels in Medical Imaging (Radiology) *Act 304* Ministry of Health Malaysia. 2013. Guidelines in Malaysia Diagnostic Reference Levels in Medical Imaging (Radiology). Putrajaya. (download https://radia.moh.gov.my/project/)
[4] ICRP 2001 Diagnostic reference levels in medical imaging: review and additional advice. *Ann ICRP* **31** 33–52
[5] IRMER 2000 Ionising Radiation (Medical Exposure) Regulations 2000 (the Regulations) and notes on good practice. P1–2
[6] Institute of Physics and Engineering in Medicine (IPEM) 2004 Guidance on the establishment and use of diagnostic reference levels for medical X-ray examinations. *IPEM Report* **88**.
[7] European Commission (EC) 1999 Guidance on Diagnostic Reference Levels (DRLs) for Medical Exposures, *Radiation Protection* **109**.
[8] ARPANSA 2008 Code of Practice for Radiation Protection in the Medical Applications of Ionizing Radiation *RPS 14*
[9] ARPANSA 2008 Safety Guide for Radiation Protection in Diagnostic and Interventional Radiology. *RPS 14.1*
[10] McCollough C H, Primak A N, Braun N, Kofler J, Yu L, and Christner J 2009 Strategies for Reducing Radiation Dose in CT *Radiol Clin North Am.* **47(1)** 27–40
[11] Karim M K A, Hashim S, Bakar K A, Muhammad H, Sabarudin A, Ang W C, Bahruddin N A 2016 Establishment of multi-slice computed tomography (MSCT) reference level in Johor, Malaysia. *J of Physics: Conference Series.* **694** 1–5

**Acknowledgements**

The authors would like to thank radiographers and all staff at Imaging Unit, AMDI USM for their help and support throughout this work.

IOP Conf. Series: Journal of Physics: Conf. Series **851** (2017) 012004 doi:10.1088/1742-6596/851/1/012004

# The effective dose result of $^{18}$F-FDG PET-CT paediatric patients

**D Hussin[1], M A Said[2], N S Ali[2], A A Tajuddin[1] and R Zainon[1]**

[1]Kluster Oncology Science and Radiology, Advanced Medical Dental Institute, Universiti Sains Malaysia (USM), Bertam, Penang, Malaysia
[2] Department of Nuclear Medicine, National Cancer Institute, No 4 Jalan P7, Presint 7, Putrajaya, Malaysia.

E-mail: dhalisa82@gmail.com

**Abstract.** Paediatric patient received high exposure from both CT and PET examination. Automatic Exposure Control (AEC) is important in CT dose reduction. This study aimed to compare the effective dose obtained from PET-CT scanner with and without the use of AEC function. In this study, 68 patients underwent PET-CT examination without the use of AEC function, while 25 patients used the AEC function during the examination. Patients involved in this study were between 2 to 15 years old with varies of malignancies and epilepsy diseases.The effective dose obtained from PET and CT examinations was calculated based on recommendation from International Commission on Radiological Protection (ICRP) Publication 106 and ICRP publication 102. The outcome of this study shows that the radiation dose was reduced up to 20% with the use of AEC function. The mean average of effective dose result obtained from PET and CT examinations without the use of AEC and AEC function were found to be as 6.67 mSv, 6.77 mSv, 6.03mSv and 4.96 mSv respectively. Where total effective dose result of PET-CT with non-AEC and AEC were found to be 13.44 mSv and 10.99 mSv respectively. Conclusion of this study is, the installation of AEC function in PET-CT machine does play important role in CT dose reduction especially for paediatric patient.

## 1. Introduction

Children are more sensitive to ionising radiation which lead to carcinogenic effect rather than adult [1]. Awareness on ionising radiation risk potential to the paediatric patient who undergo PET-CT examination had increased attention from medical lines [2,3]. The PET-CT machine become useful tools in diagnostic, staging, monitoring response to therapy and various type of malignancies [6-10]. The combination of CT and PET scanners, gives information on anatomy and metabolic information respectively. The scanned PET-CT images is more clear rather than a single scan of CT or PET scanner and it helps physician to make decision on further treatment for the patient [11]. There is a demand on the use of PET-CT scan due to its benefits. However, the disadvantage of these technologies is patients who undergo PET-CT examination will expose to dual radiation generated by both CT and PET scanners [12].

As an alternative solution to this radiation risk issue, the technique of Automatic Exposure Control (AEC) function was introduced in PET-CT imaging. It is mainly focused in reducing the CT dose exposure during the scanning procedure. AEC is based on tissue density where imaging parameters such as tube current and voltage were adjusted automatically to the patient's tissue thickness. Determination of whole-body effective dose is one of the important parameters to prove that the AEC can reduce the radiation dose to minimum level as required by the regulations. Many studies done on comparison of PET-CT effective dose between the PET-CT machine which installed with AEC function and without AEC function.

Effective dose as defined in ICRP Publication 60 [13] is referred to a quantity of weighted sum of equivalent doses to all relevant tissues and organ. CT estimation dose can be derived from other CT dose indicator such as CT Dose Index (CTDI) or Dose Length Product (DLP) [1]. On the other hand, PET dose estimation uses dose coefficient provided by International Commissional Radiological Protection (ICRP) 106 [14]. Radiopharmaceutical of $^{18}$F-FDG is well known and widely use in clinical PET imaging. It produced by the bombardment from the cyclotron [12].

## 2. Materials and Methodology

This retrospective study involved 68 paediatric patients (4 females and 54 males) that were referred to undergo PET-CT examination without the use of Automatic Exposure Control (AEC) technique. 25 patients (9 females and 16 males) underwent the PET-CT examination with the use of AEC technique. Both group of paediatric patients were scanned with the same paediatric imaging protocol. The patients age involved in this study were in the range from 2-15 years and suspected malignancies disease such as Hodgkins Lymphoma, DLBCL, cancer and epilepsy. The patients underwent a standard clinical PET-CT scan procedure in the National Cancer Institute from the beginning of the consultation until received a complete of PET-CT images.

The calculation of PET-CT effective dose includes both internal and external exposure as provided in ICRP publication 102, 103 and 106. The internal radiation exposure referred to the exposure emitted from radiopharmaceuticals $^{18}$F-FDG. The $^{18}$F-FDG is a positron emitters with beta energies of 240 keV and gamma 511 keV. The internal exposure to patients was computed based on equation (1) [12, 17].

$$D_T = A \; x \; \Gamma_T^{FDG} \quad (1)$$

where $D_T$ is the absorbed doses to a tissue or organ, T, A is the activity (MBq) of $^{18}$F-FDG administered to the patient and $\Gamma_T^{FDG}$ is dose coefficient provided by ICRP 106 for a variety of organs and tissues of the adult MIRD phantom [12,17]

The effective dose, E received by patients were evaluated based on equation (2).

$$E = \sum_T W_T \, x \, D_T = A \, X \sum_T W_T \, x \, \Gamma_T^{FDG} = A \; x \; \Gamma_T^{FDG} \quad (2)$$

where $\Gamma_T^{FDG}$ = 56uSv/MBq dose coefficient for the effective dose F18-FDG (age 2-5 years), $\Gamma_T^{FDG}$ = 37uSv/MBq dose coefficient for the effective dose F18-FDG (age 6-10 years), $\Gamma_T^{FDG}$ = 24uSv/MBq dose coefficient for the effective dose F18-FDG (age 11-15 years), $W_T$ tissue weighting factors ($\sum_T W_T=1$) given in ICRP Publication 103 [16]

Assessment of external radiation exposure to the patients was obtained from CT imaging. Based on ICRP publication 102 [18], the external radiation exposure was evaluated based on CT Dose Index (CTDI) and Dose Length Product (DLP). The CTDI is defined as dose profile for a single-slice along a line parallel to the axis of rotation divided by the nominal thickness [19-20].

While determination of external exposure to the patient is basically from the CT scan that generates the x-ray. As referred to ICRP publication 102 [18], external exposure will determine using the CT Dose Index (CTDI) and Dose Length Product (DLP) value which can obtained direct from screen computer scan.

The effective dose, E for external exposure was then calculated according to equation (3) [18]

$$E = k \; x \; DLP \quad (3)$$

where k is coefficient based on empirical weighting factor, which functional of the anatomical region scanned ($mSv.mGy^{-1}.cm^{-1}$) in ICRP 102 [18] and k=0.015 for trunk.

## 3. Results and Discussions

A summary of patient's information data and standard clinical whole-body $^{18}$F-FDG scan parameters used at National Cancer Institute, Putrajaya Malaysia was presented in Table 1. The average patient body weight and height was 30.43 kg, 38.27 kg and 130.99 cm and 149.99 cm for the case without the use of AEC and with the use of AEC function in PET-CT examination.

IOP Conf. Series: Journal of Physics: Conf. Series **851** (2017) 012004 doi:10.1088/1742-6596/851/1/012004

**Table 1.** Summary of patient's information and whole-body PET-CT scan parameters

| Case study | Gender | Body weight (kg) | Height (cm) | Tube Current (mAs) | Voltage (kV) | Pitch |
|---|---|---|---|---|---|---|
| 68 (without the use of AEC) | 14 females 54 males | 30.43 | 130.99 | 80 | 120 | 1.75 |
| 25 (with the use of AEC) | 9 females 16 males | 38.27 | 149.99 | 50-100 | 140 | 1.25 |

For machine equipped with AEC function, the CT parameter like tube current, tube voltage and pitch settings was varies between 50-100 mAs, 140 kVp and 1.25 while fixed amount of 80 mAs, 120 kVp and 1.75 for PET-CT machine without AEC function. CT parameter settings such as tube current were adjusted automatically based on patient's tissue density because of installation of AEC function.

Table 2 shows the comparison result between PET-CT effective dose using the AEC function and without the use of AEC function within age categorisation. Based on ICRP publication 106, the paediatric patients who received radiopharmaceutical $^{18}$F-FDG was categorised into 4 groups: 0-1 year, 2-5 years, 6-10 years and 11-15 years. The activity dose administered was based on patients' body weight where 5 MBq/kg and 6 MBq/kg was injected to patients who underwent the treatment with and without the use of AEC functions. The dose administered to patients was slightly different due to the use of AEC function.

**Table 2.** Comparison of PET-CT effective dose result

| Age (y) | Case Study (n) | | Weight (kg) | | Administered Activity (MBq) | | PET Effective Dose (mSv) | | CT Effective Dose (mGy) | | Total PET-CT Effective Dose | |
|---|---|---|---|---|---|---|---|---|---|---|---|---|
| | W/o AEC | With AEC | W/o AEC | With AEC | W/o AEC | With AEC | W/o AEC | With AEC | W/o AEC | With AEC | W/o AEC | With AEC |
| 0-1 | 0 | 0 | 0 | 0 | 0 | 0 | 0 | 0 | 0 | 0 | 0 | 0 |
| 2-5 | 9 | 5 | 14.56 | 21.10 | 113.74 | 105.52 | 6.37 | 5.90 | 3.50 | 2.64 | 9.87 | 8.14 |
| 6-10 | 23 | 9 | 24.57 | 27.31 | 178.22 | 136.54 | 6.59 | 5.05 | 5.23 | 3.08 | 11.82 | 8.13 |
| 11-15 | 36 | 11 | 38.14 | 55.62 | 240.67 | 286.29 | 5.78 | 6.87 | 6.69 | 7.62 | 12.87 | 14.49 |

Figure 1 shows the comparison of PET-CT whole-body effective dose to patient by age categorisation between underwent PET-CT machine equipped with AEC function and without the use of AEC function. Patient underwent a single PET-CT examination in a day.

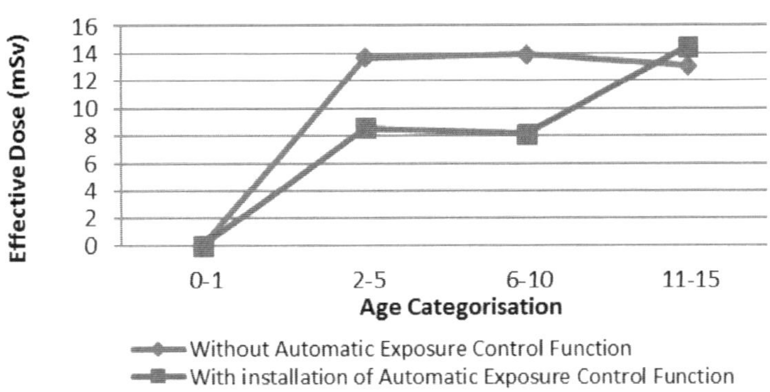


**Figure 1.** Comparison of PET-CT Whole-body Effective Dose to Paediatric patients based on Age categories between machine with AEC and without AEC function.

PET-CT effective dose at age of 2-5 years and 6-10 years was found to show differences between the machine equipped with AEC and without AEC function. However, findings in this study found that the effective dose for age group of 11-15years shows patient whom underwent PET-CT examination with AEC slightly high compared to without AEC function.

Figure 2 shows whole body PET-CT effective dose in paediatric patients. Results in this study found that the radiation dose exposure was reduced up to 20% to the patient who underwent PET-CT scan that with the use of AEC function. Installation of AEC software contributes to the external exposure (CT exposure), which related with the X-ray tube. Results from this studies in-line with Soni et al., [4] where CT exposure dose is lower than PET radiation dose due to low CT dose was used in the study. The CT dose is proportional to the tube current. Higher tube current will give more radiation dose to patients [4,6]. Thus, adjustment of tube current based on patient size will reduce the radiation dose to patients especially in paediatric patients [24,25]. Furthermore, other CT imaging parameters such as tube voltage, pitch, time per bed and rotation can be adjusted manually to obtain good images [22] rather than rely on AEC function only.

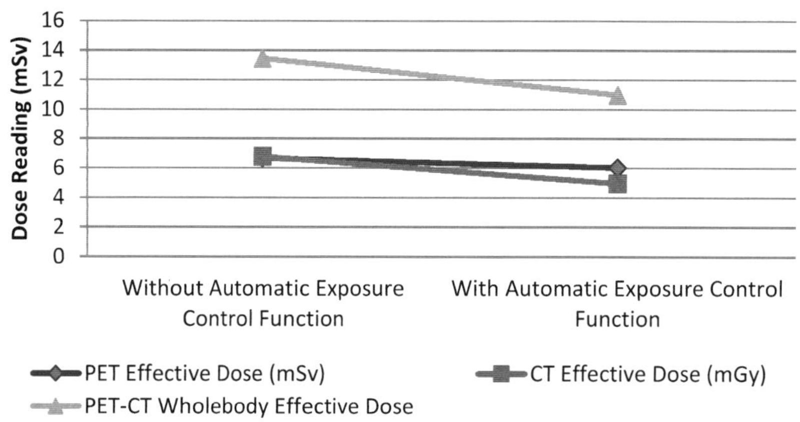


**Figure 2.** Comparison of PET-CT Whole-body Effective Dose

As for paediatric patient the suggested PET acquisition mode is in 3D compared from 2D [26]. With the use of 3D acquisition mode, the PET detector is managed to collect more data from the photons that travel in the body which effect from the collimations interaction compared to 2 D which the ability to gain data is less due to filter design. With this specialities, the administered dose activity to the patient can be minimise and this indirectly reduce the effective dose [28-30]. The use of 3D acquisition mode and OSEM reconstruction method is to ensure obtaining get a good quality image [26].

Patient's body mass index (BMI) also contributes to the radiation dose exposure. Patient with a small size and short received higher dose compared to bigger size patients. The reason of this is because of less water density in the body and the distance between organ to organ in small range and angle lead to have a lot of scattered radiation produced during the photons travelling with a short distance between each organs at the medium of less water compared to patient having big and tall figure [27].

## 4. Conclusion

The whole-body PET-CT effective dose was reduced up to 20% with the use of AEC function compared to without the use of AEC function. The AEC function is based on patient size and it related to external exposure which is CT exposure. There are other factors that can be applied to reduce the radiation dose exposure such as other CT parameter, injected administered dose, acquisition mode and scan time. Perhaps, the paediatric scan protocol shall set based on guidelines [23] that based on weight or age for the same PET-CT machine series. These will balance the unnecessary exposure especially when dealing with paediatric patient.

## 5. References

[1] Brady Z et.al. 2012 Comparison of organ dosimetry methods and effective dose calculation methods for paediatric CT *Australas Phys Eng Sci Med* **35**:117-134
[2] Marylyn J G et.al 2011 Approach to promotion and implementation of action on radiation protection for children *Radiation Protection Dosimetry* Vol. **147** No1-2 pp 137-141
[3] Brenner D J and Hall E J 2007 Computed Tomography-an increasing source of radiation Exposure *N.Engl.J.Med* **357** 2277-2284
[4] Soni C C et.al 2010 Estimated cumulative radiation dose from PET/CT in children with ma malignancies: a 5-year retrospective review *Pediatr Radiol* **40**:681–686
[5] Federman N, Feig S A 2007 PET/CT in the evaluating pediatric malignancies: a clinician's perspective *J Nucl Med* **48**:1920–1922
[6] Tatsumi M, Miller J H, Wahl R L 2007 FDG PET/CT in evaluating paediatric malignancies *J Nucl Med* **48**:1923–1931
[7] Rhodes M M, Delbeke D, Whitlock J A et al 2006 Utility of FDG-PET/CT in follow-up of children treated for Hodgkin and non-Hodgkin lymphoma *J Pediatr Hematol Oncol* **28**:300–306
[8] Miller E, Meltser U, Avrahami G et al 2006 Role of 18F-FDG PET/CT in staging and follow-up of lymphoma in pediatric and young adult patients. *J Comput Assist Tomogr* **30**:689–694
[9] Hudson M M and Donaldson S S 1999 Treatment of pediatric Hodkin's lymphoma. *Semin Hematol* **36**:313–323
[10] McCarville M B, Christie R and Daw N C 2005 PET/CT in the evaluation of childhood sarcomas. *AJR* **184**:1293–1304
[11] Townsend D W and Beyer T A 2002 Combined PET/CT scanner : the path to true image Fusion *Br.J.Radiol*.**75**,S24 S30
[12] Brix G, Lechel U et.al 2005 Radiation exposure of patients undergoing whole-body dual modality F18-FDG PET/CT examinations. *J.Nucl.Med* **46**, 608-613
[13] International Commission on Radiological Protection, ICRP, *1990* Recommendations of the International Commission on Radiological Protection. *ICRP Publication 60 Ann ICRP* 1991.**21**(1-3)
[14] International Commission on Radiological Protection, ICRP *2008* Radiation Dose to patients from radiopharmaceuticals-addendum 3 to ICRP publication 53. *ICRP Publication 106 Ann ICRP* **38**(1-2)
[15] Marguerite T H et.al *2002* MIRD dose estimate report no 19: radiation absorbed dose estimates from F18FDG *Journal of Nuclear Medicine* **43**(2):200-204
[16] International Commission on Radiological Protection, ICRP 2007 Recommendations of the International commission on Radiological Protection *ICRP Publication 103. Ann ICRP* **37**(2-4)
[17] Huang B et.al 2009 Whole-body PET/CT scanning: estimation of radiation dose and cancer risk Radiology **251**(1):p166-174
[18] International Commission on Radiological Protection, ICRP *2007* Managing Patient Dose in Multi-Detector Computed Tomography(MDCT) *ICRP Publication 102 Ann ICRP* **37**(1)
[19] Imhof H, Schinany et.al 1981 Spiral CT and radiation dose *Eur J Radiol* **47**(1):p.29-37
[20] Shope T B, Gagne R M et.al 1981 A method for describing the dose delivered by transmission x-ray computed tomography *Med Phys* **8**(4):p.488-495.
[21] Huda W et.al 2010 Organ doses to adult patients for chest CT *Med Phys* **37**(2):p.842-847
[22] Kitiwat K, Anchali K et .al 2010 The determination of patient dose from F18-FDG PET/CT examination *Radiation Protection Dosimetry* Vol. **141**, No 1, pp.50-55
[23] Frush D P 2005 Computed tomography: Important considerations for pediatric patients *Expert Rev. Med. Devices* **2**(5), 567-575.
[24] Greess H, Lutze J, Nomayr A et al 2004 Dose reduction in subsecond multislice spiral CT examination of children by online tube current modulation *Eur. Radiol* **14**(6), 995-999
[25] Greess H, Nomayr A, Wolf H et al 2002 Dose reduction in CT examination of children by an attenuation-based on-line modulation of tube current (CARE Dose) *Eur. Radiol.* **12**(6), 1571-1576

[26] Vassileva J et.al 2015 A study to establish international diagnostic reference level for paediatric computed tomography *Radiation Protection Dosimetry* pp. 1–1

[27] Qiu R, Li J, Zhang Z, Wu Z, Zeng Z, Fan J 2008 Photon SAF calculation based on the chinese mathematical phantom and comparison with the ORNL phantoms. *Health Physics* **95**: 716-724

[28] Jun Z, Katherine B, Michael K. et. al 2016 Evaluation of the Feasibility of Low Dose 18F-FDG PET in Pediatric Oncology: How Low Can It Be? *J Nucl Med* **57**(2) : 431

[29] Adam M. A, Marla S, Grace S. Phillips et. al 2011 Evaluation of Optimal Acquisition Duration or Injected Activity for Pediatric $^{18}$F-FDG PET/CT *J Nucl Med* **52**(7) :1028-1034

[30] Ming K. C, David H. M. , David W. C 2016 Determining the Minimal Required Radioactivity of $^{18}$F-FDG for Reliable Semiquantification in PET/CT Imaging: A Phantom Study *J. Nucl. Med. Technol.(*44)1: 26-30

**Acknowledgments**

The author would like to thank the cooperation given by all staff at the Nuclear Medicine Department, National Cancer Institute for the information and support. This study has been approved by the Medical Research & Ethics Committee Ministry of Health (MoH) (NMRR-14-1577-23841(IIR). and from Human Research Ethics Committee USM No : USM / JEPeM / 15090300.

IOP Conf. Series: Journal of Physics: Conf. Series **851** (2017) 012005 doi:10.1088/1742-6596/851/1/012005

# Effect of slice thickness on image noise and diagnostic content of single-source-dual energy computed tomography

**Marwan Alshipli.[1] and Norlaili A. Kabir[1]**

[1]School of Physics, Universiti Sains Malaysia, 11800, Pulau Pinang, Malaysia

E-mail: marwanayasrh@yahoo.com

**Abstract**. Computed tomography (CT) employs X-ray radiation to create cross-sectional images. Dual-energy CT acquisition includes the images acquired from an alternating voltage of X-ray tube: a low- and a high-peak kilovoltage. The main objective of this study is to determine the best slice thickness that reduces image noise with adequate diagnostic information using dual energy CT head protocol. The study used the ImageJ software and statistical analyses to aid the medical image analysis of dual-energy CT. In this study, ImageJ software and F-test were utilised as the combination methods to analyse DICOM CT images. They were used to investigate the effect of slice thickness on noise and visibility in dual-energy CT head protocol images. Catphan- 600 phantom was scanned at different slice thickness values; .6, 1, 2,3,4,5 and 6 mm, then quantitative analyses were carried out. The DECT operated in helical mode with another fixed scan parameter values. Based on F-test statistical analyses, image noise at 0.6, 1, and 2 mm were significantly different compared to the other images acquired at slice thickness of 3, 4, 5, and 6 mm. However, no significant differences of image noise were observed at 3, 4, 5, and 6 mm. As a result, better diagnostic image value, image visibility, and lower image noise in dual-energy CT head protocol was observed at a slice thickness of 3 mm.

## 1. Introduction

Computed tomography scan (CT) is a cross-sectional medical imaging technology introduced in 1972 by British engineer Godfrey Hounsfield. Computed tomography is the first radiologic modality that requires the use of computers to scan the internal organ of the body. This technology creates data depending on the different attenuation levels of the scanned area [1]. CT is classified according to the type of energy technology; the classifications are single- (SECT) and dual-energy CT (DECT). DECT provides two different attenuation levels (low and high) that construct two data sets for the same scanned area. Therefore, DECT presents more advantages to improve image quality in comparison with SECT, including the correction of the beam-hardening artifact, reduction in the radiation dose level and increase of the diagnostic content [2]. In addition, DECT developed additional techniques, such as the reduction in the number of slices and virtual non-enhanced images techniques, for reducing the radiation dose level while increasing the image quality [3].

Image noise impacts the image quality in CT images. This appears as an irregular granular pattern that degrades image quality. Image noise is based on the standard deviation of pixel intensity values or CT number for uniform anatomical region. Although DECT improves image quality, it still has approximately the same or slightly lower image noise level compared with single-energy CT [4, 5]. Despite the thinner slices increased the image noise, the image visibility of a small lesion has been improved by provide more diagnostic information. As a result, a balance between the change in slice thickness with diagnostic information and image noise must be taken into account. Therefore, a number of studies were conducted to reduce image noise in DECT protocols. The present study primarily aims to use statistical analysis combined with ImageJ program to determine the effect of slice thickness on image noise with the maximisation of diagnostic content in DECT head protocol.

### *1.1. Effect of image noise and slice thickness on DECT images*

Image noise is proportionally related to the variation in CT number values for a homogeneous area [6]. Therefore, image noise is statically defined based on standard deviation of CT number or pixel intensity values in a physical uniform region. Noise decreases as the variance (standard deviation) for pixel value decreases. In addition, CT noise is defined depending on the number of photons received by the detectors (quantum noise). Image noise decreases, when the quantity of the photons received by the detectors increase [6, 7].

Slice thickness values are determined by the operator in accordance with clinical examination requirements; its values usually fall from 1 to 10 mm. Image noise is inversely relational with the variation square root in the beam thickness (slice thickness). Thinner slice thickness increases image noise [8]. Despite thinner slices increasing image noise, the visibility of a small lesion was improved by providing more diagnostic content [9]. Therefore, a balance between the change in slice thickness with the diagnostic content and image noise should be considered. As a result, several researchers including Yao [10] and others [11] [12] compared CT image quality and diagnostic content. In addition, different CT modalities were used to assess the noise and visibility of CT images. For example, in a study conducted in 2016, cone beam computed tomography (CBCT) was used to scan two phantoms consisting of materials that mimic the mandible and its surrounding tissue. The study concluded that the lowest slice thickness increased the CT image details and structures despite the higher noise [13].

### *1.2. Use of statistical analysis to evaluate image quality in literature survey*

According to a review in 2014 [14], a statistical analysis using SPSS software was used to determine the mean and standard deviation to express the image noise (SD). The authors used independent t test samples to compare the significance of the quality of CT images between the data presented as mean ± SD. The data was collected to assess the image quality and radiation dose in multidetector CT with two groups: the first group at 120 kVp and the second group used automated attenuation-based tube potential optimisation mode (CARE kV). The t-test results determined that CARE kV reduced the radiation dose level without a significant difference in image quality between the two groups [14].

Another study by Matsumoto et al. compared the image noise in virtual monochromatic images (VMSI) with conventional 120 kVp CT image through statistical analysis. They used paired Student's statistical t test to determine the significance among the different CT images. We investigated varying impact noise levels by comparing the mean standard deviation in the regions of interest. The statistical results indicate that the noise level on the 67 to 72 keV images is significantly lower than on the 120 kVp image [15].

In 2016, Ford et al. published a paper describing the direct impact of slice thickness value on the 3D reconstruction of anatomical structures using a GE Light-Speed VCT 64 slice scanner head protocol at an initial of 0.625 mm slice thickness .The initial slice thickness values were resliced for each head of the patients and arranged from 1 to 5 mm. The statistical analysis using one-way ANOVA confirmed the significant differences in volume, surface area, and part-to-part comparison analyses from the 0.625 mm skulls. The study recommended that a maximum slice thickness of 1.25 mm can be utilised when the 3D reconstruction of anatomical features and thicknesses were greater than 1.25 mm resulted in a loss of fidelity on the representative anatomy [16].

## 2. Materials and methods

This study was conducted at the Imaging Unit, Advanced Medical and Dental Institute, Universiti Sains Malaysia, Bertam, Pulau Pinang, Malaysia. The study was performed using a Catphan 600 phantom (150 x 190 mm). As shown in Figure 1, the phantom was made from five different modules; CTP 404, 591, 528, 515 and 486. However, the fifth module (image uniformity module CTP 486) performed in this study to measure image noise. The phantom was scanned using a single-source dual-energy CT scanner (SOMATOM Definition AS, Siemens 2014, serial number 95187, Germany). After processing and reconstructing, three series of images will be produced; The first one will be with a low energy with the value (80 kVp), the second series will be a high energy with the value (140 kVp) and the third series creates by the combination of two datasets of two energies [2] [4].

CT Images were acquired at different slice thickness to determine the best slice thickness that reduces image noise with better image quality. The ImageJ Java-based image processing program software developed at the National Institutes of Health was used to analyse the DICOM CT images. F-test statistical analysis method was carried out to compare the variance significance of the obtained data. This test applied to determine the significance of variances between two groups of samples created depending on the change of the slice thickness value. The F-test distribution is formed by the ratio of two independent chi-square variables divided by their respective degrees of freedom. Therefore, this function is used to determine if two samples have different variances or not. The F value can be used to decide whether results are significant enough to reject the null hypothesis. The null hypothesis is rejected if the F value calculated from the data is greater than the critical value of the F -distribution for some desired false rejection probability.

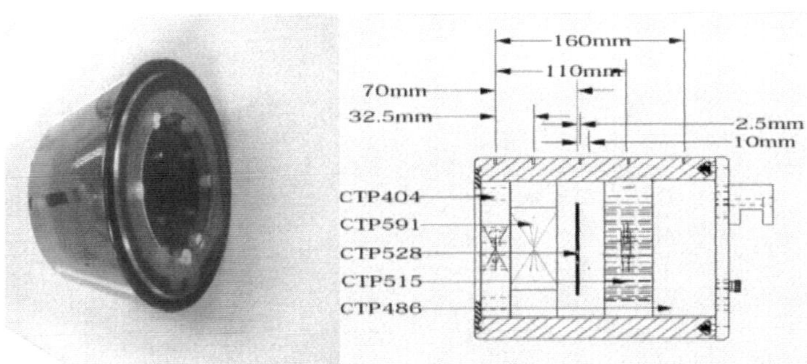


**Figure 1.** Modules and diameters of the Catphan 600 phantom.

### *2.1. Scan process*

The DECT head protocol was employed on the Catphan 600 phantom with different slice thickness values of 0.6, 1, 2, 3, 4, 5, and 6 mm. The DECT was operated in axial projection and helical scanning mode with another constant scan parameter values and fixed kVp (80 and 140). The phantom was centered, and the topogram image was produced. Afterwards, the DECT head adult protocol was applied to create different images with various slice thickness values. Table 1 shows the main parameters used in this study for DECT head protocol.

**Table 1.** DECT routine head protocol parameters.

| **Parameters** | **Head protocol dual-energy CT** |
|---|---|
| Scan mode | Spiral |
| mAs | 330 for 80 kVp<br>78 for 140 kVp |
| Pitch | 1.2 |
| Slice thickness (mm) | Range of (0.6 to 6 mm) |
| Rotation time (s) | 0.5 |
| Field of view (mm) | 217 mm |
| CARE dose type | CARE Dose4D |

### *2.2. Analysis process*
Image analysis was carried out using the ImageJ software. The pixel intensity values of the Catphan 600 phantom images at varied slice thickness were measured. The region of interest (ROI) for all DICOM images was drawn as a rectangular shape with an area approximately 1,500 mm² as shown in Figure 2. Statistical analysis was implemented using the F-test function. The F-test function was utilised to compare the statistical noise between DECT DICOM images at different slice thickness. The F-test was applied to determine the significance of variances between the two sample groups created depending on the change of the slice thickness value because variance and standard deviation are measures of how spread out a distribution is.

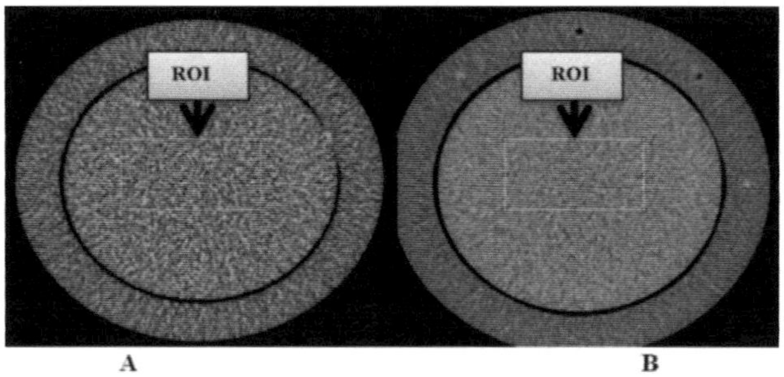


**Figure 2.** Regions of interest at 0.6 (A) and 6 mm slice thicknesses (B).

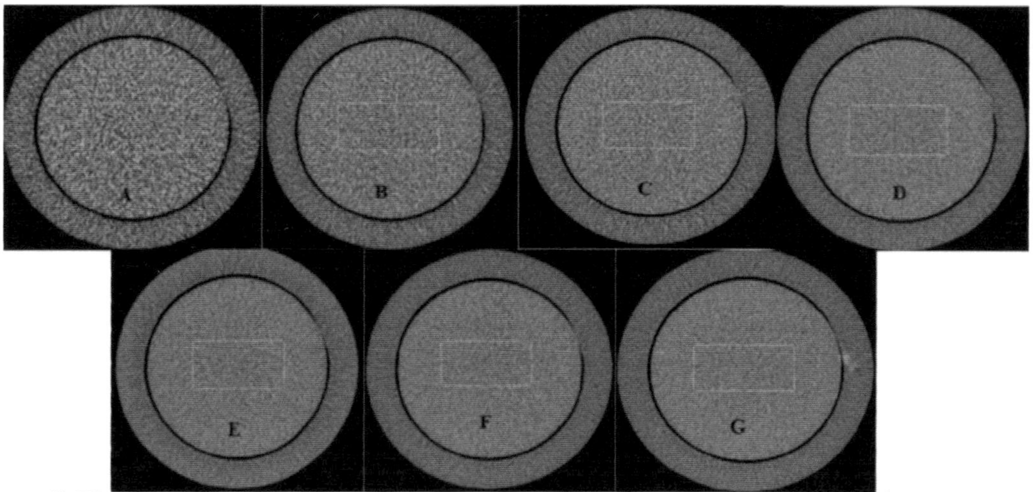


**Figure 3.** The images from A–G display the noise in the DECT routine head protocol images at different slice thickness values of 0.6, 1, 2, 3, 4, 5, and 6 mm, respectively.

## 3. Results and discussion
Statistical noise refers to the variance among the pixel intensity values (standard deviation) in the physical uniform region. Figure 3(A–G) shows the image noise in CT images using the DECT routine head protocol at different slice thickness values of 0.6, 1, 2, 3, 4, 5, and 6 mm, respectively. The results show that noise is decreased with the increase of slice thickness. The noise was measured by calculating the standard deviation (SD) in the ROI. Table 2 displays the mean and standard deviation for the ROIs using various slice thickness values at a depth of about 19 cm. Figure 4 illustrates that the increase of the slice thickness reflected on image noise decrement. Following the standard deviation analysis, the F-test statistical analysis was carried out.

Hypothetically, increment of slice thickness simultaneously decreases the image noise and the diagnostic content of the CT images. When the slice thickness is decreased, the distinction of small lesions remarkably improved owing to the decreased partial volume effects. Thus, decreasing the slice thickness value provides higher diagnostic content for the CT image. As a result, in this article, the lowest slice thickness value with a lower image noise is considered as the best value. Based on the F-test results, the significant difference exists between slice thickness values of 0.6, 1, 2, and 3 mm and non-significant values of 3, 4, 5, and 6 mm. Table 3 shows that the *F* values are less than the *F* critical value (2.168) at a range of 3 to 6 mm. The critical value is appointed on the test distribution to determine whether to reject the null hypothesis. However, the results show that image noises between the slice thickness values from 3 to 6 mm are no significantly different. Given that lower slice thickness value has more diagnostic content, the 3 mm value is considered as the best value to reduce image noise and to provide more possible diagnostic content using the DECT head protocol.

**Table 2.** Main statistical measurements for different slice thickness values using the DECT head protocol.

| Slice thickness (mm) | Mean | Standard deviation | Variance | Area $mm^2$ | Slice number |
|---|---|---|---|---|---|
| **0.6** | 79.759 | 5.379 | 28.932 | 1,500 | 91/229 |
| **1** | 78.428 | 3.690 | 13.618 | 1,500 | 318/381 |
| **2** | 77.906 | 2.715 | 7.370 | 1,500 | 96/115 |
| **3** | 78.484 | 2.333 | 5.445 | 1,500 | 64/77 |
| **4** | 77.863 | 1.860 | 3.460 | 1,500 | 49/58 |
| **5** | 79.293 | 1.821 | 3.317 | 1,500 | 39/64 |
| **6** | 78.613 | 1.694 | 2.870 | 1,500 | 33/39 |

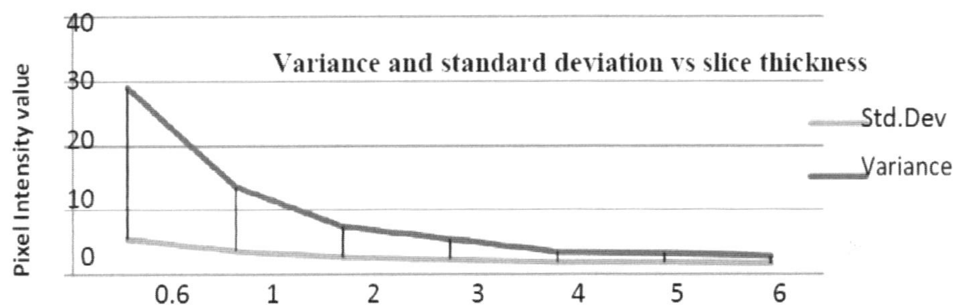


**Figure 4.** Standard deviation and variance versus slice thickness values. Standard deviation and variance are decreased with the increase of the slice thickness values especially at values less than 3 mm.

**Table 3.** F test results for slice thickness values; 3, 4, 5, and 6 mm.

| Slice thickness (mm) | Another slice thickness (mm) | *F* value | *F* critical value | Significant |
|---|---|---|---|---|
| **3** | 4 | 1.574 | 2.168 | **Non-significant** |
| | 5 | 1.642 | 2.168 | **Non-significant** |
| | 6 | 1.897 | 2.168 | **Non-significant** |
| **4** | 5 | 1.043 | 2.168 | **Non-significant** |
| | 6 | 1.206 | 2.168 | **Non-significant** |
| **5** | 6 | 1.156 | 2.168 | **Non-significant** |

Finally, as noted in Table 2 and Figure 3, the change in variance and standard deviation from 3 to 6 mm is lower than that of 0.6 to 3 mm. The effect of the increased slice thickness value from 3 to 6 mm is minimal compared with the images acquired at other slice thickness values. These results match the F-test measurements which concluded that no significant image noise exist at a range of 3 to 6 mm.

## 4. Conclusions

The imageJ analyses show that image noise decrease with the increase of slice thickness. However, the F-test results show no significant difference of the image noise of DICOM images acquired at 3, 4, 5, and 6 mm. As the diagnostic value decrease with the increased of slice thickness, the 3 mm slice thickness is considered as the best slice thickness value that reduces image noise with adequate diagnostic content in the dual-energy CT head protocol. The major limitation of this study is the single source dual energy CT was limited to fixed tube energy technique. Knowing that, the ability to change values between two different energies might help in giving better image quality

## 5. References

[1] Jacobs, R. and M. Quirynen, *Dental cone beam computed tomography: justification for use in planning oral implant placement.* Periodontology 2000, 2014. **66**(1): p. 203-213.
[2] Johnson, T.R.C., *Dual-energy CT: general principles.* AJR. American journal of roentgenology, 2012. **199**: p. 3-8.
[3] Hu, L., et al., *Radiation dose and image quality with abdominal computed tomography with automated dose-optimised tube voltage selection.* The Journal of international medical research, 2014.
[4] Matsumoto, K., et al., *Virtual monochromatic spectral imaging with fast kilovoltage switching: improved image quality as compared with that obtained with conventional 120-kVp CT.* Radiology, 2011. **259**: p. 257-262.
[5] Yu, L. and S. Leng, *Image Reconstruction Techniques.* Image Wisely, 2010: p. 1-3.
[6] Gutjahr, R., et al., *Human Imaging With Photon Counting–Based Computed Tomography at Clinical Dose Levels: Contrast-to-Noise Ratio and Cadaver Studies.* Investigative radiology, 2016. **51**(7): p. 421-429.
[7] Goldman, L.W., *Principles of CT: radiation dose and image quality.* Journal of nuclear medicine technology, 2007. **35**: p. 213-225; quiz 226-228.
[8] Lalondrelle, S., et al., *Investigating the relationship between virtual cystoscopy image quality and CT slice thickness.* 2014.
[9] Nagel, H.D., *CT parameters that influence the radiation dose*, in *Radiation dose from adult and pediatric multidetector computed tomography*. 2007, Springer. p. 51-79.
[10] Yao, Y., et al., *Image quality comparison between single energy and dual energy CT protocols for hepatic imaging.* Medical Physics, 2016. **43**(8): p. 4877-4890.
[11] Holmes, D.R., et al., *Evaluation of non-linear blending in dual-energy computed tomography.* European journal of radiology, 2008. **68**(3): p. 409-413.
[12] Chen, C.-M., et al., *Performance of adaptive iterative dose reduction 3D integrated with automatic tube current modulation in radiation dose and image noise reduction compared with filtered-back projection for 80-kVp abdominal CT: Anthropomorphic phantom and patient study.* European Journal of Radiology, 2016. **85**(9): p. 1666-1672.
[13] Katkar, R., et al., *The effect of mA, number of basis images and export slice thickness on contrast-to-noise ratio and detection of mandibular canal on cone beam computed tomography scans: an in vitro study.* Oral Surgery, Oral Medicine, Oral Pathology and Oral Radiology, 2016.
[14] Hu, L., et al., *Radiation dose and image quality with abdominal computed tomography with automated dose-optimised tube voltage selection.* Journal of International Medical Research, 2014: p. 0300060513496173.
[15] Matsumoto, K., et al., *Virtual monochromatic spectral imaging with fast kilovoltage switching: improved image quality as compared with that obtained with conventional 120-kVp CT.* Radiology, 2011. **259**(1): p. 257-262.
[16] Ford, J.M. and S.J. Decker, *Computed tomography slice thickness and its effects on three-dimensional reconstruction of anatomical structures.* Journal of Forensic Radiology and Imaging, 2016. **4**: p. 43-46.

IOP Conf. Series: Journal of Physics: Conf. Series **851** (2017) 012006 doi:10.1088/1742-6596/851/1/012006

# Adaptive iterative dose reduction (AIDR) 3D in low dose CT abdomen-pelvis: Effects on image quality and radiation exposure

**W C Ang[1], S Hashim[1,2], M K A Karim[1,3], N A Bahruddin[1], N Salehhon[1] and Y Musa[1]**

[1] Department of Physics, Faculty of Science, Universiti Teknologi Malaysia, 81310 Skudai, Johor, Malaysia
[2]Centre for Sustainable Nanomaterials (CSNano), Ibnu Sina Institute for Scientific and Industrial Research (ISI-SIR), Universiti Teknologi Malaysia, 81310 UTM Skudai, Johor, Malaysia
[3]Department of Diagnostic Imaging, National Cancer Institute, Precint 7, 62250 Putrajaya, Malaysia

E-mail: angelangweechin@yahoo.com

**Abstract**. The widespread use of computed tomography (CT) has increased the medical radiation exposure and cancer risk. We aimed to evaluate the impact of AIDR 3D in CT abdomen-pelvic examinations based on image quality and radiation dose in low dose (LD) setting compared to standard dose (STD) with filtered back projection (FBP) reconstruction. We retrospectively reviewed the images of 40 patients who underwent CT abdomen-pelvic using a 80 slice CT scanner. Group 1 patients ($n$=20, mean age 41 ± 17 years) were performed at LD with AIDR 3D reconstruction and Group 2 patients ($n$=20, mean age 52 ± 21 years) were scanned with STD using FBP reconstruction. Objective image noise was assessed by region of interest (ROI) measurements in the liver and aorta as standard deviation (SD) of the attenuation value (Hounsfield Unit, HU) while subjective image quality was evaluated by two radiologists. Statistical analysis was used to compare the scan length, CT dose index volume ($CTDI_{vol}$) and image quality of both patient groups. Although both groups have similar mean scan length, the $CTDI_{vol}$ significantly decreased by 38% in LD CT compared to STD CT ($p<0.05$). Objective and subjective image quality were statistically improved with AIDR 3D ($p<0.05$). In conclusion, AIDR 3D enables significant dose reduction of 38% with superior image quality in LD CT abdomen-pelvis.

## 1. Introduction

Radiation protection has been a prime concern due to globally increase use of computed tomography (CT) examinations in radio-diagnostic imaging. Although CT has the capability to provide more details for early detection of diseases, it has been considered as a major radiation source that contribute to cancer risk [1]. To overcome these cancer risks, improved version of hardware and software products have been developed by multiple CT manufacturers for dose reduction while maintaining optimal diagnostic image quality.

Moreover, implementation of iterative reconstruction (IR) with newly evolved algorithms was reintroduced in clinical CT imaging further facilitate better quality image with reduced radiation exposure to the patients. Several studies were performed using hybrid or full IR algorithms with iDose and iterative model reconstruction (IMR) from Philips Healthcare [2], iterative reconstruction in image space (IRIS) [3], sinogram affirmed iterative reconstruction (SAFIRE) [2], and advanced modeled iterative reconstruction (ADMIRE) [4] offered by Siemens Healthcare, adaptive statistical iterative reconstruction (ASIR) [2] and model-based iterative reconstruction (MBIR) [5] from GE Healthcare. The recent iterative reconstruction technology, adaptive iterative dose reduction three dimensional (AIDR 3D) unveiled by Toshiba Medical Systems was introduced for clinical use [6]. This advanced IR adopted a statistical noise model and a scanner model in its operation process, followed by image noise elimination which happened in the projection data in considering of photon starvation [6].

AIDR 3D is the successor to AIDR and it considerably reduced radiation dose, lowered image noise and increased signal to noise ratio (SNR) while having same diagnostic acceptability in CT chest as reported in previous literature [6]. However, potential of AIDR 3D in clinical abdominal-pelvic CT has not been addressed formally. According to our knowledge, AIDR 3D reconstruction would obtain good diagnostic image quality while low dose setting was used. Hence, the aim of our study was to compare the image quality and radiation dose between low dose (LD) scan with AIDR 3D reconstruction and standard dose (STD) with traditional filtered back projection (FBP) reconstruction in CT abdomen-pelvic examinations with the use of 80 multi-detector CT system.

## 2. Methodology

This study was carried out in a hospital located in Johor Bahru that equipped with a clinical grade CT scanner (Acquilion PRIME 80, Toshiba Medical Systems) for radiological diagnostic purpose. CT abdomen-pelvis examinations completed without contrast administration were retrieved from hospital PACS (Picture Archiving and Communication System) and anonymised prior to analysis for this study.

We retrospectively reviewed CT images performed in patients with a protocol using same tube potential of 120 kVp and different automatic tube current modulation (SURE Exposure 3D) noise index selection. According to the CT abdomen-pelvis protocol setting in the hospital, the noise index of 12.5 is applied for all 5 mm slice thickness of the soft tissue window (window width, 400 HU; window level, 40 HU) in LD scan and noise index of 10 is used in the STD scan. Other acquisition parameters are fixed: detector collimation = 100 × 0.5 mm, slice interval = 5 mm, rotation time = 0.5 s, beam pitch = 0.81 and reconstruction kernel FC18 at matrix size of 512 × 512. The CT images acquired in LD scan and reconstructed with AIDR 3D was categorized as group 1, while STD scan reconstructed using FBP was categorized as group 2.

All CT images were analysed on a dedicated medical image post-processing workstation (Myrian® 641.14.3, Intrasense). Measurement of image quality involved both objective and subjective criteria. Objective image quality was assessed by measuring the standard deviation (SD) of the attenuation value (HU) at the predefined ROIs in abdominal region as the image noise. In this study, circular 5 $mm^2$ ROIs were drawn in homogenous area at right lobe of liver and aorta in axial image located at the level of main portal vein [7]. Each ROI measurement was repeated three times and the mean value was taken. Subjective image quality evaluation involved two experienced radiologists in reading clinical CT abdomen-pelvis examinations. Image assessment criteria comprised of subjective image noise, image artifacts and overall diagnostic acceptability. Each image data set from group LD AIDR and group STD FBP was randomly displayed and reviewed directly without technical parameter, reconstruction technique and patient's information. The image quality rating is based on a qualitative five point likert scales (1=unacceptable; 2=suboptimal; 3=average; 4=good; 5=excellent) with no time restriction on viewing duration.

The radiation dose was evaluated by CT dose index volume ($CTDI_{vol}$), expressed in mGy which displayed on the CT console upon completion of the scan. The total scan length of each scan was recorded. Patient's demographic data including age, weight and height was also documented. The body mass index (BMI) of each patient was calculated as weight in kilograms (kg) divided by height in meter squared ($m^2$). Radiation dose, scan length and objective image noise, and scores for subjective image criteria obtained from LD AIDR 3D reconstruction were compared with STD FBP reconstruction. Statistical analysis was performed with commercially available statistical package SPSS version 23.0.

## 3. Results

In this study, the number of patients in LD AIDR (group 1) and STD FBP (group 2) were equal (20 patients in each group). The mean age, weight and height (± standard deviation) for group 1 patients was 41 ± 17 years, 59.03 ± 12.75 kg and 1.67 ± 0.03 m; 52 ± 21 years, 61.83 ± 13.53 kg and 1.68 ± 0.04 m for group 2 patients. Group 1 and group 2 patients had mean BMI (± standard deviation) of 20.99 ± 4.0 and 22.17 ± 4.61 respectively, ranging from minimum of 16.63 to the maximum 30.29.

### *3.1. Objective and subjective image quality*

All CT abdomen-pelvis examinations included in the present study obtained diagnostic quality images. The objective image quality in mean SD of HU measured at liver and aorta were 12.95 ± 2.03 versus 18.61 ± 2.74 and 13.65 ± 1.99 versus 18.96 ± 3.58 respectively for LD AIDR 3D versus STD FBP. Significant lower objective image noise was achieved by AIDR 3D reconstruction as compared to FBP images (all, $p<0.05$) in spite of LD scan was performed (figure 1).

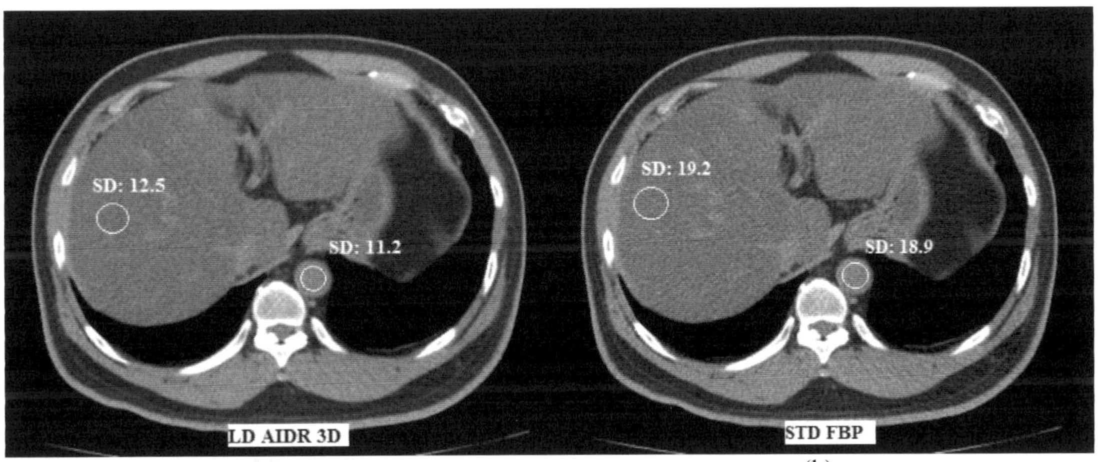


(a) (b)

**Figure 1.** Comparison of objective image quality reconstructed with (a) LD AIDR 3D and (b) STD FBP. The measured objective image noise at liver and aorta were lower in (a) with SD of 12.5 and 11.2, as compared to (b) with SD of 19.2 and 18.9, respectively. The image quality is markedly better in the LD AIDR 3D reconstruction than in the STD FBP reconstruction.

The rating results of subjective image quality by radiologists are shown in table 1. The lowest mean score ± standard deviation of 2.80 ± 0.38 was found in group 2 for subjective image noise criteria while the highest mean score ± standard deviation of 3.93 ± 0.29 was recorded in group 1 for image artifacts parameter. When compared to group 2, group 1 images scores showed significant reduced subjective image noise and artifacts by 32% and 28 % respectively. Overall diagnostic acceptability was considered good in both LD scan and STD scan with mean scores statistically higher in AIDR 3D images ($p<0.05$). The LD AIDR 3D images were superior to the STD FBP images in terms of all subjective criteria of image quality.

**Table 1.** Subjective image scores analysis.

| Subjective criteria | Group 1 LD AIDR 3D | Group 2 STD FBP | *p* |
|---|---|---|---|
| Subjective image noise | 3.70 ± 0.29 | 2.80 ± 0.38 | |
| Image artifacts | 3.93 ± 0.29 | 3.08 ± 0.44 | <0.05 |
| Overall diagnostic acceptability | 3.78 ± 0.26 | 3.03 ± 0.26 | |

*3.2. Scan length and radiation dose*

The mean scan length ± standard deviation in group 1 patients was 47.3 ± 3.44 cm (range, 40-55 cm) and 46.83 ± 4.22 cm (range, 40.5-55 cm) in group 2 patients. No significant difference in mean scan length was found between the assessed patients in LD AIDR 3D and STD FBP protocols ($p>0.05$).

The mean $CTDI_{vol}$ ± standard deviation for LD CT and STD CT were 4.03 ± 0.95 mGy (range, 3.2-7.0 mGy) and 6.45 ± 1.49 mGy (range, 5.3-11.9 mGy), respectively. There was a significant dose reduction of 38% in group 1 when compared to group 2 ($p<0.05$). Figure 2 shows the distribution of scan length and $CTDI_{vol}$ in our study with horizontal lines indicate median, whiskers minimum and maximum values.

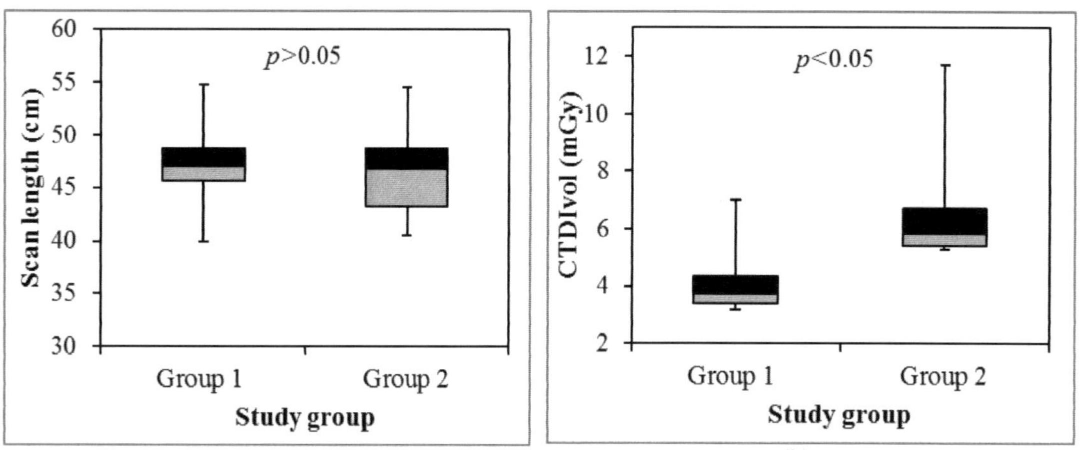


(a) (b)

**Figure 2.** (a) Scan length of CT abdomen-pelvis in group 1-LD AIDR 3D and group 2-STD FBP were almost identical with $p>0.05$. (b) $CTDI_{vol}$ was significantly higher in group 2 as compared to group 1, $p<0.05$.

## 4. Discussion

Among all iterative reconstruction technologies in the CT market, AIDR 3D is a relatively new iterative reconstruction technique introduced by Toshiba Medical Systems. Although iterative reconstruction is considered as a new development in image processing, it has been long occurred and installed in the very first CT scanners [8]. Prior major limitation of iterative reconstruction on the reconstruction time had overcome by recent expedited computer technologies. Thus, the ultimate advantage of iterative reconstruction was known to process images with less noise than the FBP. Subsequently, reduced radiation dose could also be achieved without compromising image quality as demonstrated in previous phantom and patients studies [9-12].

In this study we analysed the effects of AIDR 3D in clinical low dose abdominal-pelvic CT image quality improvement and radiation dose reduction. Our results in the present study showed that application of AIDR 3D is useful in low dose CT abdomen-pelvis with consistently superior objective and subjective image quality as compared with FBP in STD scan. In addition, ROIs measurements of objective image noise in the two regions (liver and aorta) found an average 30% reduction of image noise using AIDR 3D reconstruction compared with images reconstructed with FBP specifically. These results are in line with prior researchers' finding where AIDR 3D greatly reduces image noise in LD CT protocol [6, 13, 14]. Since AIDR 3D is available in 3 settings (mild, standard and strong), it is expected that stronger AIDR 3D setting would provide more distinct image denoising effects than the AIDR 3D standard setting used in the current study. Therefore, further study should be involve different strength of AIDR 3D setting in various CT examinations.

Apart from objective image quality improvement with AIDR 3D, group 1 (LD AIDR 3D) image quality evaluation by radiologists demonstrated excellent subjective scores throughout the image noise, image artifacts and diagnostic acceptability criteria in the present study. It is worth to mention that none of the image is graded score value of 2 or less out of 5 in AIDR 3D images regardless the subjective parameters being assessed. In contrast, five cases out of 20 patients in group 2 (STD FBP) were rated as suboptimal image when considering the image noise and image artifacts subjectively. Accordingly, the diagnostic acceptability also quarterly decreased in group 2 images with FBP reconstruction. The results of our investigation extend the feasibility of AIDR 3D in obtaining remarkable better quantitative and qualitative image quality than Juri *et al.* and Gervaise *et al.* initial experiences on LD CT compared to routine dose CT [13, 15]. The benefits of AIDR 3D reconstruction not only removing unnecessary image noise and artifacts, it also indirectly eased the radiologists in making diagnosis without superimposition of noise on the anatomy of interests. However, diagnostic acceptability in this study does not represent diagnostic accuracy in detection of diseases and anatomical details. Hence, further study on the effects of AIDR 3D on diagnostic accuracy is required.

As above-mentioned, AIDR 3D attributed image noise and artifacts reduction. Thus, it is possible to use lower level acquisition parameters and therefore reduce the dose to the patients. In view of comparison analysis between LD AIDR 3D group and STD FBP group in the current study, it confirmed that significant less radiation dose as low as 3.2 mGy could have delivered in CT abdomen-pelvis by using AIDR 3D reconstruction. This positive finding is consistent with ALARA principle where radiation exposure as low as reasonably achievable should be applied in all ionizing radiation examinations without degrading image quality.

The average $CTDI_{vol}$ in the present study reduced from 6.45 ± 1.49 mGy to 4.03 ± 0.95 mGy, reaching 38% dose saving. Our results are in good agreement with William *et al.* study where 39% dose reduction was achieved by AIDR 3D in CT coronary angiography [16]. Similarly, a phantom study carried out by Joemai *et al.* established a calculated average dose reduction of 36% in order to maintain the low contrast detectability [17]. Although some study showed half dose delivered in CT abdomen-pelvis when low-dose AIDR 3D was used, CTDIvol obtained in this study was far less than 10.3 ± 3.6 mGy as stated in Gervaise et al. study [15]. Therefore, AIDR 3D can be a great tool in optimizing radiation exposure while increasing the image quality as image quality and diagnostic accuracy is often interrelated [18].

There were few limitations in our study reported here. First, this was a single centre study and had a relatively small sample size in each group. We included these patients retrospectively in order to avoid additional irradiation to the patients that would have resulted from performing a prospective study. Accordingly, the number of patients was limited for this preliminary results. However, future studies of a larger sample size must be conducted to confirm our results. Second, we only evaluated CT abdomen-pelvis on the CT unit used for our study with specific low dose and standard dose settings at different reconstructions. This study concept could also be adopted and extended to other imaging protocols and CT units made by other manufacturers in future research.

## 5. Conclusion

Our study demonstrated that there was 38% dose reduction in LD CT protocols compared to STD CT. Moreover, by using AIDR 3D reconstruction in LD CT, considerable improvement in objective and subjective image quality which was superior to STD FBP reconstruction in spite of lower radiation dose. AIDR 3D reconstruction is a promising CT dose optimization tool that benefit clinical CT imaging by producing excellent image quality at low dose exposure to the patients.

## 6. References

[1] De Gonzalez A B, Mahesh M, Kim K P, Bhargavan M, Lewis R, Mettler F and Land C 2009 *Arch. Intern. Med.* **169** 2071-77
[2] Love A, Olsson M L, Siemund R, Stalhammar F, Bjorkman-Burtscher I M and Soderberg M 2013 *Br. J. Radiol.* **86** 20130388
[3] Bittencourt M S, Schmidt B, Seltmann M, Muschiol G, Ropers D, Daniel W G and Achenbach S 2011 *Int. J. Cardiovasc. Imaging* **27** 1081-87
[4] Gordic S, Desbiolles L, Stolzmann P, Gantner L, Leschka S, Husarik D B and Alkadhi H 2014 *Clin. Radiol.* **69** 497-504
[5] Pickhardt P J, Lubner M G, Kim D H, Tang J, Ruma J A, Del Rio A M and Chen G H 2012 *AJR. Am. J. Roentgenol.* **199** 1266
[6] Yamada Y, Jinzaki M, Hosokawa T, Tanami Y, Sugiura H, Abe T and Kuribayashi S 2012 *Eur. J. Radiol.* **81** 4185-95
[7] Kim J H, Kim M J, Kim H Y and Lee M J 2014 *Clin. Imaging* **38** 710-4
[8] Fleischmann D and Boas FE 2011 *Eur. Radiol.* **21** 510-7
[9] Hara A K, Paden R G, Silva A C, Kujak J L, Lawder H J and Pavlicek W 2009 *AJR. Am. J. Roentgenol.* **193** 764-71
[10] Sagara Y, Hara A K, Pavlicek W, Silva A C, Paden R G and Wu Q 2010 *AJR. Am. J. Roentgenol.* **195** 713-9
[11] May M S, Wüst W, Brand M, Stahl C, Allmendinger T, Schmidt B, Uder M and Lell MM 2011 *Invest. Radiol.* **46** 465-70
[12] Pontana F, Pagniez J, Flohr T, Faivre J B, Duhamel A, Remy J and Remy-Jardin M 2011 *Eur. Radiol.* **21** 627-35
[13] Juri H, Matsuki M, Itou Y, Inada Y, Nakai G, Azuma H and Narumi Y 2013 *J. Comput. Asist. Tomogr.* **37** 52-7
[14] Matsuki M, Murakami T, Juri H, Yoshikawa S and Narumi Y 2013 *Acta Radiol.* **54** 869-75
[15] Gervaise A, Osemont B, Louis M, Lecocq S, Teixeira P and Blum A 2014 *Diagn. Interv. Radiol.* **95** 47-53
[16] Williams M C, Weir N W, Mirsadraee S, Millar F, Baird A, Minns F, Uren N G, McKillop G, Bull R K, Van Beek E J and Reid J H 2013 *Clin. Radiol.* **68** 570-7
[17] Joemai R M, Veldkamp W J, Kroft L J, Hernandez-Giron I and Geleijns J 2013 *AJR. Am. J. Roentgenol.* **201** 1291-7
[18] Dewey M *et al.* 2010 *AJR. Am. J. Roentgenol.* **194** 93

**Acknowledgements**
The authors would like to express their appreciation for the support of the Research University Grant Scheme with Project No Q.J130000.2517.13H39 and MyMaster MyBrain15 Scholarship.

# Evaluation of the effect of low tube voltage on radiation dose and image quality

**Nik Norhasrina Nik Din[1], Rafidah Zainon[2] and A.T Abdul Rahman[1, 2, 3]**

[1]School of Physics and Material Studies, Faculty of Applied Sciences, Universiti Teknologi MARA, 40450 Shah Alam, Selangor, Malaysia.
[2]Radiotherapy & Oncology Department, Advanced Medical & Dental Institute, Universiti Sains Malaysia, Bertam 13200 Penang, Malaysia.
[3]Institute of Science, Universiti Teknologi MARA, 40450 Shah Alam, Selangor, Malaysia.
[4]Centre for Nuclear and Radiation Physics, Department of Physics, University of Surrey, GU2 7XH Guildford, United Kingdom

E-mail: nikhasrina@gmail.com

**Abstract**. Number of Computed Tomography (CT) examinations performed worldwide is increasing. In 2010, the FDA issued an initiative to reduce unnecessary radiation exposure from CT imaging. The aim of this study is to evaluate the effect of low tube voltage on radiation dose and image quality using CTDI phantom. The CTDI phantom was scanned with dual energy CT at 80 kV and 120 kV with the tube current from 150 mAs to 350 mAs. Pitch was 1.0 while slice thickness was 1 mm and 5 mm. Results show if mAs was increased, the SNR values also will be increased. The 5 mm slice thickness shows higher SNR value compared to 1 mm slice thickness. As the voltage and tube current increased, the amount of dose absorbed is also increased because current is proportional to photon flux.

## 1. Introduction

Number of x-ray Computed Tomography (CT) examinations performed worldwide is increasing rapidly. The study investigates the best way to reduce the radiation dose absorbed by the patient from a dual energy CT using the concept as low as reasonably achievable (ALARA). Kun Tang et al (2012) found that scanning technique using low tube voltage is better than the standard tube voltage of 120 kV because the radiation dose at the center and periphery could be reduced by 35% and 13% without affecting the low contrast detectability.

## 2. Materials and Methods

Somatom Definition AS+128 Slice CT with a single source dual energy x-ray is investigated in the study. This study used CTDI phantoms made of a cylindrical polymethyl methacrylate (PMMA) and body phantom of 320 mm in diameter.

All scans were helical with pitch of 1.0 and slice thickness either 1 mm and 5 mm. Imaging was performed with two tube voltages of 80 kV and 120 kV. Tube current used were 150, 200, 250 and 300 mAs. Estimated $CTDI_{vol}$ body adult were recorded for each scan for dose estimation. $CTDI_{vol}$ is obtained from the CT given in $CTDI_{vol}$ unit.

For the evaluation of image quality, Weasis v.1.2.7 software was used. The signal to noise ratio (SNR) was measured using the software based on the CT number, CT background and standard deviation in the images. The signal to noise ratio was calculated by using equation (1)

$$SNR = CT\ number - CT\ background/\ Standard\ Deviation \qquad (1)$$

## 3. Results

Analysis of SNR for 80 kV is shown in Figure 1 for 1 mm and 5 mm slice thickness. For slice thickness of 1 mm, the SNR is increased by 82.23% when the tube current is increased from 150 mAs to 350 mAs. However, for 5 mm slice thickness, the SNR is increased by 33.87% when the tube current is increased from 150 mAs to 350 mAs. Similar trend is seen in Figure 2 for 120 kV but higher increase of SNR was observed with mAs in the range of 300 too 300 mAs for 5 mm slice thickness.

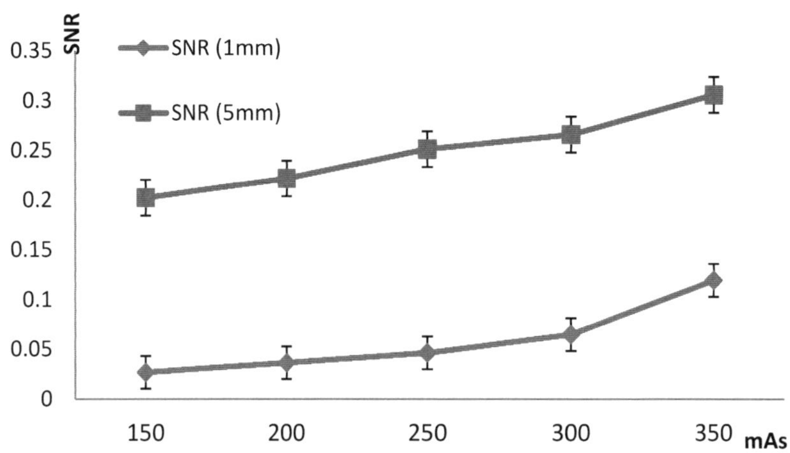


**Figure 1.** SNR for slice thickness 1 mm and 5 mm at 80 kV

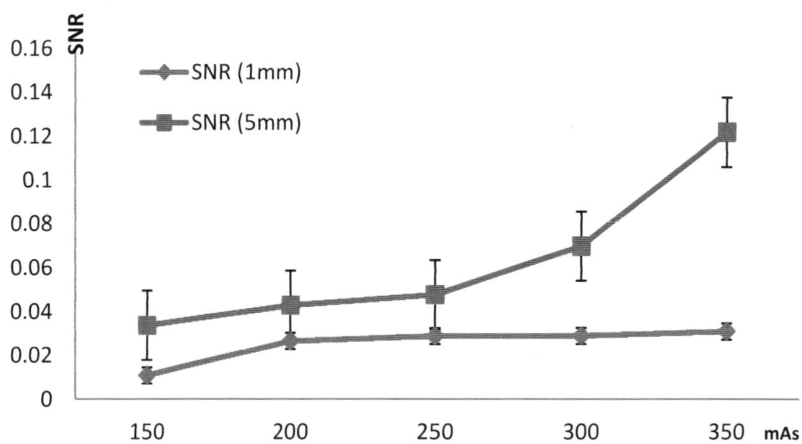


**Figure 2.** SNR for slice thickness 1 mm and 5 mm at 120 kV

Table 1 shows the $CTDI_{vol}$ for slice thickness 1 mm and 5 mm at 80 kV and 120 kV. For 80 kV and 1 mm slice thickness, the amount of $CTDI_{vol}$ was raised to 65.16% between the lowest (150 mAs) and the highest mAs (350 mAs). However, for slice thickness 5 mm, the amount of $CTDI_{vol}$ was increased to 72.31%. Large increment happened on slice thickness 1 mm between 150 mAs to 200 mAs. There is an increment of 59.40% in $CTD_{vol}$. compared to 42.56% increment for 5 mm slice thickness.

A Mann-Whitney U test shows significant difference in the slice thickness 1 mm 80 kV (Md = .028900, n = 10) and 120 kV (Md = .046100, n = 5), U = 10.5, z = -1.789, p = .074, r = 0.4619 (This could considered a large effect size using Cohen (1988)) whereas for slice thickness 5 mm Mann-Whitney U test revealed significant difference in the 80 kV (Md = 0.047600 , n = 10) and 120 kV (Md = 0.25110, n = 10), U = 0, z = -2.611, p =0.009, r = .83.Effect size , r shows large effect .

**Table 1.** The $CTDI_{vol}$ (mGy) for slice thickness 1 mm and 5 mm at 80 kV and 120 kV

| Tube Current (mAs) | 1 mm at 80 kV | 5 mm at 80 kV | 1 mm at 120 kV | 5 mm at 120 kV |
|---|---|---|---|---|
| 150 | 0.0108 | 0.0337 | 0.1687 | 0.2021 |
| 200 | 0.0266 | 0.0428 | 0.2043 | 0.2215 |
| 250 | 0.0289 | 0.0476 | 0.0463 | 0.2511 |
| 300 | 0.0290 | 0.0699 | 0.0598 | 0.2658 |
| 350 | 0.0310 | 0.1217 | 0.1090 | 0.3056 |

An independent-sample t-test was conducted to compare the dose scores for 80 kV and 120 kV. There was significant difference in scores for 80 kV (M=0.024460, SD = 0.0082637) and 120 kV (M=0.117620 SD = 0.0682396; ; t (-0.3031). The magnitude of the differences in the means (mean difference = 0.0931600, 95% . Confidence interval of the difference -0.1775593 until -0.087607) was very large effect (eta squared = 0.5345)> The guidelines (proposed by Cohen 1988, pp 284-7) for interpreting this value .01= small effect, .06 = moderate effect, .14 = large effect.

## 4. Discussion

The image noise was found correlated to the tube current. From the result, with the increased of tube current the SNR will be increased**.** This is because the amount of photon flux increased. This study used different slice thickness because of its dependence on the thickness of the CT detector. Because the scanners collimate the beam to the width of the detectors, slice thickness is equal to the collimator width. So, different slice thickness could differentiate either 5 mm slice thickness or 1 mm slice thickness were better.

The number of studies showing the utility of CT scan reported in the literature has increased, In this study, the CT dose index volume ($CTDI_{vol}$) used based on manufacturer's data for estimation of radiation dose. CTDI, expressed in terms of air kerma in milligray, was obtained directly from the computer without put any detector. $CTDI_{vol}$ is a better measurement of CT radiation dose for applications where the patient table is incremented during the scan.

The results show that it is possible to reduce radiation dose between 80 kV and 120 kV tube voltage. However, it has limitations. $CTDI_{vol}$ is an averaged dose due to homogeneous cylindrical phantom. The measurement is an approximation of patient dose. Besides that, $CTDI_{vol}$ phantom does not provide sufficiently long scatter path relative to the typical length of a human. Hence, patient dose may be underestimated with $CTDI_{vol}$.

## 5. Conclusion

As a conclusion, this research shows the low tube voltage can still provide optimal image quality. Further work will be carried out to provide guidelines for hospital in addressing in the choice of imaging parameters for dual energy CT scan.

## 6. References

[1] M.Leidholdt, J.& M.Bone, J.2002. Computed Tomography. The Essentials Physics of Medical Imaging Second Edition ed. Philadelphia,USA: Lippincott Williams & Wilkins.

[2] Bushberg, J.T./Seibert. J.A., Leidholdt,E.M. & Jr. , J.M.B. 2012. X-ray Dosimetry in Projection Imaging and Computed Tomography. The Essentials Physics of Medical Imaging. Third Edition ed. Philadelphia,USA: Lippincott Williams&Wilkins.

[3] Funama, Y., Awai, K., Nakayama, Y., Kakei, K., Nagasue, N., Shimamura, M., … Yamashita,

Y. (2005). Radiation dose reduction without degradation of low-contrast detectability at abdominal multisection CT with a low-tube voltage technique: phantom study. Radiology, 237(3), 905–910.

[4] Goodenough, D. 2012. Catphan 500 and 600 manual. *Greenwich, NY: The Phantom Laboratory*. HUDA, W., SCALZETTI, E. M. & LEVIN, G. 2000. Technique Factors and Image Quality.

[5] Goldman, L. W. (2007). Principles of CT: radiation dose and image quality. Journal of Nuclear Medicine Technology, 35(4), 213–225; quiz 226–228.

[6] Goldman, L. W. (2007). Principles of CT: radiation dose and image quality. Journal of Nuclear Medicine Technology, 35(4), 213–225; quiz 226–228.

[7] Holmberg, O., Malone, J., Rehani, M., McLean, D., & Czarwinski, R. (2010). Current issues and actions in radiation protection of patients. European Journal of Radiology, 76(1), 15–19.

[8] Jost, G., Lengsfeld, P., Voth, M., Schmid, E., & Pietsch, H. (2010). The influence of tube voltage and phantom size in computed tomography on the dose-response relationship of dicentrics in human blood samples. Physics in Medicine and Biology, 55(11), 3237–3248.

[9] Jost, G., Lengsfeld, P., Voth, M., Schmid, E., & Pietsch, H. (2010). The influence of tube voltage and phantom size in computed tomography on the dose-response relationship of dicentrics in human blood samples. Physics in Medicine and Biology, 55(11), 3237–3248.

[10] Martin, C. J., Sutton, D. G., & Sharp, P. F. (1999). Balancing patient dose and image quality. Applied Radiation and Isotopes, 50(1), 1–19.

[11] Martin, C. J., Sutton, D. G., & Sharp, P. F. (1999). Balancing patient dose and image quality. Applied Radiation and Isotopes, 50(1), 1–19.

[12] Schindera, S. T., Nelson, R. C., Yoshizumi, T., Toncheva, G., Nguyen, G., DeLong, D. M., & Szucs-Farkas, Z. (2009). Effect of Automatic Tube Current Modulation on Radiation Dose and Image Quality for Low Tube Voltage Multidetector Row CT Angiography. Phantom Study. Academic Radiology, 16(8), 997–1002.

[13] Tang, K., Wang, L., Li, R., Lin, J., Zheng, X., & Cao, G. (2012). Effect of low tube voltage on image quality, radiation dose, and low-contrast detectability at abdominal multidetector CT: Phantom study. Journal of Biomedicine and Biotechnology, 2012.

[14] Willi A. Kalender (2014) Dose in x-ray computed tomography.

IOP Conf. Series: Journal of Physics: Conf. Series **851** (2017) 012008 doi:10.1088/1742-6596/851/1/012008

# Evaluation of image quality and radiation dose using gold nanoparticles and other clinical contrast agents in dual-energy Computed Tomography (CT): CT abdomen phantom

**J Zukhi[1], D Yusob[1], A A Tajuddin[2], L Vuanghao[1] and R Zainon[1]**

[1]Oncological and Radiological Sciences Cluster, Advanced Medical and Dental Institute, Universiti Sains Malaysia, Bertam, 13200 Kepala Batas, Pulau Pinang, Malaysia.
[2]School of Physics, Universiti Sains Malaysia, Minden, 11800 Pulau Pinang, Malaysia.

E-mail: rafidahzainon@usm.my

**Abstract**. The aim of this study was to evaluate the image quality and radiation dose using commercial gold nanoparticles and clinical contrast agents in dual-energy Computed Tomography (CT). Five polymethyl methacrylate (PMMA) tubes were used in this study, where four tubes were filled with different contrast agents (barium, iodine, gadolinium, and gold nanoparticles). The fifth tube was filled with water. Two optically stimulated luminescence dosimeters (OSLD) were placed in each tube to measure the radiation dose. The tubes were placed in a fabricated adult abdominal phantom of 32 cm in diameter using PMMA. The phantom was scanned using a DECT at low energy (80 kV) and high energy (140 kV) with different pitches (0.6 mm and 1.0 mm) and different slice thickness (3.0 mm and 5.0 mm). The tube current was applied automatically using automatic exposure control (AEC) and tube current modulation recommended by the manufacturer (CARE Dose 4D, Siemens, Germany). The contrast-to-noise ratio (CNR) of each contrast agent was analyzed using Weasis software. Gold nanoparticles has highest atomic number ($Z = 79$) than barium ($Z = 56$), iodine ($Z = 53$) and gadolinium ($Z = 64$). The CNR value of each contrast agent increases when the slice thickness increases. The radiation dose obtained from this study decreases when the pitch increases. The optimal imaging parameters for gold nanoparticles and other clinical contrast agents is obtained at pitch value of 1.0 mm and slice thickness of 5.0 mm. Low noise and low radiation dose obtained at these imaging parameters. The optimal imaging parameters obtained in this study can be applied in multiple contrast agents imaging.

## 1. Introduction

Computed tomography (CT) is one of diagnostic tool that is applied in clinical and preclinical applications. This is because CT has high diagnostic efficiency which provides anatomical information of tumor and good accessibility. Conventional CT uses a wide energy spectrum and digital integrating sensor[1,2]. However, conventional CT cannot distinguish the tissues because the attenuation of x-ray is not at a specific energy but it is over the broad energy for all tissues within the voxel. The contrast resolution for soft tissues do not adequate for conventional CT [1,3]. Thus, innovation in CT imaging technology has overcome this limitation with the use of an advanced imaging technique known as a dual-energy CT (DECT) imaging.

In this study, the DECT with single source was used with applied energies of 80 kV and 140 kV. These two tube voltages will be fast switched to obtain the CT images. Different types of tissues that have similar density with different composition of element are able to distinguish by DECT. This is because DECT can differentiate the tissues according to their atomic number and electron density differences[4].

The higher atomic number of material are more preferable for contrast agent[1,5–7]. This is because the beam hardening which the effect of photon starvation is less worse than the contrast agent with low atomic number[2]. The use of multiple contrast agents simultaneously are possible in diagnostic imaging which may reduce the drug dose, diagnosis time, radiation dose, and improve specificity and sensitivity. Nevertheless, the conventional CT is unable to differentiate between the multiple contrast agents with wide range of concentrations and atomic numbers when they use simultaneously[8–10]. This limitation can be overcome by using DECT as two different energies are applied which can differentiate the contrast agents with different atomic numbers.

Thus, DECT with multiple clinical contrast agents and new contrast agent were used in this study. Barium (Ba), iodine (I), and gadolinium (Gd) are the clinical contrast agents that were used in this study The atomic numbers (Z) of Ba, I and Gd are 56, 53, and 64 respectively. A new contrast agent, gold nanoparticles (AuNPs) have higher atomic number (Z=79) than current used clinical contrast agents. Gold nanoparticles have all the properties and features of the contrast element. The circulation time for gold is up to 15 hours which is longer than iodine (less than 10 minutes). Gold also easy to fabricate into nanoparticles with variety shapes and sizes. The contrast-to-noise ratio (CNR) of CT image increases with the presence of gold at certain dose or the dose will be decreased without diminishing the CNR when using high atomic number contrast agent.

In the previous study, there is no measurement on the radiation dose obtained from multiple contrast agents scanning. Most of the studies only evaluated CT image quality when the contrast agents were applied. Thus, in this study, the evaluation of radiation dose using gold nanoparticles and other clinical contrast agents in DECT was proposed by applying different pitch and slice thickness. The optimum imaging parameters obtained from this work were proposed when applying barium, iodine, gadolinium and gold nanoparticles for DECT. The optimum imaging parameter is obtained when the radiation dose is lowest at an optimal image quality. Thus, the dose-weighted contrast-to-noise ratios (CNRD) were calculated to determine the optimum imaging parameters when using multiple contrast agents simultaneously in DECT.

## 2. Materials and methods

### *2.1. Phantom*

A fabricated polymethyl methacrylate (PMMA) phantom with diameter size of 32 cm as shown in figure 1 was used in this study to mimic an adult human abdomen. This phantom contains five cylindrical tubes which four of them filled with barium (EZ-CAT), iodine (IOMERON 350), gadolinium (MultiHance), and gold nanoparticles (SIGMA ALDRICH) while one of the tubes filled with water. The concentration of these contrast agents are similar with clinical used which are 4.9187 % w/v (4.6 % w/w), 350 mg/ml and 1.220 g/mL for barium (Ba), iodine (I), and gadolinium (Gd), respectively. The concentration of commercial gold nanoparticles (AuNPs) was used ($2.15 \times 10^{-7}$ g/ml). The volume of each contrast agent used in this study was 100 ml. There is a slot in the middle of each contrast agent tube to be inserted with the optically stimulated luminescence dosimeter (OSLD). Two OSLDs were inserted in each tube to measure the radiation dose and the average dose was computed. These contrast agents were scanned simultaneously using single-source CT scan (Somatom Definition; Siemens AG, Wittelsbacherplatz Muenchen, Germany).

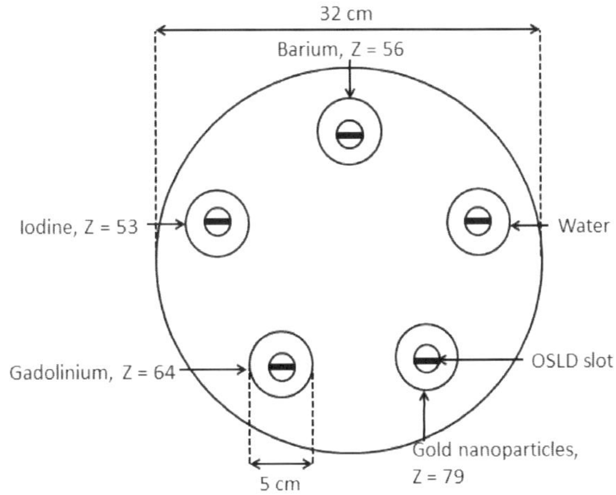


**Figure 1.** The schematic diagram of a CT abdomen phantom.

### 2.2. *Scanning Parameters and Image Analysing*

The phantom filled with water and contrast agents was scanned at 80 kV and 140 kV for DE scanning. The tube current was set automatically using CARE Dose 4D software. The pitches were set at 0.6 mm, and 1.0 mm and the slice thickness were set at 3.0 mm, and 5.0 mm. CT scanner used in this study is single source. Thus, three set images were obtained from DE scanning; where two images were acquired from 80 kV and 140 kV, and the third one was obtained from fused image.

The images obtained were analyzed using Weasis and Matlab softwares. The contrast-to-noise ratio (CNR) was calculated based on equation (1).

$$CNR = \frac{|S_A - S_B|}{\sigma_B} \quad (1)$$

where $S_A$, $S_B$, and $\sigma_B$ are mean value of contrast agent (HU), mean value of background (HU), and standard deviation of background respectively. Dose-normalised contrast-to-noise ratio dose (CNRD) was calculated to determine the optimum scanning parameters for gold nanoparticles and other clinical contrast agents using DECT[11]. CNRD was calculated using equation (2).

$$CNRD = \frac{CNR}{(Dose)^{1/2}} \quad (2)$$

## 3. Results and Discussion

### 3.1. *Contrast-to-Noise Ratio*

There are three set of images were obtained from each scanning as the single source DECT (80 kV and 140 kV) was used in this study. These images obtained were at 80 kV, 140 kV and fused image. The fused image from DECT was obtained after the scanning is assumed to be similar with the images that taken at energy of 120 kV in single-energy CT (SECT)[12,13]. All the images obtained were analyzed using Weasis and Matlab softwares to measure the CT number of the contrast agents. Then, the CNR of the contrast agent was calculated. Three regions-of-interest (ROIs) were chosen for contrast agents and background.

The CNR values obtained from DECT images were shown in figure 2. The highest CNR was obtained using DECT imaging. This is due to reduction of image noise during image reconstruction of the fused image using sharp convolution Kernel. Moreover, the image obtained was the output of quick removal of beam hardening errors as each dataset pair is obtained at the same point along the route through the patient[14,15]. The applied tube voltage of 140 kV shows higher CNR compared to CNR obtained at tube voltage of 80 kV for iodine, gadolinium and gold nanoparticles. This is because more radiation dose were exposed and less image noise produced at higher tube voltage applied[11]. However, tube voltage of 80 kV shows higher CNR than 140 kV for barium as shown in figure 2. This is because barium has higher attenuation at low tube voltage. Therefore, the contrast of the image was brighter when the attenuation of x-ray beam is high. In general, iodine gives high CNR at DECT imaging compared to other clinical contrast agents. This is due to high concentration of iodine was used which gave more attenuation occurred in iodine. The x-ray attenuation was affected by concentration and atomic number. The pattern of CNR against tube voltage for 0.6 mm pitch and 3 mm slice thickness was similar to figure 2.

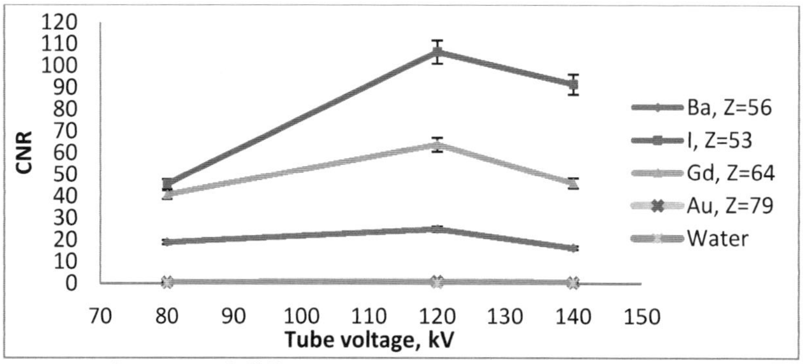


**Figure 2.** The CNR of different contrast agent against tube voltage at pitch of 1 mm and slice thickness of 5 mm.

Figure 3 shows the CNR of fused image for all contrast agents. The CNR was evaluated with different slice thickness at pitch of 1 mm. The CNR of all contrast agents increase when the slice thickness increases. The beam width and the voxel size changed when changing the slice thickness[16]. Bigger beam width and voxel size were produced when applying large slice thickness which captured more photons. These captured photons produced the contrast of the image. The gold nanoparticles shows the lowest CNR compared to other contrast agents as shown in figure 3. The CNR of gold nanoparticles was up to 0.6 slightly similar to water. This finding contradicts from other studies due to lower concentration of the gold nanoparticles used. The concentration of gold nanoparticles from other studies[10,17,18] were between 2.0 mg/ml to 13.6 mg/ml while this study used 2.15 x $10^{-7}$ g/ml. This low concentration reduced the x-ray attenuating atom per agent as low particles number found in gold nanoparticles[7]. Thus, the CNR for gold nanoparticles is lowest than other contrast agents as its concentration is lowest. The CNR pattern of all contrast agents when applying 0.6 mm pitch also increased with slice thickness.

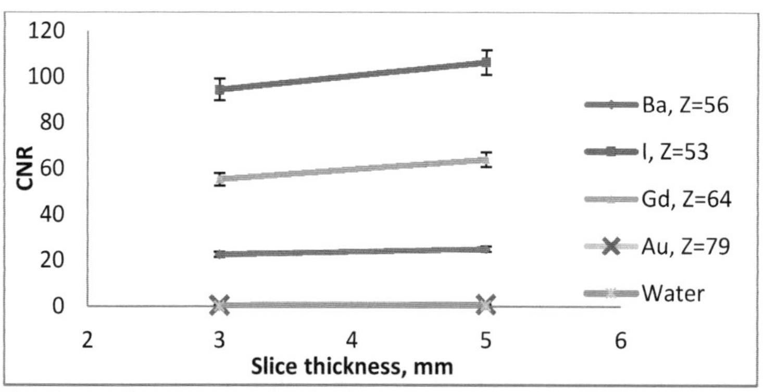


**Figure 3.** The CNR of contrast agent obtained at 80/140 kV DECT against 1 mm pitch.

### *3.2. Radiation Dose*

The average measured radiation dose obtained from OSLD was shown in figure 4. The accumulated radiation dose from 80 kV and 140 kV imaging in DECT is corresponding to the radiation dose from the fused image. From figure 4, the radiation dose of the contrast agents and water decreases with the increasing of the pitch. In multislice CT (MSCT), pitch is defined as table distance traveled in one 360° gantry rotation divided by total thickness of all simultaneously acquired slices. The radiation exposure to the contrast agents were reduced when high table speed applied. The measured radiation dose using OSLD decreases when pitch and slice thickness increases. Christner et al[19] also found a large percentage of dose reduction for high pitch while Srivastava et al[20] found that the dose reduced with increasing the slice thickness.

The use of CARE Dose 4D for tube current allows optimal tube current to give the adequate radiation dose and image quality. The highest radiation dose was obtained with Ba followed by gold nanoparticles and gadolinium. The low concentration of Ba causes less x-ray photon absorbed and allows more x-ray photon to pass through the Ba which measured by OSLD. Although that the concentration of the gold nanoparticles is lower than the barium, it shows a slightly lower radiation dose than barium. This is due to higher atomic number of gold nanoparticles compared to barium. The concentration and atomic number of the contrast agent affects the attenuation of the x-ray. The lowest radiation dose was obtained in this study with the use of iodine. The same pattern of radiation dose was obtained in this study with the use of slice thickness of 3 mm and 5 mm. In clinical application, saline solution is injected after contrast agent. The saline is not used in this study as the phantom is fixed with the region of interest.

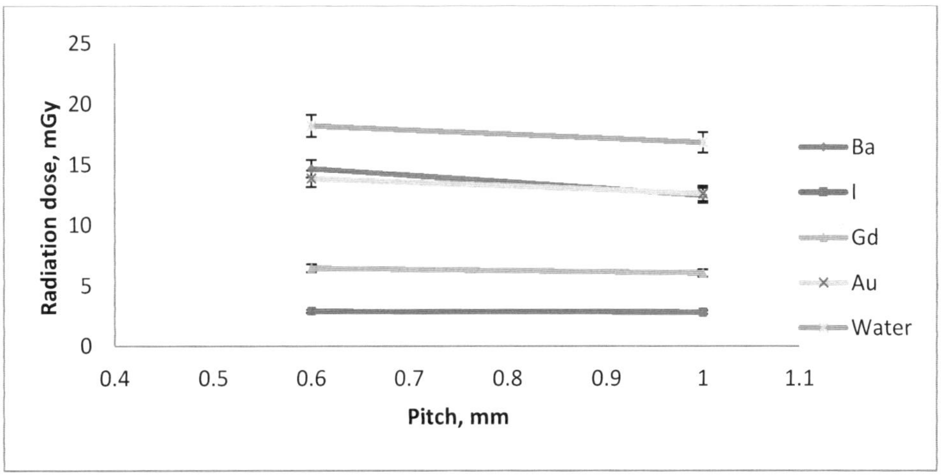


**Figure 4.** The graph of radiation dose against pitch at slice thickness of 5 mm for 80/140 kV DECT image.

The optimum scanning parameters for gold nanoparticles and other clinical contrast agents in DECT were determined by calculating CNRD as shown in equation 2. The maximum CNRD represents the minimum radiation dose that measured for a given image quality[11]. The values of CNRD of all contrast agents and water were listed in table 1. The maximum values of CNRD were obtained from high pitch of 1 mm and slice thickness of 5 mm for barium, gadolinium, and gold nanoparticles. Iodine shows maximum CNRD at pitch of 0.6 mm and slice thickness of 5 mm. The maximum CNRD for barium, iodine, gadolinium, and gold nanoparticles were 7.08, 64.30, 26.00, and 0.23 respectively.

**Table 1.** The CNRD of gold nanoparticles and other clinical contrast agents using DECT for different pitch and slice thickness.

| Pitch (mm) | Slice thickness (mm) | Ba | I | Gd | Au | Water |
|---|---|---|---|---|---|---|
| **0.6** | 3 | 6.25 | 61.50 | 18.49 | 0.15 | 0.06 |
| | 5 | 6.93 | 64.30 | 25.58 | 0.10 | 0.02 |
| **1** | 3 | 5.48 | 33.32 | 25.21 | 0.15 | 0.01 |
| | 5 | 7.08 | 63.67 | 26.01 | 0.23 | 0.07 |

The overall optimum imaging parameters for multi contrast agent imaging obtained from this study were at 1 mm pitch and 5 mm slice thickness. The maximum slice thickness for DECT imaging was up to 10 mm. The slice thickness of 5 mm is an acceptable slice thickness that has less diminished in image resolution.

## 4. Conclusion

The image quality and radiation dose using commercial gold nanoparticles and clinical contrast agents in DECT were evaluated in this study. The attenuation of barium and iodine are higher when applying low tube voltage because both contrast agents have lower atomic number compared to gadolinium and gold nanoparticles. The high concentration of iodine used causes more attenuation at tube voltage 120 kV. The tube voltage, slice thickness, and pitch affect the image quality and radiation dose to patient. Thus, the use of higher tube voltage and slice thickness gives high contrast-to-noise ratio (CNR). The DECT gave highest CNR on the fused image. Low radiation dose was obtained with the use of high pitch. Thus, pitch of 1 mm and slice thickness of 5 mm were recommended as the optimum scanning parameters for simultaneous multiple contrast agents imaging with DECT.

## 5. References

[1] He P, Wei B, Feng P, Chen M and Mi D 2013 Material discrimination based on K-edge characteristics. *Comput. Math. Methods Med.* **2013**

[2] He P, Wei B, Cong W and Wang G 2012 Optimization of K-edge imaging with spectral CT. *Med. Phys.* **39** 6572

[3] Anderson N G, Butler A P, Scott N J A, Cook N J, Butzer J S, Schleich N, Firsching M, Grasset R, de Ruiter N, Campbell M et al. 2010 Spectroscopic (multi-energy) CT distinguishes iodine and barium contrast material in MICE. *Eur. Radiol.* **20** 2126-2134

[4] Johnson T R C, Krauβ B, Sedlmair M, Grasruck M, Bruder H, Morhard D, Fink C, Weckbach S, Lenhard M, Schidt B, Flohr T, Reiser M F and Becker C R 2007 Material differentiation by dual energy CT : initial experience. *Eur. Radiol.* **17** 1510-1517

[5] Bateman C J, Rajendran K, de Ruiter N J A, Butler A P, Butler P H and Renaud P F 2015 The Hidden K-edge Signal in K-edge Imaging. *Phys. Med.* arXiv:1506.04223

[6] Meng B 2016 Image reconstruction for X-ray K-edge imaging with a photon counting detector *Proc. SPIE Int. Soc. Opt. Eng.* vol 9212, ed S R Stock pp 0-8

[7] Lusic H and Grinstaff M W 2014 X-Ray Computed Tomography Contrast Agents. *Chem. Rev.* **113** 1641-1666

[8] Zhiliang C, Ajian A Z, James Z H, Viadimir R M and Andrew T 2013 Multifunctional Nanoperticles: Cost versus benefit of adding targeting and imaging capabilities. *Science* **338** 903-910

[9] Shilo M, Reuveni T, Motiei M and Popovtzer R 2012 Nanoparticles as computed tomography contrast agents : current status and future perspectives. *Nanomedicine* **7** 257-269

[10] Ghadiri H, Ay M R, Shiran M B, Soltanian-Zadeh H and Zaidi H 2013 K-edge ratio method for identification of multiple nanoparticulate contrast agents by spectral CT imaging. *Br. J. Radiol.* **86**

[11] Lee S, Kim H, Kim D and Lee C 2010 Evaluation of Dose Reduction and Image Quality in

Pediatric Multi-detector CT. *J. Korean Phys. Soc.* **57** 183-191
[12] Siemens A G 2008 *Dual Energy CT.* (Germany: Siemens) p 10
[13] Marugami N, Kitano S, Takahama J, Takahashi A, Miura S, Kichikawa K and Kashihara/JP 2010 Dual energy CT of hepatocellular carcinoma : Can blending of 80 and 140 kV datasets improve image quality ? *Eur. Radiol.* 1-11
[14] Silva A C, Morse B G, Hara AK, Paden R G, Hongo N and Pavlicek W 2011 Dual-energy (spectral) CT: applications in abdominal imaging. *Radiographics.* **31** 1031-1046
[15] Yu L, Leng S and Mccollough C H 2012 Dual-Energy CT – Based Monochromatic Imaging. *AJR Am. J. Roentgenol.* **199** 9-15
[16] Goldman L W 2007 Principles of CT : Radiation Dose and Image Quality *. *J. Nucl. Med. Technol.* **35** 213-226
[17] Ducote J L, Alivov Y and Molloi S 2011 Imaging of nanoparticles with dual-energy computed tomography. *Phys Med Biol.* **56** 2031-2044
[18] Mullner M, Schlattl H, Hoeschen C and Dietrich O 2015 Feasibility of spectral CT imaging for the detection of liver lesions with gold-based contrast agents – a simulation study. *Phys. Med.* 1-8
[19] Christner J A, Zavaletta V A, Eusemann C D, Walz-flannigan A I and Mccollough C H 2010 Dose Reduction in Helical CT: Dynamically Adjustable z -Axis X-Ray Beam Collimation. **194** 49-55
[20] Srivastava S P, Cheng C and Das I J 2016 The effect of slice thickness on target and organs at risk volumes , dosimetric coverage and radiobiological impact in IMRT planning. *Clin. Transl. Oncol.* **18** 469-479

**Acknowledgements**
The authors would like to acknowledge the financial support from Ministry of Higher Education through Fundamental Research Grant Scheme (FRGS).

# Evaluation of efficacy of metal artefact reduction technique using contrast media in Computed Tomography

**Diana Yusob[1], Jihan Zukhi[1], Abd Aziz Tajuddin[2] and Rafidah Zainon[1]**

[1]Oncological and Radiological Sciences Cluster, Advanced Medical and Dental Institute, Universiti Sains Malaysia, 13200 Bertam, Kepala Batas, Pulau Pinang, Malaysia.
[2] School of Physics, Universiti Sains Malaysia, 11800 Pulau Pinang, Malaysia.

E-mail: rafidahzainon@usm.my

**Abstract.** The aim of this study was to evaluate the efficacy of metal artefact reduction using contrasts media in Computed Tomography (CT) imaging. A water-based abdomen phantom of diameter 32 cm (adult body size) was fabricated using polymethyl methacrylate (PMMA) material. Three different contrast agents (iodine, barium and gadolinium) were filled in small PMMA tubes and placed inside a water-based PMMA adult abdomen phantom. The orthopedic metal screw was placed in each small PMMA tube separately. These two types of orthopedic metal screw (stainless steel and titanium alloy) were scanned separately. The orthopedic metal crews were scanned with single-energy CT at 120 kV and dual-energy CT at fast kV-switching between 80 kV and 140 kV. The scan modes were set automatically using the current modulation care4Dose setting and the scans were set at different pitch and slice thickness. The use of the contrast media technique on orthopedic metal screws were optimised by using pitch = 0.60 mm, and slice thickness = 5.0 mm. The use contrast media can reduce the metal streaking artefacts on CT image, enhance the CT images surrounding the implants, and it has potential use in improving diagnostic performance in patients with severe metallic artefacts. These results are valuable for imaging protocol optimisation in clinical applications.

## 1. Introduction

Computed Tomography (CT) is an imaging modality that combines multiple x-ray projections taken from different angles and produce detailed image views of all types of body tissues. However, one of the CT imaging limitations is the difficulty in evaluating the image with the artefacts arising from metallic implants in patient's body [1]. Metal implants such as shrapnel, surgical clips, pacers, joint prostheses, wires, pedicle screw placement, stenting, and so on usually undergo follow-up CT imaging [2-4]. These metallic implants lead to severe streak and shadow artefacts in CT images that superimpose the structures of interest and deteriorate image quality, which is important for diagnostic imaging [4, 5]. This is because the metallic objects is a high density material, which is strongly scattered to the transmitted x-ray beam during CT examination. It results in the production of severe image artefacts due to the lack of the output data projection in CT images [6]. Therefore, it is difficult for radiologists to interpret the CT images with metallic implants due to the artefacts problem. This situation hampers the diagnostic interpretability of the implants themselves, metallic-bone interfaces, and adjacent soft tissue itself. These diagnostic criteria are important to prevent cracking or loosening and sufficient verification with adequate protection of the implant [7] and ruling out of hematoma, evaluation of blood vessels near implants or inflammation in the adjacent soft tissue. However, due to the occurrence of metal artefacts, the diagnosis on CT images remains challenging with many cases rendered uninterpretable, even with hard convolution kernels and widened CT density ranges [1].

Various metal artefact reduction techniques have been developed to reduce metal artefacts, and most commonly used reconstruction algorithm for CT are filtered back projection (FBP), interpolation methods, and iterative reconstruction methods. These methods involve in identification and replacement of the metal-corrupted data in CT images. However, the limitations are most of the techniques do not remove all sources of streaking artefacts, cause loss of spatial resolution in images and introduce the secondary artefacts that may lead to misinterpretation [8-10]. Other than reconstruction technique, the use of contrast media is also proposed as metal artefact reduction technique [11-13]. Contrast media are a group of medical drugs used to improve the visibility of internal organs and structures in imaging techniques. Contrast media commonly used to visualise vessels, tissues, organs, urinary tract and so on. They are helpful in distinguishing between the normal and the pathological areas. The use of contrast media is usually safe, and adverse effects are usually mild and limited. Side effects range from contrast media range from a mild inconvenience, such as itching, delayed allergic reactions, anaphylactic reactions, and cutaneous reactions, to life-threatening emergencies [14]. Lambert et al. (2015) study on the complementary of contrast media using bismuth, tantalum, and tungsten for patients with metal implants in dual-energy CT. They found that metal artefact severity was reduced in high-keV images across changes in the metal artefact types [11]. Other study done by Pjontek et al. (2015) found that contrast media and metal artefact reduction software (MARS) in angiographic CT is significantly improves image quality in patient after coiling or clipping [12]. Saake et al. (2012) use intravenous contrast media in angiographic CT follow-up imaging of intracranial flow diverting device. They found that result of digital subtraction angiography and intravenous contrast media images scored identically in all patients with stent deployment [13].

Therefore, the purpose of this study was to evaluate the efficacy of the metal artefact reduction method using clinical contrast media in abdomen CT scan to produce high image quality.

## 2. Materials and Methodology

### *2.1. Phantom Design*

A water-based abdomen phantom of diameter 32 cm was designed and fabricated using polymethyl methacrylate (PMMA) materials. The PMMA consists of five small tubes with diameter of 5 cm. The contrast media were filled in these small tubes. The small tubes were mounted in a circular pattern in abdomen phantom, which was then filled with water. A 100 ml of three clinical contrast materials (iodine, barium and gadolinium) with a clinical concentration of 350 mg/ml, 49 mg/ml, and 340 mg/ml were used respectively. Orthopedics metal screw represents the prototype of a high attenuation material that causes streak artefacts in CT imaging. Two type of metal implants materials with 6 mm diameter were used in separate phantom configuration to produce different artifact types.

The first configuration for non-contrast based phantom, used 316-grade stainless steel orthopedic screws, attached along the small PMMA tubes. The second configuration used 5-grade titanium alloy orthopedic screws. While for contrast based phantom, the first configuration used 316-grade stainless steel orthopedic screws, attached along the small PMMA tubes. The small PMMA tubes contain contrast media and orthopedic screw material each scan. The other remaining two tubes were filled with water with orthopedic screws. The second configuration used 5-grade titanium alloy orthopedic screws. The phantom configuration was shown in Figure 1.

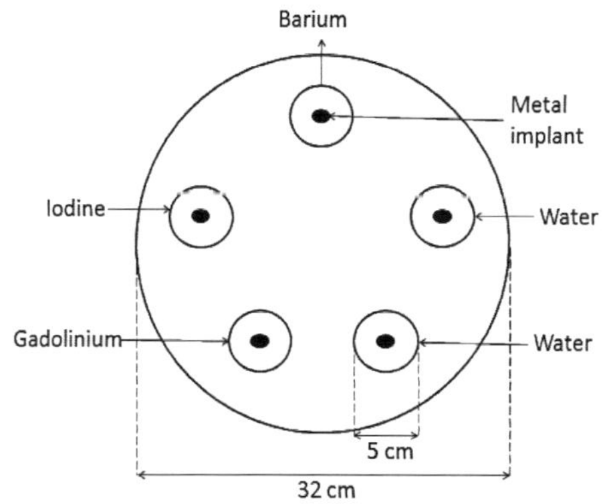


**Figure 1.** Axial image of the CT abdomen phantom.

*2.2. CT scan parameters*

All CT examinations were performed using CT scanner SOMATOM Definition AS, Siemens Healthcare, Forchheim, Germany (128 slices system) for abdomen routine scan. Table 1 shows a summary of CT parameters used in this study.

**Table 1.** Summary of CT scan parameters

| | Single-energy CT | Dual-energy CT |
|---|---|---|
| Tube voltage (kV) | 120 | Fast kV-switching between 80 and 140 |
| Tube current (mAs) | Tube current modulation (care4Dose setting) | |
| Pitch (mm) | 0.35, 0.60, 1.20 | |
| Slice Thickness (mm) | 1.0, 3.0, 5.0 | |

*2.3. Image Analysis*

The signal-to-noise ratio (SNR) before and after applying metal artefacts reduction technique using contrast media were evaluated. The analysis of SNR was performed by drawing the five region-of-interest (ROI) using circular area of 50 $mm^2$ at the surrounding of each small PMMA tubes. The location of ROIs was standardized in the image analysis. The SNR was calculated with respect to the background (water). An average from five ROIs and background were used to calculate the SNR. The SNRs were calculated by dividing average of mean CT numbers by standard deviation of the background as shown in equation 1 [15].

$$\mathrm{SNR} = \frac{S_0\,(HU)}{\sigma_0\,(HU)} \qquad (1)$$

where, $S_0$ is the signal intensities, and $\sigma_0$ is the standard deviation of the background.

*2.4. Statistical Analysis*

The qualitative scales of image quality were reported based on SNR measurement. The students't-test was performed to determine the significant differences between the results. A p-value < 0.05 was accepted as statistically significant.

## 3. Results and Discussions

*3.1. Single-energy and dual-energy CT*

Figure 2 shows SNR against slice thickness for non-contrast based phantom and contrast based phantom in single-energy and dual-energy CT. SNR for single-energy CT and dual-energy CT for non-contrast based phantom is not significantly different with p-value of 0.547. The contrast based phantom is also not significantly different with p-value of 0.982. Same findings was found by Exhibit et al. (2016) stated that image quality of blended images of dual-energy CT is similar to the conventional single-energy 120 kV acquisitions [16].

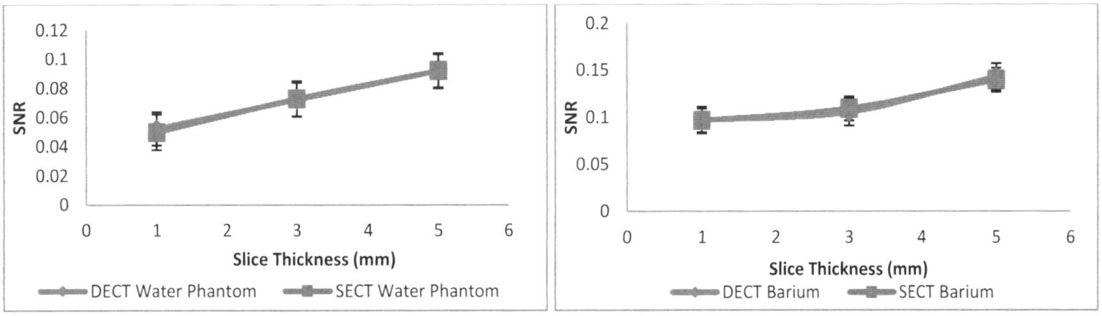


(a) (b)

**Figure 2.** SNR against slice thickness for (a) non-contrast based phantom and (b) contrast based phantom.

*3.2. Stainless steel and titanium alloy*

The severity of streaking artefacts on CT images varied with the presence of orthopedics screws depending on the material type. The streak artefacts are present between the orthopedic screw and elsewhere in the phantom (background). However, the streaking artefacts are more pronounced with the use of stainless steel screw compared to titanium alloy screw as shown in Figure 3.

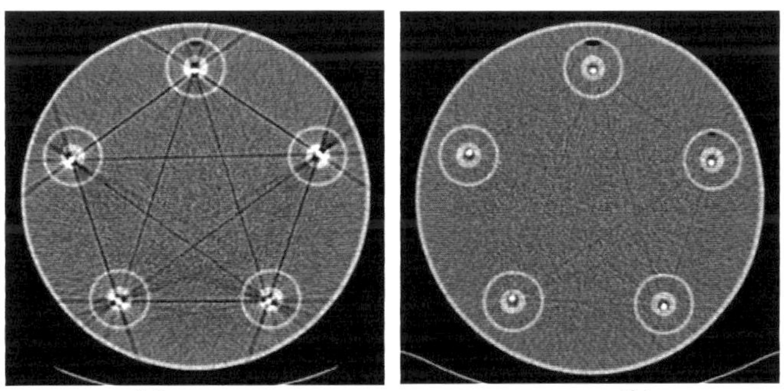

(a) (b)

**Figure 3.** CT images with metallic artefacts in the presence of (a) stainless steel and (b) titanium alloy.

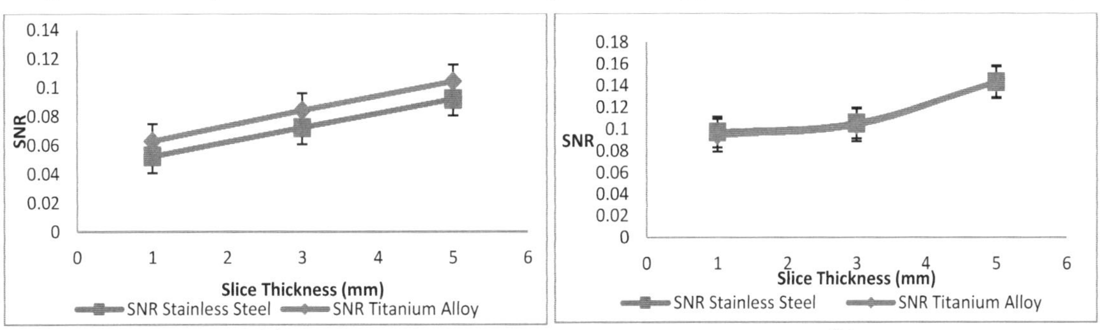


(a) (b)

**Figure 4.** The SNR against slice thickness for (a) non-contrast based phantom and (b) contrast based phantom.

Figure 4 (a) shows the SNR against slice thickness for titanium alloy and stainless steel in non-contrast based phantom. Titanium alloy shows higher SNR values compared to stainless steel (p-value = 0.002). The SNR is inversely proportional with the noise in the CT images [15]. The higher the SNR, the lower the noise appears in the images. Higher noise in CT images was obtained with the use of stainless steel. This is due to the higher of atomic number of stainless steel compared to titanium alloy. The atomic number of stainless steel ($Z$=26.7) is higher than the titanium alloy ($Z$=21.4). Thus, it causes higher scattered x-ray radiation.

Figure 4 (b) shows the SNR against slice thickness for titanium alloy and stainless steel in contrast based phantom. SNR of stainless steel and titanium alloy is not significantly difference with p-value of 0.204. This is because the x-ray attenuation of contrast media is higher compared to the orthopedic screw [11, 17]. Thus, the SNR value calculated for stainless steel and titanium alloy for contrast based phantom is not significantly difference.

However, the SNR values obtained with non-contrast based phantom appear higher when compared to SNR values of contrast based phantom. This shows that the CT images with contrast based phantom have less image noise compared to non-contrast based phantom. Furthermore, the use of contrast media helps to reduce metallic artefact, increase edge detection of soft tissue and enhances the image quality at the area of implants.

### *3.3. CT data acquisition*

Pitch and slice thickness that used will also affect the quality of the CT image obtained. Figure 5 shows that the SNR results obtained at different pitch. The SNR values increase with the increment of slice thickness. The SNR decreases when the pitch increases as shown in Figure 6.

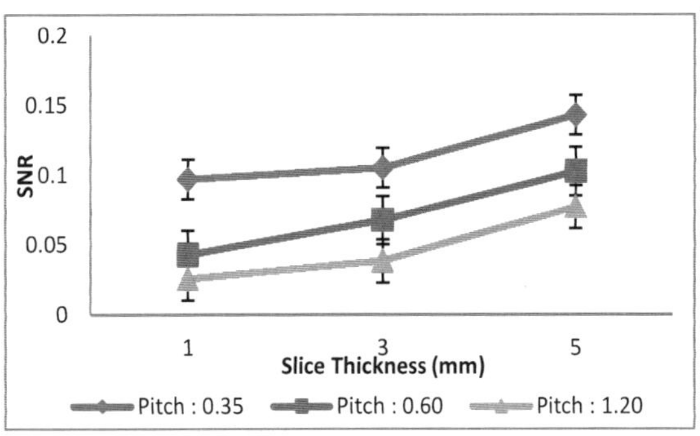


**Figure 5.** SNR against slice thickness.

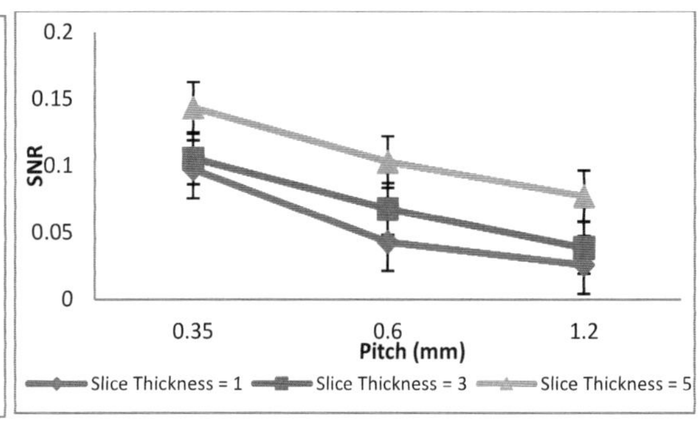


**Figure 6.** SNR against pitch.

Figure 5 shows the SNR value increases when the slice thickness increases. Thinner slice thickness reduces the size of the voxels. Smaller voxels absorbs or captures less number of x-ray photons. Thus, this will increase image noise; due to greater spatial resolution. Although thicker slice thickness have lower spatial resolution, they also have lower noise in CT images. However, if the slice thickness is too large, the CT images obtained will be affected by partial volume effect (PVE). The PVE occurs when different tissues or objects are represented by the same voxel. PVE is also known as partial volume averaging [19]. Thus, slice thickness of 5 mm shows optimum parameters for orthopedic implant with higher image quality with higher SNR value.

Figure 6 shows that the SNR value decreases when pitch increases. This is due to high noise in the images [18]. Pitch is defined as table distance traveled in one 360° gantry rotation divided by beam collimation. However, pitch affects both image quality and patient dose. Increasing the pitch value increases the speed of moving patient through the x-ray beam and reduces the time required to cover an anatomical area. Higher pitch of 1.20 mm results in decreased patient dose, but, it also decreased the image quality (lowest SNR value). This is because there are gaps in the x-ray beams where the tissue is not irradiated. However, lowest pitch (0.35 mm) resulted in better image quality, but it gave high patient dose due to x-ray beam overlap and it causes a volume of tissue is irradiated more than once per scan.

Thus, pitch of 0.60 mm shows an optimum parameters with higher image quality (higher SNR value), and give optimum patient dose. The higher pitch and lower slice thickness settings resulted in the degradation of contrast and diagnostic image quality of the CT images. The use of thinner slice thickness settings resulted in a noisy image due to the greater spatial resolution. Therefore, the SNR values increase as slice thickness increases, and decrease as pitch increases. Thus, the use contrast media was able to enhance the CT images and reduce artefacts from the metallic implants, where the contrast-based materials provided higher x-ray attenuation that caused the high atomic number of iodine ($Z$=53), barium ($Z$=56) and gadolinium ($Z$=64).

*3.4. Comparison for contrast media and non-contrast media*

Figure 7 shows a graph of SNR against slice thickness for non-contrast based phantom and contrast based phantom. The graphs shows that the higher SNR value was obtained with the use of contrast based phantom compared to non-contrast based phantom (p-value = 0.015). Murakami et al. (2010) stated that the SNR value is inversely proportional with the noise in the images [15]. Lower SNR value have higher noise appears in CT images. This explained that contrast based phantom has lower noise and streaky image than non-contrast based phantom in CT images. This is because the x-ray attenuation of contrast media is higher compared to the orthopedic screw [11, 17].

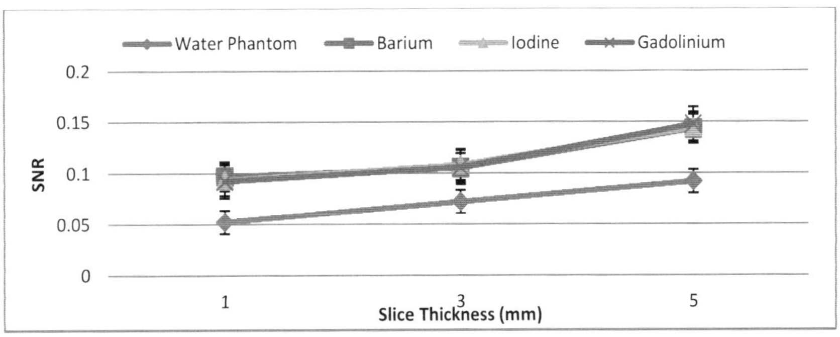


**Figure 7.** SNR against slice thickness in contrast based phantom and non-contrast based phantom.

## 4. Conclusion

In conclusion, our phantom study of water-based abdomen phantom in single-energy and dual-energy CT, we found that excessively low slice thickness and high pitch settings resulted in degradation in diagnostic image quality. Although an increase in pitch setting leads to decrease radiation dose towards patient, however these will an increase noise in CT images. Attenuation of contrast material may cause an increase in backround noise in addition to the beam hardening artefacts. Thus, our results suggest that it is desirable to use pitch of 0.60 mm, and slice thickness of 5.0 mm. The use of clinical contrast media is able to reduce metallic artefacts from implants (19 to 46%), and improved image quality in the area of metallic implants. Furthermore, the contrast media also improves the relative convergence in the area of metallic implants (edges and attenuation values) than the image without contrast media. The result of this study can be translated to clinical imaging, and contrast administration can be done by intravenous injection for iodine and gadolinium, while administrated orally for barium.

## 5. References

[1] Lee M J, Kim S, Lee S A, Song H T, Huh Y M, Kim D H, Han S H and Suh J 2007 Overcoming artifacts from metallic orthopedic implants at high-field-strength MR imaging and multi- detector CT *Radiographics* **27** 791–803

[2] Campi A, Ramzi N, Molyneux A J, Summers P E, Kerr R S, Sneade M, Yarnold J A, Rischmiller J and Byrne J V 2007 Retreatment of ruptured cerebral aneurysms in patients randomized by coiling or clipping in the International Subarachnoid Aneurysm Trial (ISAT) *Stroke* **38** 1538–44

[3] Schulze R, Heil U, Gross D, Bruellmann D D, Dranischnikow E, Schwanecke U and Schoemer E 2011 Artifacts in CBCT: a review *Dentomaxillofac. Radiol.* **40** 265–273

[4] Yu L, Li H, Mueller J, Kofler J M, Liu X, Primak A N, Fletcher J G, Guimaraes L S, Macedo T and McCollough C H 2009 Metal artifact reduction from reformatted projections for hip prostheses in multislice helical computed tomography:techniques and initial clinical results *Invest. Radiol.* **44** 691–696

[5] Zhang X, Wang J and Xing L 2011 Metal artifact reduction in x-ray computed tomography (CT) by constrained optimization *Med. Phys.* **38** 701–711

[6] Robert M, Michael G, and David H 2004 Artifact analysis and reconstruction improvement in helical cardiac cone beam CT *IEEE Tran. Med. Imaging.* **23( 9)** 1150–1164

[7] White L M and Buckwalter K A 2002 Technical considerations: CT and MR imaging in the postoperative orthopedic patient *Semin. Musculoskelet. Radiol.* **6** 5–17

[8] Watzke O and Kalender W A 2004 A pragmatic approach to metal artifact reduction in CT: Merging of metal artifact reduced images *Eur. Radiol.* **14** 849-856

[9] Fleischmann D and Boas F E 2011 Computed tomography—old ideas and new technology *Eur. Radiol.* **21** 510–517

[10] Willemink M J, Leiner T, De Jong P A, De Heer L M, Nievelstein R A J, Schilham A M R and Budde R P J 2013 Iterative reconstruction techniques for computed tomography part 2: Initial results in dose reduction and image quality *Eur. Radiol.* **23(6)** 1632–1642

[11] Lambert JW, Edic P M, FitzGerald P F, Torres A S and Yeh B M 2015 Complementary contrast media for metal artifact reduction in dual-energy computed tomography *J. Med. Imaging* **2(3)** 033503 1-8

[12] Pjontek R, Onenkoprulu B, Scholz B, Kyriakou Y, Schubert G A, Nikoubashman O, Othman A, Wiesmann M and Brockmann M A 2015 Metal artifact reduction for flat panel detector intravenous CT angiography in patients with intracranial metallic implants after endovascular and surgical treatment *J. Neurointerv. Surg.* **8** 824–829

[13] Saake M, Struffert T, Goelitz P, Ott S, Seifert F, Ganslandt O and Doerfler A 2012 Angiographic CT with intravenous contrast agent application for monitoring of intracranial flow diverting stents *Neuroradiology* **54** 727–735

[14] Andreucci M, Solomon R and Tasanarong A 2014 Side Effects of Radiographic Contrast Media: Pathogenesis, Risk Factors, and Prevention *BioMed Research International* **741018** 1-20

[15] Murakami Y, Kakeda S, Kamada K, Ohnari N, Nishimura J, Ogawa M, Otsubo K, Morishita Y and Korogi Y 2010 Patient Safety 64-Section Multidetector 3D CT Angiography: Evaluation with a Vascular Phantom with Superimposed Bone Skull Structures *Am. J. Neuroradiol.* **31(4)** 620-5

[16] Exhibit E, Zhang C and Nicolaou S 2016 Dual-energy computed tomography (DECT): Review of key principles and utility in common clinical settings *Electronic Presentation Online System* ECR2016/C-1624 1–17

[17] Courtney A Coursey, Rendon C Nelson, Daniel T Boll, Erik K Paulson, Lisa M Ho, Amy M Neville, Daniele Marin, Rajan T Gupta and Sebastian T Schindera 2010 Dual-Energy Multidetector CT:

IOP Conf. Series: Journal of Physics: Conf. Series **851** (2017) 012009 doi:10.1088/1742-6596/851/1/012009

How Does It Work, What Can It Tell Us, and When Can We Use It in Abdominopelvic Imaging? *Ragiographics* **30(4)** 1037-1052
[18] M Lewis, K Reid and A P Toms 2013 Reducing the effects of metal artifact using high keV monoenergetic reconstruction of dual energy CT (DECT) in hip replacements *Skeletal. Radiol.* **2(2)** 275-82
[19] J F Barrett and N Keat 2004 Artifacts in CT:recognition and avoidance *Radiographics* **24(6)** 1679–91

**Acknowledgment**
The authors would like to thank the financial assistance from Ministry of Higher Education Malaysia through Fundamental Research Grant Scheme (FRGS).

# Development of a thresholding algorithm for calcium classification at multiple CT energies

**Ng, LY. [1], Alssabbagh, M.[1], Tajuddin, A. A.[1], [2], Shuaib, I.L.[1], Zainon, R.[1]**

[1] Advanced Medical and Dental Institute, Universiti Sains Malaysia, 13200 Kepala Batas, Pulau Pinang, Malaysia
[2] School of Physics, Universiti Sains Malaysia, 11800 Pulau Pinang, Malaysia

E-mail: rafidahzainon@usm.my

**Abstract.** The objective of this study was to develop a thresholding method for calcium classification with different concentration using single-energy computed tomography. Five different concentrations of calcium chloride were filled in PMMA tubes and placed inside a water-filled PMMA phantom (diameter 10 cm). The phantom was scanned at 70, 80, 100, 120 and 140 kV using a SECT. CARE DOSE 4D was used and the slice thickness was set to 1 mm for all energies. ImageJ software inspired by the National Institute of Health (NIH) was used to measure the CT numbers for each calcium concentration from the CT images. The results were compared with a developed algorithm for verification. The percentage differences between the measured CT numbers obtained from the developed algorithm and the ImageJ show similar results. The multi-thresholding algorithm was found to be able to distinguish different concentrations of calcium chloride. However, it was unable to detect low concentrations of calcium chloride and iron (III) nitrate with CT numbers between 25 HU and 65 HU. The developed thresholding method used in this study may help to differentiate between calcium plaques and other types of plaques in blood vessels as it is proven to have a good ability to detect the high concentration of the calcium chloride. However, the algorithm needs to be improved to solve the limitations of detecting calcium chloride solution which has a similar CT number with iron (III) nitrate solution.

## 1. Introduction

According to the review done by George and Matthew, the ranking for atherosclerotic cardiovascular diseases in causing of death is the top which takes up about thirty percent of all deaths in the worlds [1]. There was about fifty percent of contracted acute coronary syndrome in the patients who did not show any symptoms. Thus, there is importance in identifying atherosclerosis in the early stage. Calcium is one of the compositions which appears in plaque calcification in the coronary vessels. The formation of atherosclerotic plaques and the total coronary plaque burden in coronary correlated closely with calcification (Hong, Bae, and Pilgram 2003)[2]. CT is a non-invasive method to diagnose and help in the prognosis of coronary artery calcification. The CT numbers can be varied for different types of plaques and the attenuation rate of calcium is high compared to other types of plaque. Thus, the efficacy and sensitivity for detection of calcification plaque in coronary are high [3-6]. A lot of studies were performed in order to detect the calcification and identify the composition of the plaque forming in the coronary. The Agatston score system which was introduced by Dr. Authur Agatston has collaborated in cardiac CT routine and has widely been mentioned in many studies in order to characterize the calcified plaque. The Agatston score is a semi-automated tool; it is used to compute the calcification score in coronary arteries and will characterize the calcified tissue if the threshold value is greater than 120 HU for the images obtained at the 120 kV in single energy [7]. The Agatston score only analyses the calcification level and is unable to distinguish between the calcification plaque and iron deposit plaque as the CT number of these two types of plaque was too close.

Computed tomography was invented by Sir Godfrey Newbold Hounsfield from London in the early 1970's [8]. The x-ray tube rotates 360 degrees along to the gantry rapidly to produce 2D images of the patients. Detectors were installed around the gantry to measure transmitted radiation, and then these readings were used to reconstruct the final images. In the 1980's, the slip ring technology was invented. It allows the x-ray tube to continuously rotate along the gantry and the spiral scan was developed in computed tomography at that time. Slip ring technology solved the problem of acquiring a single slice of CT image with scanners in the past. Single-energy computed tomography (SECT) scanners only use a single x-ray tube to perform the scanning at a fixed tube voltage.

Elementary composition of each material was different and the attenuation coefficient of the photons was affected by this. The absorption and scatter of the photons resulted in the reducing of the intensity of the x-ray when it penetrates through a matter. When the x-ray energy increase the photoelectric effect will decrease and the campton scattering effect will increase. These will affect the attenuation coefficient of the materials. Generally, different materials will have different attenuation values or CT numbers and it is represented on grayscale [9]. Thus, when an image is acquired from conventional CT, the images will show the useful structural information of objects or materials.

Nowadays, a lot of the diagnostic processes are performed using a computer which is also known as a computer-aided diagnosis (CAD). The use of this system to analyse the image automatically has significantly improved the data handling process and reduced human errors. The diagnosis results obtained from the system are fast and reliable when collaborated with high-performance computers. There were several problems faced when performing segmentation in CT images, for example, the uniformity of the intensity, artefact and the similarity attenuation value of the soft tissue. Thus, in medical image processing, there were no common algorithms especially when applying the automatic segmentation technique to evaluate the images [10].

In many medical image processing studies, thresholding, edge detection, K-mean and others are commonly used while performing segmentation [10-13]. The thresholding technique is the fastest and easiest method compared with other methods [10, 13]. The threshold level can be determined either with the global thresholding method or local thresholding method. For global thresholding methods, the suitable threshold value will be defined according to the observer's preferences. The threshold value can be defined using the trial and error method or through observation of the histogram modes of that particular image, to differentiate between that targeted object and the surroundings. Multiple thresholds were set for multiple window regions in the local thresholding method [13,14]. This technique is used to separate the object and divide them into different areas in the image according to the color, brightness and contrast [10]. The multi-thresholding method had been used in this study for the image processing section to distinguish the various concentrations of calcium chloride solution, iron (III) nitrate and sunflower oil.

## 2. Materials and Methods

A total of eight PMMA tubes were filled with the solutions in this study. All five tubes were filled up with calcium chloride anhydrous grade of analytical reagents (Grade AR) with concentration of 20.98 mg/ml, 70.89 mg/ml, 152.10 mg/ml, 204.93 mg/ml and 356.75 mg/ml, respectively. The sixth and seventh tubes were filled up with Iron (III) nitrate with a concentration of 51.71 mg/ml and 96.85 mg/ml respectively. The eighth tube was filled up with sunflower oil. All tubes were held by a holder and placed in a cylindrical-shaped PMMA container. The diameter and the height of the PMMA container are 10 cm by 10 cm (Figure1). Difference concentration of calcium chloride and iron (III) nitrate is to stimulate the difference level of calcification in the plaque. The PMMA container was filled up with water and various materials were scanned in single energy mode in this study.

The 128-detector single-source dual-energy CT (Somatom Definition; Siemens AG, Wittelsbacherplatz Muenchen, Germany) was used in this study. This scanner is capable of producing single-energy and dual-energy images. Five different tube voltages were set (70 kV, 80 kV, 100 kV, 120 kV and 140 kV) for single-energy scanning. In this study, the current tube was automatically set using the CARE DOSE 4D. The automatic current modulation software from Siemens reduces the dose and maintains the quality of the images concurrently. The slice width was set to 1 mm for all the energy levels.

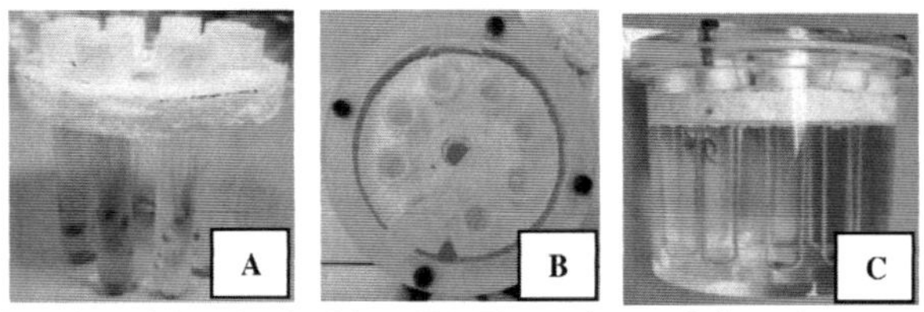


**Figure 1.** (A) PMMA tube placed in the holder (B) Top view of the phantom (C) Side View of the phantom after filled with water

The output DICOM images from SECT were analysed using a Java-based open source software which is ImageJ 1.5g, developed by the National Institute of Health (NIH) [15]. The area, mean (CT number) and standard deviation of the selected region of interest (ROI) of the each material were then measured by ImageJ. The results were exported and saved in an Excel file.

The DICOM format CT images acquired from the scanned images were also imported into the developed system. MATLAB® software v. R2014a was used to develop the new algorithm. The algorithm of the developed system was applied to the image in order to compute the ROI to obtain the value of the area, the mean CT number and the standard deviation for that selected ROI. The location and the size of the area of the ROI placed in develop system were the same with the ROI placed in the ImageJ. The DICOM images which imported to the developed system were in pixel value. Thus, a conversion from pixel value to CT number was needed. Equation 1 was used to convert the pixel value of the images to CT number for each selected ROI of each material with different solution concentration.

$$CT\ numbers\ (HU) = Pix \times Rslope + RIntercept \quad (1)$$

The CT numbers (HU) obtained from the developed software using the pixel value (Pix) multiply with the RescaleSlope (*Rslope*) and add with the RescaleIntercept (*RIntercept*). The value of the Rescale slope and the Rescale Intercept could differ if the images are obtained from different systems. The value of the Rescale Slope and Rescale Intercept can be found in the metadata in the DICOM file [16]. The results converted from pixel value to CT numbers of the ROI of each material in the images were stored in an array format. The information stored in the array form will be used to compute the mean value of the CT numbers and the standard deviation value for each ROI from each material.

DICOM images were imported to the developed software where initially the dimension was in pixel-based units and not in length-based units. A conversion was required to convert the pixel-based units to length-based units. Equation 2 was applied to convert the pixel-based units' value for the area of ROI to the length-based units which are in millimetres (mm). The size of the area of the ROI was used to make the selection at the same location and area on DICOM images in ImageJ software.

$$Area\ of\ ROI\ (mm^2) = (PixLength \times PS) \times (PixWidth \times PS) \quad (2)$$

From Equation 2, the *PixLength* and the *PixWidth* refer to the length and width of the rectangular shape, respectively. The pixel spacing (PS) refers to the vertical and the horizontal distance between two pixels which are adjacent to the unit of millimetre (mm). Basically, the distance between 2 pixels

IOP Conf. Series: Journal of Physics: Conf. Series **851** (2017) 012010 doi:10.1088/1742-6596/851/1/012010

was measured from centre to centre. The algorithm was evaluated using the same images which were analysed by ImageJ. The location and area of the ROI placed on each material in the images were same. The mean and the standard deviation was then computed and stored in an array format. The results from the ImageJ and the new system were compared and the percentage difference was calculated. The purpose of comparison between ImageJ and the developed software is to ensure the conversion value of the developed system was correct in the first place before applying these values to the segmentation section.

The segmentation process of the new algorithm is working based on a multi-thresholding technique. The minimum and maximum threshold value for each material were set based on the mean CT numbers and the standard deviations obtained from the new algorithm. This system was able to segment only one material at a time. The system will perform the segmentation process after the user inserts the value of the tube voltage and selects the type of material desired for segmentation, after the input of the DICOM image. The system converts the pixel value to a CT number for the DICOM image and compares the CT number of every pixel in the images with the threshold range predetermined in the system. The CT numbers within the range will be segmented and displayed in a binary image.

## 3. Results and discussion

The developed software was used to evaluate the CT numbers of the DICOM images and the results were compared with the ones acquired from the ImageJ. The comparison and verification processes were carried out to ensure that the conversion was performed correctly before proceeding to the segmentation stage. ImageJ software was chosen for the verification process. While obtaining the CT number and standard deviation with ImageJ, the ROI size and location were placed carefully in the same slice number. In this study, the ROI with an area of 0.41 $cm^2$ was used. The average of the CT number and standard deviation were computed for every material for single energy images, for all five energy levels. The results of the CT number for each material for all energy levels acquired from the developed software and ImageJ were illustrated in figure 2.

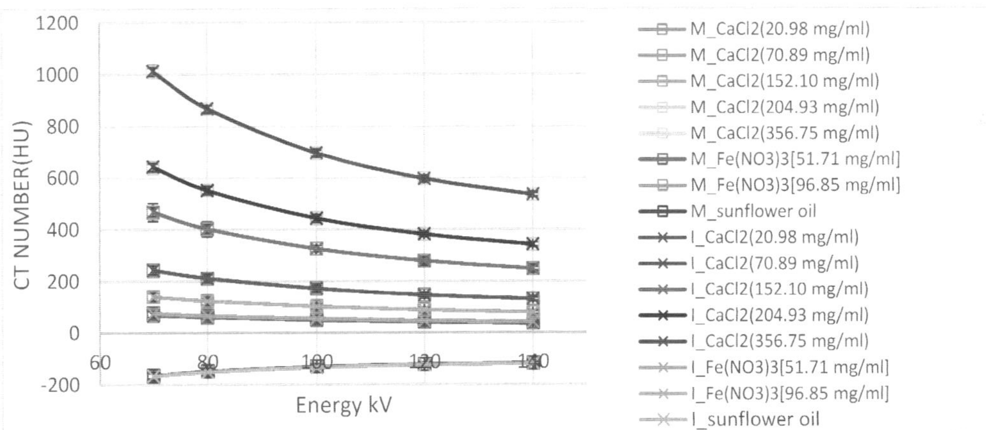


**Figure 2.** The obtained CT number from the new algorithm compared with the CT numbers obtained from ImageJ at different energy levels in SECT.

Figure 3 illustrates the process of segmentation for the developed software. After the user imported a DICOM image into the system, the system would require the user to manually input the specific energy level (70, 80, 100, 120 and 140) for the image which was imported and the material (from 1 to 8) which the user intends to segment. Then, the system would convert the DICOM image in pixel value to CT number and compare the value of the CT number across the entire pixel in the image. If the CT number of that particular pixel falls within the ranges set, the system will label and segment it. The system will only able to segment one material at a time.

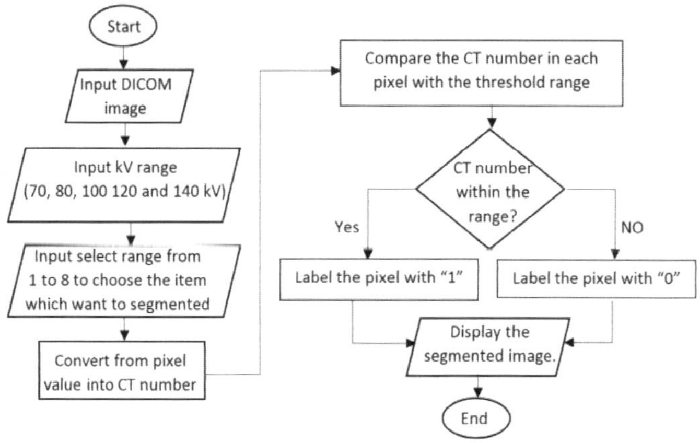


**Figure 3:** Process flow for the image segmentation algorithms.

The algorithm was tested with the CT images from all five energy levels for all materials. Figure 4 shows the set of results of the segmented CT images of various concentrations of calcium chloride in various energy levels. The algorithm in the system was unable to distinguish between calcium chloride (20.98 mg/ml) and iron (III) nitrate (51.71 mg/ml) in all energy levels (figure 4(G) and 4(H)). Although these two were a different material, the CT number of these two materials at these concentrations produced a very similar CT number. Thus, the system was unable to differentiate them. In contrast, the algorithm showed a promising result to distinguish the calcium chloride solution with high concentrations for all energy levels (figure 4(A) to 4(F)). The measured CT numbers for these high concentrations can be distinguished very easily by the thresholding segmentation technique.

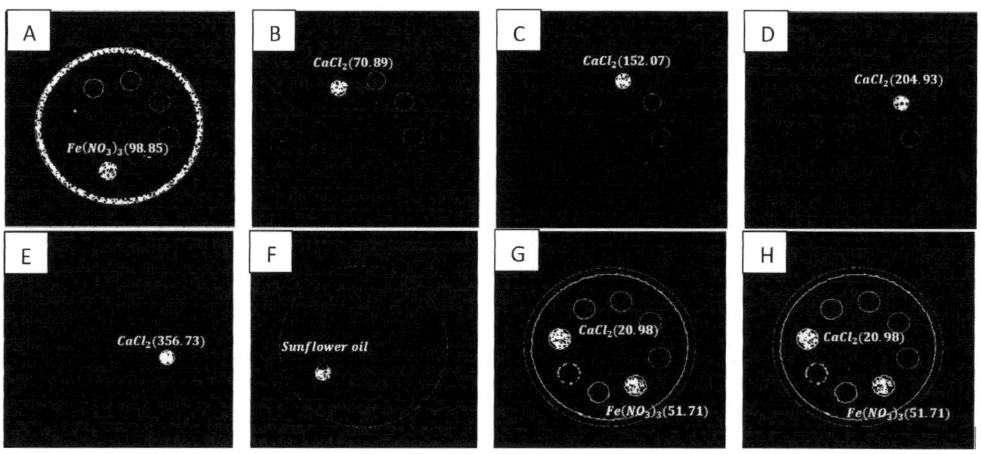


**Figure 4:** The output images after applying the thresholding algorithm on the single energy images.

## 4. Conclusion

The results of the applied multi-thresholding technique algorithm to distinguish different types of materials with various concentrations are promising across all energy in single energy scanned. This technique might be helpful in distinguishing different types of plaques which form in blood vessels. Identifying the type of plaque is very important in clinical studies. The clinical doctor will provide a proper and better treatment plan for a patient if the system is able to identify the type of plaque.

Although there are two materials consisting of different elementary compositions, the CT number obtained will sometimes be very close or similar to each other such as it happened it in this study. In low concentration of both calcium chloride (20.977 mg/ml) and iron (III) nitrate (51.706 mg/ml), their

CT number were very similar to each other. In clinical studies, there is a limitation of differentiating the calcium deposit plaque with iron deposit plaque as the CT numbers are too close such as in this study. Therefore, other techniques would be required to collaborate with this technique in order to differentiate between the materials which consist of very close CT numbers. We can further improve the system by including the material decomposition technique into our algorithm. To perform the material decomposition technique, dual energy images would be needed as these images are able to provide extra information than single energy images. This extra information could be used to further distinguish between those materials.

## 5. Reference

[1] Youssef, George, and Matthew J Budoff. 2012. "Coronary Artery Calcium Scoring, What Is Answered and What Questions Remain." *Cardiovascular diagnosis and therapy* **2**(2): 94–105.
[2] Hong, Cheng, Kyongtae T Bae, and Thomas K Pilgram. 2003. "Coronary Artery Calcium: Accuracy and Reproducibility of Measurements with Multi-Detector Row CT--Assessment of Effects of Different Thresholds and Quantification Methods." *Radiology* **227**(3): 795–801.
[3] Henzler, Thomas et al. 2011. "Attenuation-Based Characterization of Coronary Atherosclerotic Plaque: Comparison of Dual Source and Dual Energy CT with Single-Source CT and Histopathology." *European Journal of Radiology* **80**(1): 54–59.
[4] Motoyama, Sadako et al. 2009. "Computed Tomographic Angiography Characteristics of Atherosclerotic Plaques Subsequently Resulting in Acute Coronary Syndrome." *Journal of the American College of Cardiology* **54**(1): 49–57.
[5] Zainon, R. et al. 2012. "Spectral CT of Carotid Atherosclerotic Plaque: Comparison with Histology." *European Radiology* 22(12): 2581–88.
[6] Kristanto, Wisnumurti et al. 2013. "A Meta Analysis and Hierarchical Classification of HU-Based Atherosclerotic Plaque Characterization Criteria." *PLoS ONE* **8**(9): 1–13.
[7] Yamak, D et al. 2013. "Coronary Calcium Quantification Using Contrast-Enhanced Dual-Energy Computed Tomography Scans." *Journal of Applied Clinical Medical Physics* **14**(3): 4014.
[8] G.N, Hounsfield. 1973. "Computerized Transverse Axial scanning (tomography): Part I Description of System." *British Journal of Radiology* **46**(552): 1016–22.
[9] Alvarez, Robert E, and Albert Macovski. 1976. "Energy-Selective Reconstructions in X-Ray Computerized Tomography." *Physics in medicine and biology* **21**(5): 733–44.
[10] Sharma, Neeraj, and Lalit Arggarwal. 2010. "Automated Medical Image Segmentation Techniques." *Journal of medical physics* **35**(2): 3–14.
[11] Demirkaya, O, M H Asyali, and P K Sahoo. 2008. *Image Processing with MATLAB: Applications in Medicine and Biology*. CRC Press.
[12] Kamdi, Shilpa, and R K Krishna. 2011. "Image Segmentation and Region Growing Algorithm." *International Journal of Computer Technology and Electronics Engineering* **2**(1): 103–7
[13] Rais, Naveed Bin, M Shehzad Hanif, and A Taj. 2004. "Adaptive Thresholding Technique for Document Image Analysis."
[14] Gonzales. 2014. "Digital Image Proccesing Using MATLAB." *Igarss* ***2014*** (1): 404–17.
[15] Rasband, Wayne. Image J. [internet] [cited 2016 May 10]. Available from: https://imagej.nih.gov/ij/index.html
[16] Kamath, Srijit et al. 2011. "An Image Quality Comparison Study between XVI and OBI CBCT Systems." *Journal of applied clinical medical physics / American College of Medical Physics* **12**(2): 3435.

## Acknowledgments

We would like to acknowledge funding from the Ministry of High Education Malaysia via its Fundamental Research Grant Scheme (FRGS) for supporting this work.

# Investigation of the characteristics of Automatic Exposure Control (AEC) of a Computed Tomography (CT) scanner by utilising cylindrical and anthropomorphic phantoms

**W E P Rulaidi, M S N Huri & K H Ng**

Department of Biomedical Imaging, Faculty of Medicine, University of Malaya, 50603 Kuala Lumpur, Malaysia.

E-mail: shahrun@ummc.edu.my

**Abstract**. One method to optimise the use of x-rays in CT and hence a reduction in patient dose is the application of automatic exposure control (AEC). This study measured the effective mAs, image noise and volume CT dose index ($CTDI_{vol}$) as the result of changing the AEC index on a Siemens Somatom Definition 64 slices dual source CT scanner. The scans were performed on four phantoms of different geometries, namely the 16 and 32 cm cylindrical CTDI phantoms and two anthropomorphic phantoms, RANDO (20 cm effective diameter) and ATOM (19.8 cm effective diameter). Results showed that the effective mAs increased with increasing tube potential (kVp) and Quality Reference mAs (QRM), therefore increasing $CTDI_{vol}$ while reducing image noise. Meanwhile, no changes of radiation dose and image noise were observed when the pitch was increased. However, for the largest phantom (32 cm effective diameter), a constant effective mAs was found between 120 and 140 kVp. The same trend was also found with increasing QRM from 300 mAs to 400 mAs suggesting a certain limitation of the AEC has been reached. In conclusion, this study showed that AEC is affected by kVp and QRM but not by pitch selection. Further work is required to quantify the characteristics of the AEC system in relation to the mentioned parameters for better optimisation.

## 1. Introduction

Computed Tomography (CT) has been used over 45 years and has undergone a lot of improvements. Consequently, its development boost the number of CT examinations, which resulted in the highest collective dose among medical imaging modalities [1-2]. Previously, in order to reduce radiation dose, a fixed tube current was utilised where the user manipulates the tube current based on patient size. However, the human body is not homogenous along the direction of scan. This resulted in different dose absorption in a patient. Therefore, CT manufacturers introduced an automatic exposure control (AEC) system to reduce the radiation dose accordingly to patient sizes and variation in thickness to produce a better image quality [2-5].

The modulation of tube current can be influenced by patient positioning [6-8], size, attenuation properties in different regions of the body, and technical factors such as tube potential (kVp), Quality reference mAs (QRM) and pitch [9-11]. QRM is the technical factor used by Siemens AEC system

(CARE Dose4D) to represent the required image quality. The aim of this study was to investigate the effect of AEC system on the radiation dose delivered and image noise based on various phantom geometries. In addition, the influences of kVp, QRM, and pitch to the modulated effective mAs in AEC system were also studied.

## 2. Methods and materials

Four phantoms including a 32 cm and 16 cm Computed Tomography Dose Index (CTDI) phantom, adult female ATOM (CIRS, Norfolk, Virginia, USA) and RANDO (model ART-300A, Radiology Support Devices, Inc., Long Beach, CA, USA) phantom were scanned by using Siemens AEC system (CARE Dose4D) of a 64 slices dual source CT scanner (Somatom Definition AS, Siemens, Erlangen, Germany). In this study, only the abdominal region was investigated. The size of the phantoms is represented by effective diameter given by the following equation [1]:

$$\text{effective diameter} = \sqrt{\text{Lateral}\times\text{Anteriorposterior}} \quad (1)$$

in which the size of 16 cm and 32 cm CTDI, ATOM and RANDO phantom were 16 cm, 32 cm, 19.8 cm and 20 cm respectively.

The parameter used during scanning was shown in Table 1. The entire phantom had been scanned by varying the kVp (80 kVp, 100 kVp, 120 kVp, 140 kVp), Quality Reference mAs (QRM) (50 mAs, 100 mAs, 210 mAs, 300 mAs, and 400 mAs) and pitch (0.85 and 1.5). For pitch study, only the RANDO phantom was used. Three central slices were used for the measurement. Radiation dose was represented by $CTDI_{vol}$ and image quality was represented by image noise. Image noise evaluation was made by using RadiAnt DICOM viewer (Medixant, Poznan, Poland) and Syngo Fastview (Siemens AG, Berlin and Munchen). Four identical 3cm$^2$ region of interests (ROIs) were drawn on the central slice image and the average of the standard deviations (SD) of the four ROIs was considered to represent the image noise (refer to Figure 1).

**Table 1.** Parameters used during CT scanning.

| Parameters | Adult Protocol | Parameters | Adult Protocol |
|---|---|---|---|
| Pitch factor | 0.85 | Recon slice | 0.75 mm |
| Direction of scan | Craniocaudal | Recon kernel | B20f smooth |
| Slice thickness | 0.75 mm | Recon window | abdomen |
| Scan time | 6.09 s | Recon increment | 0.7 mm |
| Rotation time | 0.5 s | | |

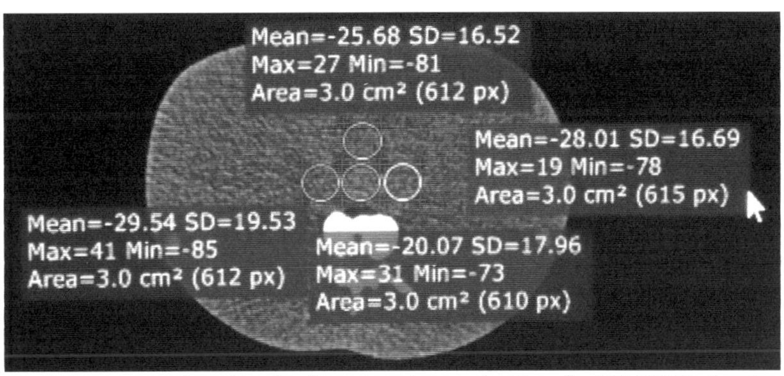


**Figure 1.** A screen-capture showing the image noise evaluation for RANDO phantom. The same technique was implemented for the other phantoms.

IOP Conf. Series: Journal of Physics: Conf. Series **851** (2017) 012011 doi:10.1088/1742-6596/851/1/012011

## 3. Results and Discussions

Similar trends were observed for measurements made on the 16 cm CTDI phantom, RANDO and ATOM phantom. For the simplicity of presentation, only the measurements of RANDO phantom was shown in Figure 2a, 3a and 4a. Figure 2b, 3b and 4b showed a different trend when using 32 cm CTDI phantom. In Figure 5, only 120 kVp was shown due to the similar trend observed for 80 kVp, 100 kVp and 140 kVp.

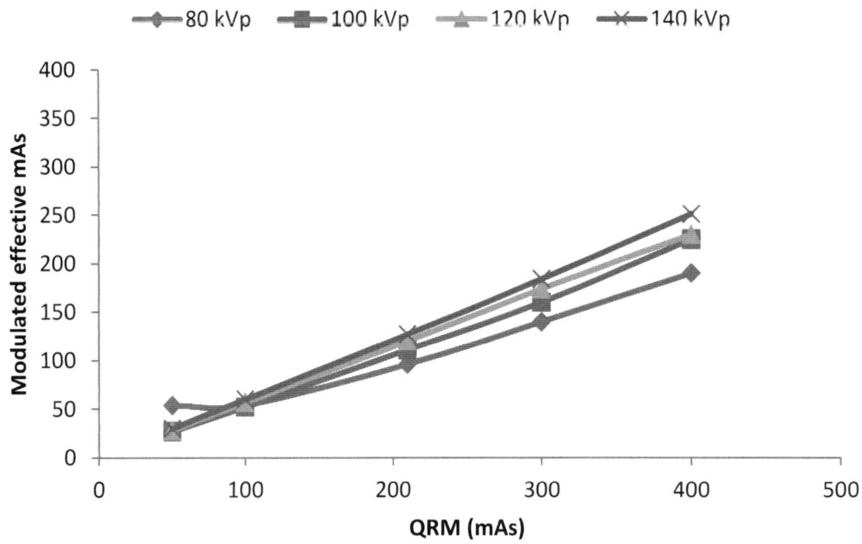


(a)

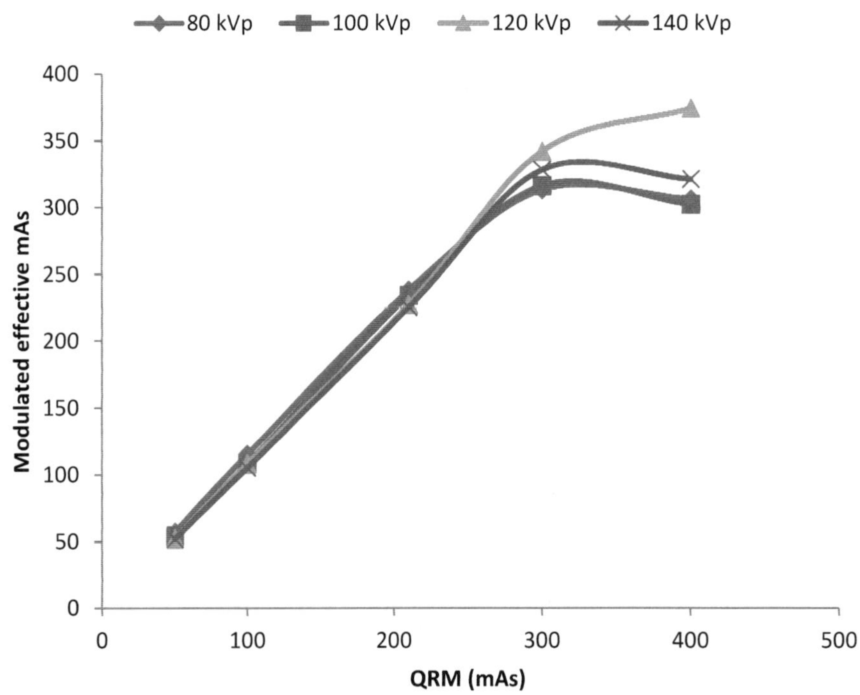


(b)

**Figure 2.** Modulation of effective mAs for varying QRM and kVp by using (a) RANDO and (b) 32 cm CTDI phantom.

IOP Conf. Series: Journal of Physics: Conf. Series **851** (2017) 012011 doi:10.1088/1742-6596/851/1/012011

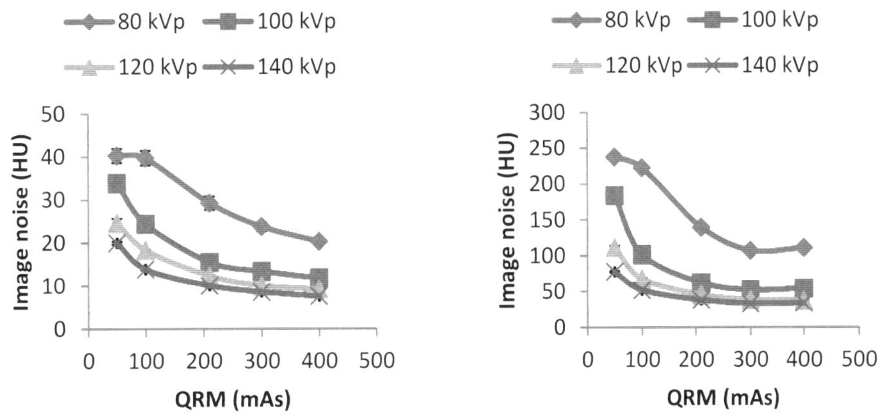


(a) (b)

**Figure 3.** Image noise for varying QRM and kVp by using (a) RANDO and (b) 32 cm CTDI phantom.

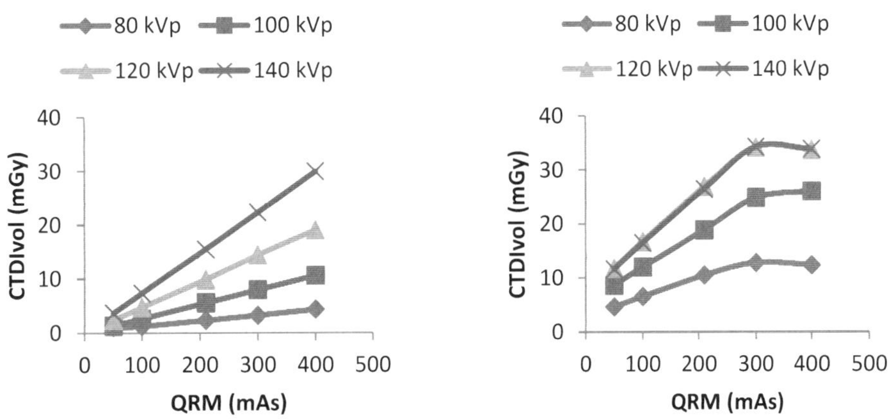


(a) (b)

**Figure 4.** Radiation dose for varying QRM and kVp by using (a) RANDO and (b) 32 cm CTDI phantom.

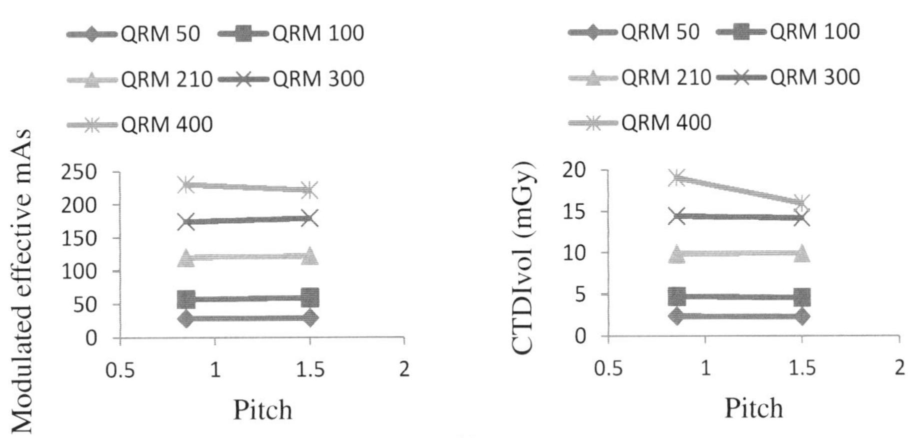


(a) (b)

**Figure 5**. Influence of pitch for (a) modulation of effective mAs (b) radiation dose for RANDO phantom with 120 kVp.

### *3.1. Influence of QRM on effective mAs, image noise and radiation dose*

On the influence of QRM to effective mAs, when the QRM was increased the effective mAs also increased. QRM is the technical factor used by the Siemens AEC system (CARE Dose4D) to represent the required image noise. In order to achieve the prescribed image noise, the AEC system modulates the effective mAs.

In general, increased QRM represents lower image noise, which is achieved by increasing the effective mAs and subsequently increasing the radiation dose. Figure 2, 3 and 4 shows the effect of modulating the QRM on the effective mAs, image noise and radiation dose respectively.

This trend was observed for the measurement using the RANDO phantoms (shown in Figure 2a, 3a and 4a). In contrast, a slightly different trend was observed for measurements made in the 32 cm CTDI phantom (shown in Figure 2b, 3b and 4b). Using the 32 cm phantom, increasing QRM increases the effective mAs (refer to Figure 2b). This resulted in increased radiation dose (Figure 4b) and decreased image noise (refer to Figure 3b). However, when increasing QRM 300 mAs to 400 mAs, the changes in the effective mAs was different for 80 kVp, 100 kVp and 140 kVp compared to the 120 kVp (refer to Figure 2b). This variation was due to the AEC system pursuing the prescribed image noise. In addition, it was also observed that increasing from QRM 300 mAs to 400 mAs for every kVp used, the radiation dose was constant (refer to Figure 4b) as well as for the image noise (refer to Figure 3b). This result showed that the AEC system was limiting the dose delivered when QRM 400 was used for 32 cm effective diameter phantom.

### *3.2. Influence of kVp on effective mAs, image noise and radiation dose*

On the influence of kVp to effective mAs, when the kVp was increased the effective mAs also increased. This trend was observed for the measurement using the 16 cm CTDI phantom, RANDO and ATOM phantoms (shown in Figure 2a). This resulted in an increase of radiation dose and a decreased image noise (shown in Figure 3a and 4a). In contrast, a slightly different trend was observed for measurements made in the 32 cm CTDI phantom (shown in Figure 2b, 3b and 4b). Using the 32 cm phantom, increasing kVp decreases almost constantly the effective mAs (refer to Figure 2b). This variation was due to the AEC system pursuing the prescribed image noise. However it resulted in increased radiation dose (Figure 4b) and decreased image noise (refer to Figure 3).

The previous result showed that increasing effective mAs resulted in an increase of radiation dose and a decrease in image noise. However, based on the influence of kVp to effective mAs for 32 cm diameter phantom, it showed that a decrease in effective mAs resulted in an increase of radiation dose and a decrease of image noise (refer to Figure 3b and 4b). In this case, it is due to higher kVp increasing the ability of photons to penetrate the phantom and therefore more photons were being collected. This reduced the image noise and from the exposure of high energy photons, radiation dose was increased. In addition, this result also showed that the relations of image noise as an inverse function of the square root of mAs was not obeyed for 32 cm phantom.

It was also observed that increasing from 120 kVp to 140 kVp for every QRM used, the radiation dose was constant (refer to Figure 4b). This result showed that the AEC system was limiting the dose delivered when 140 kVp was used for the 32 cm diameter phantom.

### *3.3. Influence of pitch on effective mAs and radiation dose*

On the influence of pitch to effective mAs and radiation dose, no significant change was observed. CARE Dose4D uses effective mAs setting in which the effective mAs will be constant although there is a change in pitch, consequently the mA will be increased as the pitch increased (i.e. for a 1 s rotation time and 240 effective mAs, mA is 240 mA with pitch of 1 but uses 300 mA with pitch of 1.25) and it is governed by the following equation [8]:

$$effective\ mAs = \frac{mAs}{pitch} \quad \text{.........} \ (2)$$

However, QRM 400 mAs showed that an increase in pitch resulted in a decrease in radiation dose (refer to Figure 5b). This result shows that the AEC system was limiting the dose delivered when QRM 400 mAs was used for the 32 cm diameter phantom.

## 4. Conclusions

This study showed that tube potential and selection of QRM affects the radiation dose and image noise for small and medium size patients (16 – 20 cm effective diameter), represented by 16 cm CTDI, RANDO and ATOM phantoms. However, for large size patients, such as represented by the 32 cm diameter CTDI phantom, a similar trend was observed for QRM settings ≤ 300 mAs. Beyond QRM 300 mAs, the modulation of the effective mAs appears to be restricted by the CT internal system, shown as a plateauing of the effective mAs and subsequently limiting the radiation dose delivered to patients. This effect is particularly prominent for higher tube potentials (≥120 kVp).

Pitch factor is often used as a scanning parameter that can reduce the radiation dose. However, our study showed that when the AEC system is switched on, changing the pitch from 0.85 to 1.5 (i.e. from overscanning to underscanning the patients), the effective mAs was kept constant. As a result, no radiation dose reduction would be expected.

Overall, this study showed the importance of understanding the characteristics of specific AEC system as it can influence the radiation dose delivered to the patient.

## 5. References

[1] Smith-Bindman R, Lipson J, Marcus R, Kim K P, Mahesh M, Gould R, de Gonzalez A B and Miglioretti D L 2009 Radiation dose associated with common computed tomography examination and the associated lifetime attributable risk of cancer *Arch Intern Med.* **169** 2078-89

[2] Kaasalainen T 2015 *Academic dissertation: Optimizing computer tomography examinations by using anthropomorphic phantoms and MOSFET dosimeters* University of Helsinki, Department of Physics, Helsinki

[3] Soderberg M and Gunnarsson M 2010 Automatic exposure control in computed tomography-an evaluation of systems from different manufacturers *Acta Radiol.* **51** 625-34

[4] Rizzo S, Kalra M, Schmidt B, Dalal T, Suess C, Flohr T, Blake M and Saini S 2006 Comparison of angular and combined automatic tube current modulation techniques with constant tube current CT of abdomen and pelvis *AJR Am J Roentgenol* **186** 673-9

[5] Papadakis A E, Perisinakis K and Damilakis J 2008 Automatic exposure control in pediatric and adult multidetector CT examinations: a phantom study on dose reduction and image quality *Medical Physics* **32** 4567-76

[6] Singh S, Kalra M K, Thrall J H and Mahesh M 2011 Automatic exposure control in CT: application and limitations *J Am Coll Radiol.* **8** 446-9

[7] Gudjonsdottir J, Svensson J R, Campling S, Brennan P C and Jonsdottir B 2009 Efficient use of automatic exposure control systems in computed tomography requires correct patient positioning *Acta Radiol.* **50** 1035-41

[8] Siemens AG 2002-2004 *SOMATOM sensation 64 application guide*

[9] Keat N 2005 *MHRA evaluation report 05016 CT scanner automatic exposure control systems*

[10] Muramatsu Y, Ikeda S, Osawa K, Sekine R, Niwa N, Terada M, Keat N and Miyazaki S 2007 Performance evaluation for CT-AEC systems *Nippon Hoshasen Gijutsu Gakkai zasshi* **63** 534-45

[11] Lee C H, Goo J M, Lee H J, Ye S J, Park C M, Chun E J and Im J G 2008 Radiation dose modulation techniques in the multidetector CT era: from basics to practice *RadioGraphics* **28** 1451-9

[12] AAPM Task Group 204 2001 Size-specific dose estimates (SSDE) in paediatric and adult body CT examination *American Association of Physicist in Medicine*

**Acknowledgement**

This project is funded by a PPP grant (PO041-2015B) from University of Malaya.

IOP Conf. Series: Journal of Physics: Conf. Series **851** (2017) 012012 doi:10.1088/1742-6596/851/1/012012

# Modelling the Siemens SOMATOM Sensation 64 Multi-Slice CT (MSCT) Scanner

**A T Mohd Amin and A A Abd Rahni**

Department of Electrical, Electronics & Systems Engineering, National University of Malaysia, Malaysia

E-mail: ashrani@ukm.edu.my

**Abstract**. Reconstructing large volumetric 3D images with minimal radiation dosage exposure with reduced scanning time has been one of the main objectives in the advancement of CT development. One of its advancement is the introduction of multi-slice arc detector geometry from a cone-beam source in third generation scanners. In solving this complex geometry, apart from the known vast computations in CT image reconstruction due to large CT images, iterative reconstruction methods are preferred compared to analytic methods due to its flexibility in image reconstruction. A scanner of interest that has this type of geometry is the *Siemens SOMATOM Sensation 64 Multi-Slice CT (MSCT) Scanner*, which has a total of 32 slices with 672 detector elements on each slice. In this paper, the scanner projection is modelled via the intersecting lengths between each ray (exhibited from the source to the detector elements) with the scanned image voxels, which are evaluated using the classical *Siddon's* algorithm to generate the system matrix, $H$. This is a prerequisite to perform various iterative reconstruction methods, which involves solving the inverse problem arising from the linear equation: $\mathbf{S} = H \cdot \mathbf{I}$; where $\mathbf{S}$ is the projections produced from the image, $\mathbf{I}$. Due to the 'cone-beam geometry' along the $z$-axis, the effective field-of-view (FOV) with voxel dimensions $(0.4 \times 0.4 \times 0.4)$ $mm^3$ is $512 \times 512 \times 32$ voxels. The scanner model is demonstrated by reconstructing an image from simulated projections using the analytic *Feldkamp-Davis-Kress* (FDK) method against basic iterative image reconstruction methods.

## 1. Introduction

Third generation CT scanners are currently widely being used clinically [1]. The manufacturers use a simple acquisition principle [2] with an x-ray focal spot source that rotates on a circular path around the isocentre of the gantry, and a cylindrical detector located on the opposite side with its centre axis positioned on the focal spot. This set up known as the 'rotate/rotate' geometry [3] allows the rotating gantry to be mounted with both the x-ray tube and detector revolving around the patient.

Obtaining volumetric 3D images has been one of the objectives in the advancement in CT imaging. From the utilisation of subsequent axial scans in 'step-and-shoot'/sequential mode, to spiral/helical imaging, the technology has evolved with the introduction of multi-slice detector modules that allows volumetric acquisition even during axial scanning [4]. This is useful, especially in dynamic volume imaging for volume perfusion studies [3].

Introduced in 2004, the Siemens SOMATOM Sensation 64 Multi-Slice CT (MSCT) Scanner is a scanner model that allows an acquisition of 64-slice simultaneously with an adaptive 32-row array detector design with a double z-sampling technique [3] using a periodic motion of the focal spot in the

IOP Conf. Series: Journal of Physics: Conf. Series **851** (2017) 012012 doi:10.1088/1742-6596/851/1/012012

z-direction [5], which is also known as the z-flying focal spot (z-ffs). Nevertheless, its potential can still be enhanced. This paper provides the development of modelling the scanner as a prerequisite to perform various iterative reconstruction methods. Preliminary image reconstruction results are also presented to validate the model, using standard reconstruction techniques.

## 2. Methodology

The methodology adopted to model the MSCT scanner is based on a previous study in modelling a PET/CT scanner [6], divided into four aspects. Firstly, the geometry of the scanner gantry is modelled to simulate the rays transmitted from the source to the multi-slice 2D detector arrays. Secondly, the geometry of the phantom image is modelled to establish the field-of-view (FOV) across the axial and sagittal planes to accommodate the cone-beam geometry of the scanner. Thirdly, the intersection lengths between the rays and each of the image voxels are traced to generate the system matrix that models the scanner. The last aspect of the methodology is performing image reconstruction based on the model to validate it. This part involves reconstruction using the analytic Feldkamp-Davis-Kress (FDK) [8] method to compare with the iterative Algebraic Reconstruction Technique (ART) [7] method. The model of the scanner, and image reconstruction were performed in MATLAB.

### *2.1. Scanner geometry*

The two main components of the Siemens SOMATOM Sensation 64 MSCT scanner are the source and detector. The distance between the source and detector is 1040 mm, and the distance between the source and isocentre is 570 mm.

The source uses the Siemens STRATON X-ray tube technology [5] that allows a user to choose between acquiring 32 or 64-slice during scanning. This is based on the availability of the z-flying focal spot (z-ffs) option that utilizes a periodic motion of the focal spot in the longitudinal z-direction. Nevertheless, the scope of this paper is limited to modelling 32-slice with the z-ffs option switched off [2] for simplicity of the model.

The detector [3] is made of scintillator ceramics known as Ultra-Fast Ceramic (UFC), and is arranged in 2D modules. Each module consists of 40 × 16 detector pixels along the longitudinal and azimuthal directions (i.e. z-axis and x-y plane). Along the longitudinal direction, the detectors are arranged in an adaptive array with 32 central rows and 4 outer rows on both sides to accommodate for different collimated slice widths. In this model, only the central rows with a 0.6 mm slice width measured at the isocentre is considered. Thus, the total length at the isocentre is 32 × 0.6 mm = 19.2 mm. Taking the ratio between the distances of the source to the isocentre and detector locations, the actual detector width is evaluated as 35.03 mm, with each slice width approximately 1.09 mm. This produces a cone angle, $\Gamma \approx 1.93°$ along the z-axis direction.

Across the x-y plane, 42 detector modules are arranged in an arc with its centre axis positioned at the location of the focal spot source. Each module is separated by an equal-angle, $\Delta\varphi \approx 1.24°$ at the source with the length of each module in the azimuthal direction is 22.51 mm, given that the fan angle, $\Phi \approx 52°$ [2]. Figure 1 depicts an illustration of the fan-beam and cone-beam geometries.

The basis of CT is the detection of attenuated x-rays transmitted from the source by each of the scintillator detector elements. The capability of a scanner depends on the number of x-rays that the detector can capture, thus is reflected by the number of detector elements available. For each longitudinal row, the number of detector elements is 42 × 16 = 672 elements across the azimuthal direction. Therefore, for this MSCT scanner, the total number of detector elements is 672 × 32 = 21,504 elements.

### *2.2. Image geometry*

A realistic 3D head phantom with dimensions: 512×512×512 voxels is used, as shown in figure 2. Assuming each voxel length is 0.4 mm, the total volumetric dimension lengths would be 204.8 mm each. This is acceptable since the MSCT scanner geometry [5] has an aperture length of 700 mm, and a range of reconstruction field-of-view (FOV) between 50 – 500 mm. With these dimensions, the head phantom would entirely be covered by the fan-beam geometry with a FOV of 204.8×204.8 mm$^2$ on the axial plane

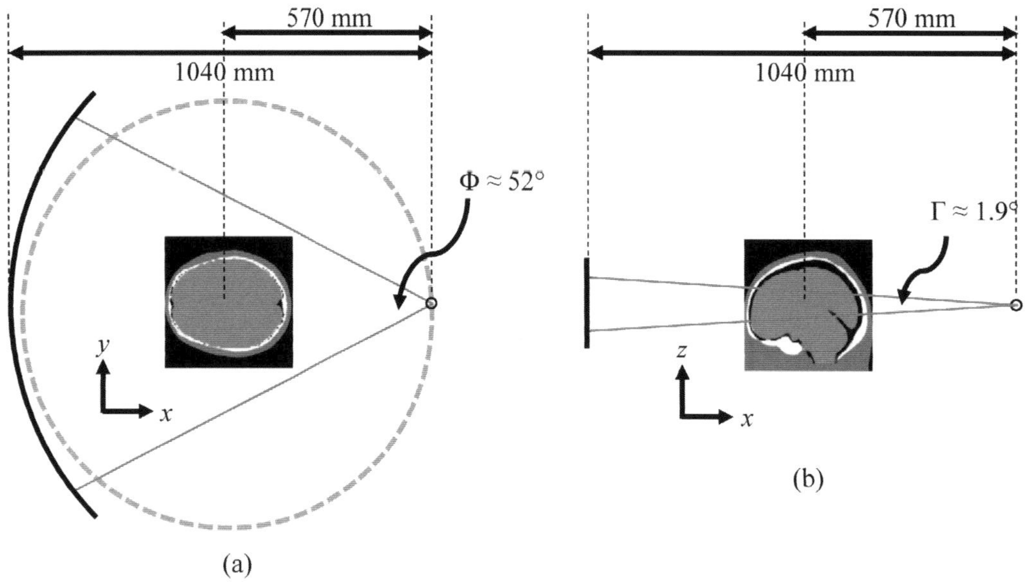


**Figure 1.** The source-detector configuration at view angle θ = 270° depicting the: (a) fan-beam, Φ ≈ 52°, and (b) cone-beam, Γ ≈ 1.9° geometries from axial and sagittal planes respectively.

Multi-slice detectors have proved to be advantageous due to its simplicity and robustness especially in reducing the radiation dose during scanning [9]. However, the cone beam angle from this geometry has its limitations such as image artefacts arising due to several factors, such as 'axial truncation' that limits the FOV, since the object is not entirely covered in the longitudinal direction. Discarding the slices that are not entirely covered by the cone angle, only 32 pixels along the z-axis are considered. This leads to a FOV of 204.8×12.30 mm$^2$ on the sagittal plane. Therefore, the head phantom's 3D FOV is 512×512×32 voxels that correspond to 204.8×204.8×12.30 mm$^3$.

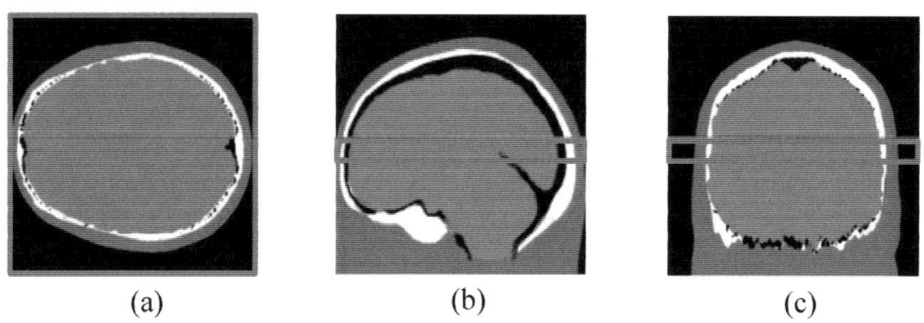


**Figure 2.** Realistic head phantom images used from different plane of views: (a) axial, (b) sagittal, and (c) coronal planes. The field-of-view (FOV) of the scanner is evaluated as 512×512×32 voxels as highlighted.

### *2.3. Ray tracing*

Transmitted x-rays that penetrate a scanned object are attenuated to a degree depending on the type of tissue as well as its thickness located within the path of each ray. In this model, the radiological path of a ray from the source through the image voxel arrays onto the detector elements, are evaluated by tracing the intersection lengths of each ray with the voxels using Siddon's algorithm [10]. Assigning

IOP Conf. Series: Journal of Physics: Conf. Series **851** (2017) 012012 doi:10.1088/1742-6596/851/1/012012

each intersection lengths parametrically with a designated coordinate based on the ray and voxel number will eventually generate the system matrix, $H$ that models the scanner for one view.

Each view will produce a projection that contains all of the generated data [11] by the transmitted x-ray from that source location. The collection of projections from all views can also be referred to as the sonogram, **S**. Minimizing the total number of projections, i.e. number of views used in a CT scan is important [12] as it will affect the radiation dose. The number of required projections is highly dependent on the scan protocols that varies for different organs [13]. A total of 360 scan views with a 1° increment around the head phantom in a circular trajectory are used in this model, with the system matrix for each view is generated and stored individually to enable different number of views to be simulated repeatedly.

The total size of the system matrix depends on the total number of rays × total number of image voxels. For 360 views, the total number of rays, $P = 360 \times 21{,}504$ rays = 7,741,440 rays, and the total number of image voxels = 512×512×32 = 8,388,608 voxels. Therefore, the total size of the system matrix for this model, $H = [7{,}741{,}440 \times 8{,}388{,}608]$.

### 2.4. *Image reconstruction*

Two methods of reconstruction are considered here which are: analytic and iterative reconstruction methods. Each of the methods is briefly described below.

*2.4.1. Analytic reconstruction.* It is known that 3D cone-beam reconstruction is a challenging problem to be solved analytically. A popular method in solving this geometry is the Feldkamp-Davis-Kress (FDK) method [8]. This method re-writes the Radon transform used in the popular Filtered Back Projection (FBP) approach in solving 2D fan-beam reconstruction into a convolution and backprojection form that is extended to the 3D cone-beam case [14].

Nevertheless, analytic reconstruction methods are well-known to be customized for idealized models that assume the data samplings are of continuous in measurements when in reality, the projections are actually discretely sampled onto the detector module elements. In practical CT applications, the projections are generated from a finite set of measurements depending on the number of detected rays, which causes the reconstruction to be ill posed.

*2.4.2. Iterative reconstruction.* Unlike the analytic approach, iterative reconstruction methods treat the image reconstruction as a discrete problem from the start [11]. It focuses on solving the linear system equation representation of the problem: $\mathbf{S} = H \cdot \mathbf{I}$, with **S** being the vector representation of projections generated from the image vector representation, **I**, and $H$ is the system matrix of the scanner. The Algebraic Reconstruction Technique (ART) is used [14] to iteratively reconstruct the image (**I**) using a uniform image ($\mathbf{I}_0$) as an initial guess and recursively computing a sequence of estimates ($\mathbf{I}_k$) subsequently until a maximum number of iterations, $k$, is satisfied. Based on Kaczmarz's algorithm, the inductive step in ART for determining the next image estimate ($\mathbf{I}_k$) is shown in equation (1) below:

$$\mathbf{I}_{k+1} = \mathbf{I}_k + \frac{S_n - \mathbf{h}_n^{\mathrm{T}} \cdot \mathbf{I}_k}{\|\mathbf{h}_n\|^2} \cdot \mathbf{h}_n \tag{1}$$

where $\mathbf{h}_n$ refers to the $n$-th row of the system matrix, $H$ and $n = k \bmod P + 1$.

## 3. Results and discussion

The evaluation results of the MSCT scanner model are presented based on the results from the head phantom image reconstruction using the two methods described in the previous section. Each of the reconstructed images is quantitatively assessed using the normalized root mean squared difference (NRMSD) value. Although the reconstruction involves a 3D head phantom with dimensions 512×512×32 voxels as shown in figure 2, only the 2D central slice of the axial plane is presented here.

*3.1. Analytic reconstruction*

Figure 3(a) shows the synthetic projection of the central slice, generated using the system matrix, *H*, with 360 views as described in section 2.3. Figure 3(b) shows the reconstructed image using the analytic FDK method with the NRSMD = 19.1%. Compared to the original head phantom image shown in figure 2(a), it can be seen that there are numerous artefacts and distortions, due to the limited number of views.

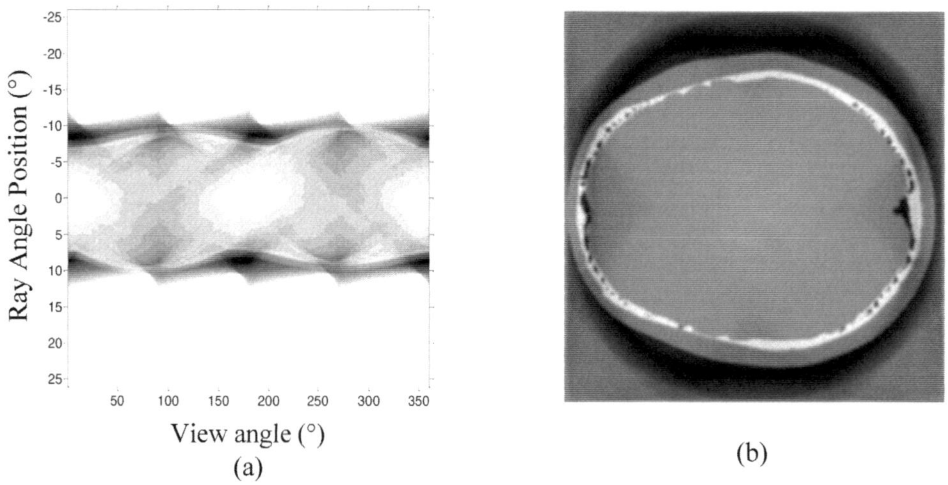


**Figure 3.** (a) A synthetic 360-view projection of the central slice, generated using the obtained system matrix, *H*, from the head phantom image used in figure 2, and (b) an axial slice of the reconstructed image using the analytic FDK method, with NRMSD = 19.1%.

*3.2. Iterative reconstruction*

Using the iterative ART method based on Kaczmarz's algorithm, a maximum number of 20 iterations were used to reconstruct the same projection shown in figure 3(a), along with the intermediary iterations at iteration 5, 10 and 15 as presented in figure 4(a). The NRMSD values for all of the reconstructed images using this method are recorded and compiled as shown in figure 4(b).

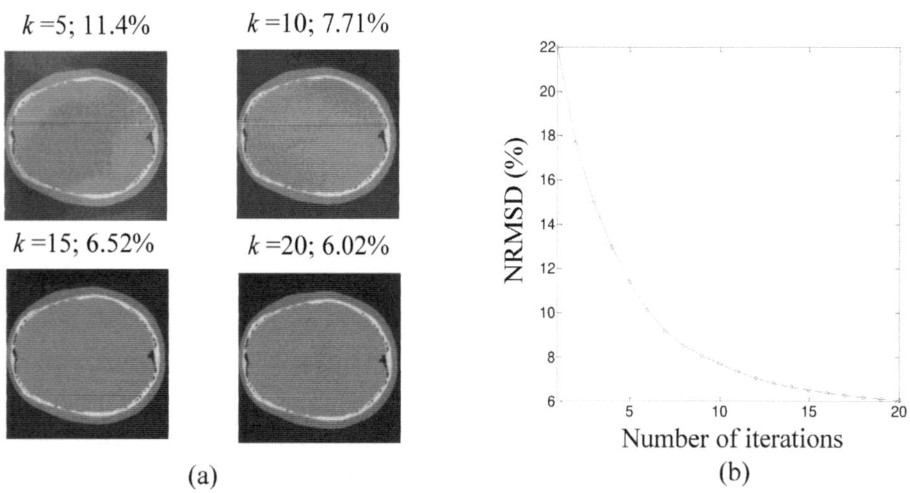


**Figure 4.** (a) Reconstructed images using the iterative ART method, at different values of iterations with their respective NRMSD values. (b) Decrement of the NRMSD value as the number of iterations is increased.

It can be seen that as the iterations progress, the percentage NRMSD between the reconstructed and actual images using the iterative ART method reduces greatly, with a higher accuracy compared to the analytic FDK method. This agrees with the fact that for a sparse number of views, iterative reconstruction methods are able to produce better quality images, indirectly minimising the need to high radiation dose during imaging.

## 4. Conclusion

The Siemens SOMATOM Sensation 64 Multi-Slice CT (MSCT) scanner was successfully modelled, with its system matrix for 360 views, $H$ generated.

This will encourage more medical imaging related researches to be carried out with clinical images available from the scanner. An improvement on the model would be to incorporate the FFS option into this model, allowing more image slice acquisition for a single view position. Apart from that is the incorporation of noise into this model.

## 5. References

[1] Kachelrieß M 2013 Interesting detector shapes for third generation CT scanners *Med. Phys* **40**
[2] Kachelrieß M, Knaup M, Penßel C and Kalender W A 2006 Flying focal spot (FFS) in cone-beam CT *IEEE Trans. Nucl. Sci.* **53** 1238–1247
[3] Ulzheimer S and Flohr T 2009 Multislice CT : Current Technology and Future Developments *Multislice CT* (Springer) Berlin Heidelberg 9–23
[4] Flohr T and Ohnesorge B 2007 Multi-slice CT technology *Multi-slice and Dual-source CT in Cardiac Imaging: Principles - Protocols - Indications - Outlook* (Springer) 41–69
[5] The ImPACT Group 2009 *Comparative specifications: 64 slice CT scanners* Centre for Evidence Based Purchasing
[6] Nayyeri F, Abd. Rahni A A and Ab Aziz A 2015 Modelling the GE Discovery 690 PET/CT scanner *2015 IEEE International Conference on Signal and Image Processing Applications (ICSIPA)* 160–164
[7] Gordon R, Bender R, and Herman G T 1970 Algebraic Reconstruction Techniques (ART) for three-dimensional electron microscopy and X-ray photography *J. Theor. Biol.*, **29** 471–481
[8] Feldkamp L A, Davis L C, and Kress J W 1984 Practical cone-beam algorithm *J. Opt. Soc. Am. A* **1** 612
[9] Hsieh J, Nett B, Yu Z, Sauer K, Thibault J and Bouman C 2013 Recent advances in CT image reconstruction *Curr. Radiol. Rep.* **1** 39–51
[10] Siddon R L 1985 Fast calculation of the exact radiological path for a three-dimensional CT array *Med. Phys.* **12** 252–255
[11] Feeman T G 2010 *The Mathematics of Medical Imaging* (Springer).
[12] Zheng Z and Mueller K 2011 Identifying sets of favorable projections for few-view low-dose cone-beam CT scanning *11th Int. Meet. Fully Three-Dimensional Image Reconstr. Radiol. Nucl. Med.*, **1** 2–5
[13] Bredenhöller C and Feuerlein U 2005 *SOMATOM Sensation 64 Application Guide*
[14] Fessler J A 2014 Fundamentals of CT Reconstruction in 2D and 3D *X-Ray and Ultrasound Imaging (Comprehensive Biomedical Physics vol 2)* ed A Brahme (Netherlands: Elsevier) 2014 pp 263 -95

IOP Conf. Series: Journal of Physics: Conf. Series **851** (2017) 012013 doi:10.1088/1742-6596/851/1/012013

# Pain when love is near

**S Tamam[1], A H Ahmad[2], M E Aziz[1], W A Kamil[1]**

[1]Department of Radiology, [2]Department of Physiology,
School of Medical Sciences, Universiti Sains Malaysia, 16150 Kubang Kerian, Kelantan, Malaysia

E-mail: sofina_tamam@yahoo.com

**Abstract:** The aim of the study is to investigate brain responses to acute laser pain when a loved one is nearby. Laser pain stimuli at individual pain threshold were delivered using Th:YAG laser to 17 female participants. The participants were categorised into two groups, *Love Hurts* or *Love Heals*, according to their responses to pain stimulation during the presence of their loved ones. fMRI brain activation was obtained using 3 T Philips Achieva MRI scanner utilising blocked design paradigm comprising 15 blocks of stimulation phase and 15 blocks of no stimulation. fMRI images were analysed using statistical parametric mapping (SPM) focusing on random effects (RFX) analysis. We found that both groups activated pain-related areas such as the thalamus, secondary somatosensory cortex, insula and cingulate cortex. However, *Love Hurts* showed more activity in thalamus, parahippocampal gyrus and hippocampus; while *Love Heals* showed more activity in the entire part of cingulate cortex during the presence of their loved ones. In conclusion, there may be specific brain regions responsible for modulation of pain due to the presence of a loved one thus manifesting as *Love Hurts* or *Love Heals*.

## 1. Introduction

Pain has diverse dimensions that renders its perception subjective. A similar type of pain may be interpreted and perceived differently by different individuals depending on many modulatory factors including cognition, psychological factors such as emotions and stress, and the threat value of the pain itself [1, 2, 3]. Previous neuroimaging studies have discovered several areas that are consistently activated during pain, the pain-related brain regions, such as secondary somatosensory cortex, insula, anterior cingulate cortex, thalamus, amygdala, ventrolateral prefrontal cortex and orbitofrontal cortex [4, 5, 6].

Pain can be psychologically influenced by the presence of partner [7], friends [8] or any significant other [9]. A common example is the different pain behaviour showed by children in the presence or absence of their parents during medical checkup or treatment. Presence of a parent seems to increase or decrease their pain behavior [10]. In adults, this modulation of pain by loved ones is seen in a study on participants who were shown pictures of their romantic partner while receiving pain [11]. The pain perceived were felt less strongly when compared to the pain felt when pictures of acquaintance were shown.

To our knowledge, no imaging study has been done to see the effect of the presence of a loved one nearby on pain threshold and brain activations. Using youths as the target group, our work aims to investigate neuronal activation due to pain when a loved one is nearby. The study utilised fMRI to capture brain activity and laser pulse as the acute pain stimulation.

## 2. Method

### 2.1. Participants

Seventeen (17) right handed female participants with age ranging between 18 and 25 years old (mean age 20.59; SD 2.85 years) participated in this study. They were all MRI compliant, healthy with no history of brain injuries, mental disorders or any serious illnesses, and gave informed consent. During the study, the participant needed to bring her loved one who could be a parent/family member, partner or best friend. This study was approved by the Human Research Ethics Committee of Universiti Sains Malaysia (USM) (Approval No.: USMKK/PPP/JEPeM [257.4.(3.16)]) allowing all volunteers, their loved ones and the researcher who was delivering the pain stimulation to stay in the scan room.

### 2.2. *Study design*

This is a within-subjects study involving two conditions known as Alone condition and Support condition. Alone condition describes the situation where the participant was alone while receiving pain stimulation, while the Support condition represents the situation where the participant was accompanied by a loved one. Based on the difference in pain threshold between the two conditions, the participants were divided into two groups named as Love Hurts, or feeling more pain when a loved one was nearby, and Love Heals, feeling less pain when a loved one was nearby.

### 2.3. *Pain Stimulation*

The determination of pain threshold for each participant was done outside the scanner. The Th:YAG laser was used to deliver the pain stimuli, generated from THEMIS laser device available in the Department of Neurosciences, Hospital Universiti Sains Malaysia (HUSM). The laser with a wavelength of 1.96 μm, working at a fixed distance of 12.7 cm, produced 5mm diameter spot on the skin. This narrow laser beam results in a pinprick pain sensation on the dorsum of the hand [12]. The dorsum of hand was selected as target area due to its easy access for pain delivery by a laser applicator which was connected to a controller device via a fiber optic cable. Several points were drawn on the dorsum of hand to mark the location of stimulation (Figure 1). Every time after the stimulation was given, the laser beam was slightly moved to a different point to avoid sensitisation and habituation effect.

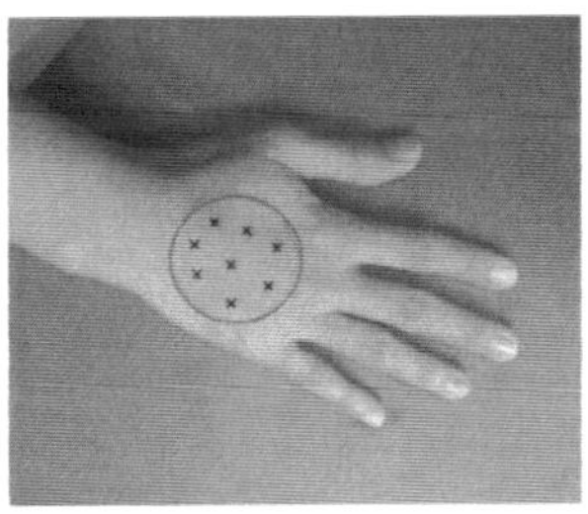

**Figure 1.** The area to receive laser heat stimulus drawn on the dorsum of the participant's right hand.

### 2.4. *Thresholding procedure*

Laser energies were delivered in an ascending and descending manner and the participants were instructed to respond verbally when they started to feel pain. The laser stimuli was started at the lowest energy level, 200 mJ and increased by 20 mJ increment until pain was reported. This process was repeated 15 times to achieve consistency. The descending series was started at the energy where pain threshold was reported. The energy was then decreased by 20 mJ until the participant reported that there was no pain felt. Similar to ascending, this process was performed 15 times before calculating the mean value of the pain. The value of laser energy when the participant started to feel pain was recorded as pain threshold.

### 2.5. *Study procedure*

Participant was placed comfortably in the MRI gantry. Prior to scanning, the participants, loved ones and researcher were ensured that they were eligible to be in MRI room. They were cautioned on MR hazards and the precautions to take. Before entering the room, any ferrous materials on their body or in pocket were removed as well as make up and jewelleries. All of them were given blanket, earplugs and MR compatible electrostatic headphone to protect their hearing. In the Alone condition, the participant was alone in MRI gantry, while in the Support condition, the participant was accompanied by the loved one, who stayed next to the MRI gantry. In Support condition, the participant was aware of the presence of the loved one. However, they were not allowed to talk or touch each other to prevent unnecessary brain activation.

The movement made from the participant's reaction may result in movement artefacts. Participants were instructed not to move during the scanning. Head motion was also restricted using head restraint. The participants had already been introduced with the painful stimuli during thresholding process, hence the stimulation during scanning was not totally unexpected and not likely to produce much motion.

### 2.6. *Functional MRI Scanning*

All participants underwent the standard procedure of fMRI scanning. An fMRI paradigm utilising a block design which consisted of 30 blocks with 15 blocks of stimulation and 15 blocks of no stimulation was arranged alternately (Figure 2). One block lasted 18 seconds in duration. In the stimulation phase, the random energy including the mean pain threshold of participant were given in pseudorandom order to prevent the participant from predicting what energy they will receive. In each block of stimulation, participant received two laser pulses also given at random time, so that the participant could not predict when they would receive the stimuli.

The brain images were obtained from 3 T scanner (Philips Achieva) equipped with a 32-channel SENSE head coil for pulse transmission and signal reception. The echo-planar imaging (EPI) utilised the following parameters: TR/TE/slice/flip angle/FOV = 2000 ms/35 ms/4 mm slices/90°/220×220 mm [13] for functional images. This produced 9 scans in each block of activity and rest, thus resulting in a total measurement of 270 scans throughout a whole run. T1-weighted scan was performed with the following parameters: TR/TE/slice/FOV = 9 ms/4 ms/4 mm slices/240×240 mm [13] for high resolution anatomical image. Total time for functional and structural scan was about 11 minutes for each participant.

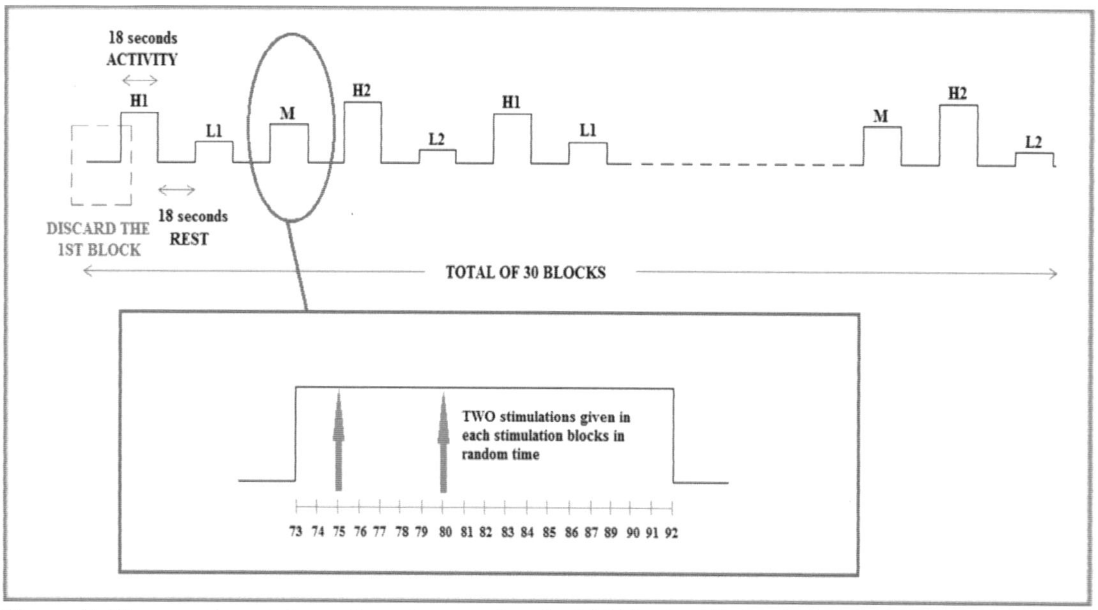


**Figure 2.** The experimental paradigm that consisted of 15 blocks of stimulation and 15 blocks of rest with 18s duration for each block. Two laser stimulations were given in each stimulation block at pseudorandom times. M is the mean pain threshold of participant; H1 and H2 are two random higher energy than the mean value; and; L1 and L2 are two random lower energy than the mean value. The first block is discarded to prevent the sequence error.

### *2.7. Data Analysis*

Functional data sets were analysed using SPM8 (Wellcome Department of Imaging Neurosciences, Institute of Neurology, University College of London, UK) and implemented in MATLAB 7.4 R2007a (Mathworks Inc., MA, USA). The pre-processing steps involved realignment, normalisation and smoothing to reduce artefacts from movement, improve the signal-to-noise ratio (SNR) and corrected into standard stereotaxic space of the Montreal Neurological Institute (MNI) template [14,15]. The smoothing used an 8-mm full-width-at-half-maximum (FWHM) Gaussian kernel [16]. Activated voxels were identified using the general linear model (GLM) with hemodynamic response function for both conditions (Alone and Support).

The analysis was focused on second-level random-effect analysis (RFX). Two-sample t-tests including all participants in each divided groups for each contrast of interest were calculated, providing a statistical parametric map of the T-statistic. The inferences were corrected (at $\alpha_{FWE}$ =0.05) and an extent threshold of $k > 10$ voxels were used to identify significant activity changes for pain across participants in each condition. The full factorial of 2 by 2 ANOVA was not suitable to use since the sample is small that it would produce no significant voxels at extent threshold of $k > 10$ voxels.

## 3. Results

### *3.1 Pain thresholding*

Pain thresholding revealed responses of resulting higher pain threshold and lower pain threshold in the Alone and Support con.dition. The variation in the responses then led to the division of the participants into two groups; participants who had lower pain threshold in the presence of the loved one (referred to as Love Hurts, n=10), and those with higher pain threshold (referred to as Love Hurts; n=7) in the same condition i.e. Support condition.

### *3.2. Effect of Interest*

The pattern of activation due to acute laser pain for each participant was unique and different from each other. Pain stimulation activated pain-related areas across all 17 participants as shown in Figure 3. In Alone condition, activations were in anterior cingulate cortex (ACC), hippocampus and ventrolateral prefrontal cortex (VLPFC), while in Support condition, the secondary somatosensory cortex (SII), insula and VLPFC were activated.

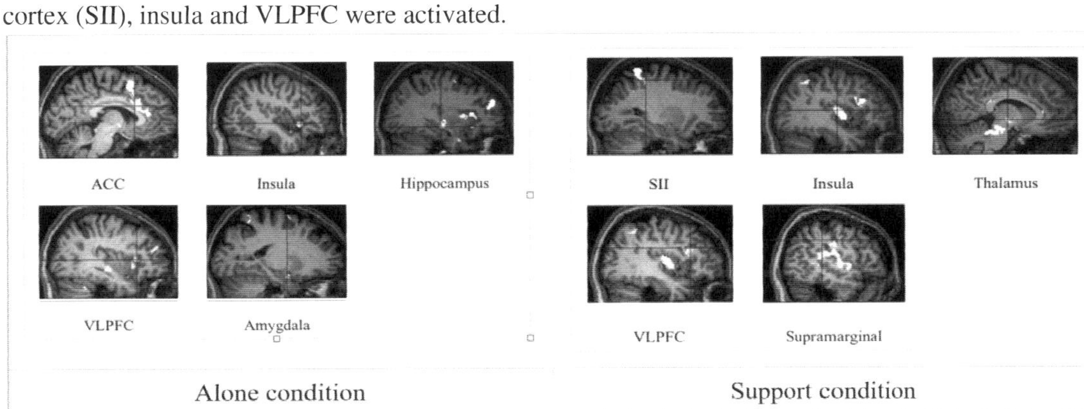


**Figure 3.** Statistical parametric map of random effect analysis (RFX) of 17 participants. In Alone condition, the anterior cingulate cortex (ACC), insula, hippocampus, ventrolateral prefrontal cortex (VLPFC) and amygdala were activated when participant was alone while receiving pain stimuli. While in Support condition, the secondary somatosensory (SII), insula, thalamus, VLPFC and supramarginal gyrus were activated in the presence of the loved one. Images obtained from FWE corrected p-value at significance level $\alpha$ =0.05, in sagittal orientation. The crosshair is the location of labelled area.

### *3.3. RFX Analyses*

To compare brain activations between Love Hurts and Love Heals groups, two-sample t-test was performed in second-level SPM analysis. Table 1 showed the comparison of brain activation between both groups.

In the group Love Hurts, feeling more pain in the presence of a loved one (Support condition) activated the left thalamus, left parahippocampal gyrus and left hippocampus, while the Love Heals group, having a loved one nearby caused less pain and activated mainly the left anterior, middle and posterior cingulate cortex. On the other hand, being alone while receiving pain activated the left insula in the Love Hurts group in accompaniment with less pain. In the Love Heals group, feeling more pain in the absence of a loved one nearby activated the left superior frontal (Table 1 and Figure 4).

**Table 1**. The brain activation contrast between groups during Alone and Support condition at corrected value of FWE, p=0.05.

| Condition | Contrast comparison | Coordinate | T-value | Activated Area |
|---|---|---|---|---|
| Alone | Love Hurts - Love Heals | -40 -12 10 | 5.13 | L - Insula |
| | | -24 -32 -8 | 4.97 | L - Hippocampus |
| | | -6 -46 18 | 3.92 | L - PCC |
| | | -10 -12 44 | 3.91 | L - MCC |
| | | 2 -2 46 | 3.78 | R - SMA |
| | Love Heals - Love Hurts | -16 -22 -18 | 4.25 | L _ Parahippocampal |
| | | -22 56 26 | 4.25 | L - Sup Frontal |
| | | -6 8 64 | 4.03 | L - SMA |
| Support | Love Hurts - Love Heals | -14 -24 10 | 4.82 | L - Thalamus |
| | | -32 -26 -20 | 4.72 | L - Parahippocampal |
| | | 16 -8 54 | 4.57 | R - SMA |
| | | -34 -22 -18 | 4.43 | L - Hippocampus |
| | Love Heals - Love Heals | 0 2 38 | 4.97 | L - ACC |
| | | 58 -24 52 | 4.42 | R - SII |
| | | 64 -18 44 | 4.24 | R - Supramarginal |
| | | -2 -32 30 | 4.09 | L - PCC |
| | | 0 -3 36 | 3.86 | L - MCC |

FWE = family wise error correction, L = left, R = right, SMA = supplementary motor area, ACC = anterior cingulate cortex, MCC = middle cingulate cortex, PCC = posterior cingulate cortex,

IOP Conf. Series: Journal of Physics: Conf. Series **851** (2017) 012013 doi:10.1088/1742-6596/851/1/012013

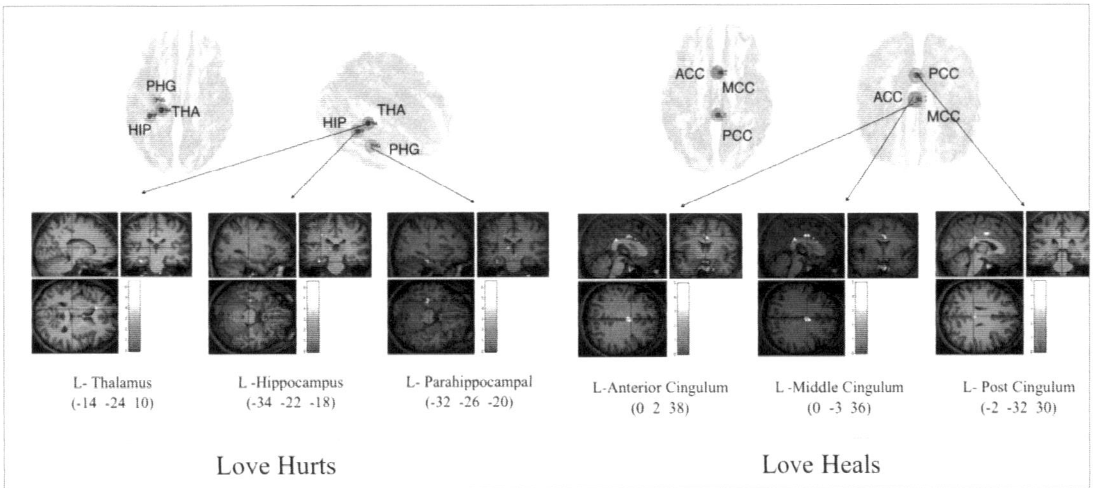


**Figure 4.** Brain activations of Love Hurts and Love Heals groups during laser pain stimulation during Support condition (having a loved one nearby). The areas and peak coordinates of thalamus, parahippocampal gyrus and hippocampus which appeared in Love Hurts represent the reaction 'feeling more pain' when a loved one was present, while the areas and coordinates of anterior, middle and posterior cingulate which appeared in Love Heals represent the reaction 'feeling less pain' when a loved one was present.

## 4. Discussion

In this study, pain stimulation activated pain-related brain regions consistent with previous pain imaging studies. The brain areas that were activated have significant function in pain processing. For instance, the thalamus acts as a relay station which is the gateway for the transmission of pain signals originating from the peripheries to the cerebral cortex[6]; while SII process the sensory-discriminative aspect of pain [17] where it gives information about the location and intensity of stimulation [6].

This study shows that feeling more pain in the presence of a loved one, the Love Hurts phenomenon, seems to be related with activations in thalamus, parahippocampal gyrus and hippocampus. On the other hand, reduction of pain when a loved one is nearby, or Love Heals, is accompanied by activations in all parts of cingulate cortex. The relevant functions of thalamus and parahippocampal gyrus in pain processing may contribute to the Love Hurts response while the hippocampus may be involved in 'tuning' the sensitivity of brain regions involved in pain processing in a context-dependent manner [10]. In Love Heals, the cingulate cortex play important role in processing cognitive factors such as attention, expectation, and anticipation, therefore is likely to be involved in processing the expectation towards the loved one [15]; as well as in processing positive emotions in relation to the loved one [18].

The activation produced by Love Hurts in Alone condition and Love Heals in Support condition represented an interaction of pain reduction; while the Love Hurts in Support condition and the Love Heals in alone condition represented the interaction of pain increment. Our results show that MCC and PCC may play a role in pain reduction consistent with their function in processing the affective-cognitive aspect of pain[5], as well as in emotional and attentional modulation of pain [4, 5]. The parahippocampal gyrus, activated in the interaction of pain increasing, has been shown to be related to expectation of high intensity stimulus [5].

## 5. Conclusion

We conclude that there may be specific brain regions responsible for modulation of pain due to the presence of a loved one thus manifesting as *Love Hurts* or *Love Heals.*

## 6. References

[1] Ahmad, A. H. & Abdul Aziz, C.B. (2014). The Brain in Pain. *Malaysian Journal of Medical Sciences.* Special Issue-Neuroscience, 21, 46-54.

[2] Ahmad, A. H. & Zakaria R. (2015). Pain in Times of Stress. *Malaysian Journal of Medical Sciences.* Special Issue-Neuroscience, 22, 50-59.

[3] Bingel, U. & Tracey, I. (2008). Imaging CNS Modulation of Pain in Humans. *Physiology.* 23, 371-380.

[4] Opisov, M. H., Dssor, G. D. and Porreca, F. (2010). Central modulation of pain. *Journal of Clinical Investigation.* 120 (11), 3379 - 3787.

[5] Wiech, K. and Tracey, I. (2013). Pain, decision and actions: A motivational perspective. *Frontiers in Neuroscience.* 46, 1 - 12.

[6] Vogt, B. A. (2015). Pain processing, cingulate cortex and the medial pain system. Retrieved December 2015, http://www.cingulumneurosciences.org/pain/shtml.

[7] Hurter, S., Paloyelis, Y., Williams, A. C. and Fotopoulu, A. (2014). Partners' empathy increases pain ratings: Effects of perceived empathy and attachment style on pain report and display. *Journal of Pain,* 15(9), 934 - 944

[8] McClelland, L. E. and McCubbin, J. A. (2008). Social influence and pain response in women and men. *J. Behav. Med,* 31, 413 - 420.

[9] Montoya, P., Larbig, W., Braun, C., Preissl, H. and Birbauner, N. (2004). Influence of social support and emotion context on pain processing and magnetic brain responses in Fibromyalgia. *Arthritis & Rheumatism,* 50(12), 4035 - 4044.

[10] McMurtry, C. M., McGrath, P. J., Chambers, C. T. (2006). Reassurance can hurt: parental behavior and painful medical procedures. *Journal of Pediatrics* 148(4), 560-561.

[11] Younger J, Aron A, Parke S, Chatterjee N, Mackey S (2010) Viewing Pictures of a Romantic Partner Reduces Experimental Pain: Involvement of Neural Reward Systems. *PLoS ONE,* 5(10), e13309.

[12] Agostino, R., Cruccu, G., Iannetti, G., Romaniello, A., Truini, A., & Manfredi, M. (2000). Topographical distribution of pinprick and warmth thresholds to CO 2 laser stimulation on the human skin. *Neurosicence Letter, 285,* 115–118.

[13] Noll-Hussong, M., Otti, A., Laeer, L., Wohlschlaeger, A., Zimmer, C., Lahmann, C., Guendel, H.(2010). Aftermath of sexual abuse history on adult patients suffering from chronic functional pain syndromes: an fMRI pilot study. *Journal of Psychosomatic Research, 68*(5), 483–7.

[14] Ashburner, J., Chen, C., Moran, R., Henson, R., Glauche, V., & Phillips, C. (2011). *SPM8 Manual The FIL Methods Group ( and honorary members )* (pp. 1–90).

[15] Cheng, Y., Chen, C., Lin, C. P., Chou, K. H. and Decety, J. (2010). Love hurts: An fMRI study. *Neuroimage.* 51, 923 - 929.

[16] Ahmad Nazlim, Y., Mazlyfarina. M., Aini, A. H., Kamil W. A., Mohd Harith, H. and Nurul Zafirah, Z. (2010). Functional specialisation and effective connectivity in cerebral motor cortices: An fMRI study on seven righthanded female subjects. *Malaysian Journal of Medicine and Health Sciences.* 6(2), 71 – 92.

[17] Craig, A. D. (2003). Pain mechanism: Labeled line versus convergence in central processing. *Ann. Rev. Neurosci.* 26, 1 - 30.

[18] Fredrikcson, B. L. and Cohn, M. A. (2008). Positive emotions. In Lewis, M., Haviland-Jonesm, J. M and Barrett, L. F. (2008). Handbook of emotion, 3rd Ed. *New York, Guilford Press.* 777 - 796.

## Acknowledgement

This study was supported by the grant of RUI (1001/PPSP/812130).

IOP Conf. Series: Journal of Physics: Conf. Series **851** (2017) 012014 doi:10.1088/1742-6596/851/1/012014

# Breast cancer cell targeted MR molecular imaging probe: Anti-MUC1 antibody-based magnetic nanoparticles

**P Moradi Khaniabadi[1*], A. M. S.A Majid[2], M Asif[2], B Moradi Khaniabadi[3], D Shahbazi-Gahrouei[4] and M. S Jaafar[1]**

[1]School of Physics, University Science Malaysia, 11800, Pulau Penang, Malaysia
[2]EMAN Research and Testing Laboratory, School of Pharmaceutical Sciences, Universiti Sains Malaysia, Penang 11800, Malaysia
[3]Child Growth and Development Research Centre, Research Institute for Primordial Prevention of Non-Communicable Disease, Isfahan University of Medical Sciences, Isfahan, Iran
[4]Department of Medical Physics, School of Medicine, Isfahan University of Medical Science, Isfahan, Iran

E-mail: med.physics64@yahoo.com

**Abstract**. Effective and specific diagnostic imaging techniques are important in early-stage breast cancer treatment. The objective of this study was to develop a specific breast cancer contrast agent for magnetic resonance imaging (MRI). In so doing, superparamagnetic iron oxide nanoparticles (SPIONs) were conjugated to C595 monoclonal antibody using EDC chemistry to produce nanoprobe with high relaxivity and narrow size (87.4±0.7 nm). To test the developed nanoprobe *in vitro*, assessments including Cell toxicity, targeting efficacy, cellular binding, and MR imaging were carried out. The results indicated that after 6 hrs incubation with MCF-7 cells at 200 to 25 μg Fe/ml doses, 76% to 16% $T_2$ reduction was obtained. The presence of iron localised in MCF-7 cells measured by atomic absorption spectroscopy (AAS) was about 9.95±0.09 ppm iron/cell at higher doses of nanoprobe. Moreover, a linear relationship between iron concentration of nontoxic SPION-C595 and $T_2$ relaxation times was observed. This study also revealed that developed nanoprobe might be used as a specific negative contrast agent for detecting breast cancer.

## 1. Introduction

The MUC1 mucin, a high molecular type transmembrane glycoprotein, is an effective inhibitor for cell-cell and cell-extracellular matrix interactions [1, 2]. Mucin is present at apical surface of epithelial cells and underglycosilated and over-expressed adenocarcinomas cells such as breast cancer [3], ovary cancer [4], prostate cancer [5], and lung cancer [6]. A common characteristic of mucin is a repeated domain which includes (a) an N-terminal extracellular domain, (b) a single membrane spanning domain, and (c) a C-terminal cytoplasmic domain with dense O-linked glycosacharid [7]. Moreover, mucin expression in normal epithelial can be dynamic varying in response to steroid hormone or cytokine influence [8]. Besides, expression of MUC1 on normal tissue is usually at a lower level and characterised by different receptor of immune epitopes. Following malignant transformation, mucin often becomes highly expressed, loses its apical restriction, and displays aberrant glycosylation and altered mRNA splice variation. In adenocarcinomas, the major change that occur is the loss of the ordered architecture of the gland, leading to no demarcation of apical and basolateral epithelial cell surface, and as a result mucin is expressed on all around the surface of the cells. As a result, the overexpression of mucin protein generally upregulates where it undergoes changes in glycosylation resulting in the exposure of the core protein tandem repeat region which is a potential target for antibody mediated therapy. Mucin modification in glycosylation and overexpression makes the tumour-associated mucin distinguishable from normal cells [9, 10]. Moreover, it is reported that mucin is expressed in 90% of breast cancers [11]. Generally, the overexpression of mucin and distribution on cell surface are assumed to influence the biological behaviour of the tumour cells during malignant transformation and tumour progression suggesting that protein may be important for maintenance or generation of the tumour. The receptor of immune epitopes to characterise the mucin on malignant cells are BC2(APDTR), HMFG1(PDTR(PA)), HMFG2 (DTR), SM3(PDTRP) and C595 (RPAP)[12-16]. Molecular imaging (MI) is one of the non-invasive imaging methods which occurs at cellular and even molecular level. MI also provides information about receptor's expressions and enzyme activities. The imaging modalities that guarantee MI are PET, SPECT, MRI, and ultrasound. The largest percentage of molecular imaging modalities depend on contrast agents [17]. The recommended agents used for molecular imaging purposes either have a specific goal to bind to the receptors over-expressed at a specific site or are cleaved by enzymes overexpressed at the location of the target.

Cancer, unregulated cell splitting, is one of the most important health issues of the present time and a leading cause of death among populations. Techniques to minimise the morbidity and mortality of this massive health issue include the recognition and modification of the risk factors, early diagnosis and treatment, and improved treatment strategies. EDC (1-ethyl-3- (3Dimethylaminopropyl) carbodiimide hydrochloride) allows for the formation of the direct component as the product of the cross-linkage reacts. An important point of interest in antibodies (Abs) is their antigen-binding region which is extremely specific and different among Abs. The EDC belongs to the groups of carbodiimides or zero-length crosslinking agents. Carbodiimides are widely used for to form amide linkage between carboxylate. In addition, EDC is utilised to interact with biomolecules, by forming an amine-reactive O-acylisourea intermediate. A possible improvement of the present biochemical reaction means that after the cross-linking and amination of the carboxyl coat, NHS esters could be linked to the amine groups on the NP which will react with the amine-reactive O-acylisourea intermediate of the antibody [18-20]. The aim of this study is to determine the efficacy of SPION-C595 as an MR imaging contrast agent targeted MUC1-overexpressing breast cancer under *in vitro* conditions.

## 2. Materials and methods

### *2.1. Chemical*

All chemicals were purchased by Sigma (USA). Nanomag-D-spio 20 nm nanoparticles (surface COOH) as well as miniMACS separator in addition to C595 mab were purchased from Miltenyi Biotech GmHb, Germany.

### *2.2. Cell*

MCF-7 breast cancer cells and EA.hy926 normal cells (as control), obtained from ATCC (USA), were routinely cultured in pre-warmed DMEM (37 ºC in water bath) supplemented with 10% of fetal bovine serum (FBS) antibiotics (100 IU/ml penicillin and 100 µg/ml streptomycin), 1% v/v essential amino acids, and 2 mM L-glutamine. The cells were grown in culture flasks with media (15 ml media/175 $cm^3$) and incubated at 37 ºC, in a humidified atmosphere of 95%/5% air/$CO_2$. Also, the cells were harvested after reaching approximately 80% confluence where they were presented as monolayer.

### *2.3 Conjugation of C595 monoclonal antibody with SPION*

Fabrication and the characterisation of the nanoprobe (SPIONs-C595) were performed as described in previous studies [21]. In brief, a 500 µl of the nanomag®-D-spio was mixed with 3 µmol N-ethyl-N-(3-

dimethyl aminopropyl) carbodiimide hydrochloride (EDC) and 10 μmol N- hydroxyl succinimide (NHS) in 125 μl 0.5M 2-(N-morpholino) ethanesulfonic acid (MES) buffer. The suspension was mixed for 90 min at room temperature by shaker. Then, the suspension was washed twice with l ml Phosphate Buffered Saline (PBS), pH=7.4 through MS column. After that, a 100 μl MUC1 (C595) mab was added to activate nanoparticles and the suspension was shook for 3 hrs. After 3 hrs the suspension was washed twice with 1 ml PBS through MS column. Then, the 30 μl glycine was added to quench the reaction of the final mixture. Finally, the suspension was passed through the high gradient field for last three times washing with PBS. The amount of immobilised monoclonal antibody (mAb) was calculated by Bradford method.

### *2.4. Cell viability test*

EDC method was utilised to fabricate the nanoparobe. Effect of SPION-C595 on viability of cells was evaluated using MTT assay. Cells from 70-80% confluent cultures were harvested by tripsinization and re-suspended in 5 ml fresh medium. The number of cells was determined and the cell count was adjusted to $5\times10^4$/ml. Subsequently, 100 μl of cell suspension was seeded in each well and cells were incubated overnight for attachment. The nanoprobe were diluted in cell culture medium to obtain 6 times the final concentration and 100 μl was added to each well. Moreover, the treated cells were incubated for 2 to 48 hrs. Then, a 20 μl of MTT solution was added to the cold cultures medium in each well, and 200 μl of culture medium containing the MTT reagent incubated for 3 to 5 hrs. At the end of the incubation, the supernant was aspirated carefully and the water insoluble formazan salt was solubilised in 100 μl DMSO per well. After 10 min incubation at 37°C, optical density of the violet was measured by a microplate reader at a primary wavelength of 570 nm and a reference wavelength of 620 nm [22].

### *2.5. Determination of efficacy of nanoprobe*

The specific and cellular uptake of C595 functionalised and non-functionalised NPs on MCF-7 cells were examined by Prussian blue staining. Also, cells were seeded at a density of $2\times10^5$ cells per well. The cells were incubated overnight at 37 °C in a humidified atmosphere of 95%/5% air/$CO_2$. When the flask was confluent, the monolayer MCF-7 cells were exposed to SPION-C595 and nanomag®-D-spio at 100, and 25 μg Fe/ml of final concentrations for 4 hrs. After the incubation time, the medium of the cells was removed and the cells were washed with PBS and fixed in 4% paraformaldehyde (PFA) for 20 minutes inside an incubator, followed by washing the cells three times. The fixed cells were incubated with 10% potassium ferrocyanide for 5 min and 10% potassium ferrocyanide in 10% hydrochloric acid for 30 min at room temperature (25 °C). The cells were counterstained with nuclear fast red for 5 min after two times washing with PBS. Finally, the fixed cells were washed three times with PBS and were sent for taking image by microscope.

### *2.6. MR in vitro imaging*

For imaging, the MCF-7 cells were grown to confluence in 6-well plates. Different concentration levels (25, 50, 100, 200 μg Fe/ml) of SPION-C595 were prepared in fresh media. Then, breast cancer cells were treated with prepared nanoprobes for 6 hrs incubation time at 37 °C, followed by 3 times washing with PBS, and detaching the cells by the 150 μl trypsin per well. To prepare two million cells for each MR imaging samples, after tyrosination cells were centrifuged and adjusted by counting cell method. The cells were suspended in 250 μl 1% agarose gel in PBS and were transferred into 1.5 ml centrifuge tubes for imaging. Moreover, untreated cells in 1% agarose were considered as negative control (T) while distilled water and (E) 1% agarose were considered as positive controls. The Eppendorf tube box holder was filled with 1% agarose gel to eliminate the background noise. All relaxation times were obtained on a using a 1.5T clinical MRI machine (GE Healthcare, Wisconsin, USA). $T_1$ and $T_2$ relaxation time measurements were modifications of the method of previous studies [24, 25]. Furthermore, to determine the slope of the regression line, the relaxivities were calculated by plotting $R_1$ and $R_2$ values versus different concentrations of SPION-C595.

## 3. Results and Discussion

### *3.1. Cell viability test*

As shown in Table 1, SPION-C595 did not exhibit any toxicity toward the cells as the nanoprobe was localised in the cytoplasm. Furthermore, prolonging the incubation of MCF-7 cell with SPION-C595 did not affect either cell morphology or proliferation ability. The results also revealed that there was no toxicity upon conjugation, even at the highest concentration of components (Figure 1). On the other hand, there were statistically significant differences between the concentrations among those within the same incubation time ($P < 0.05$ for all) [23, 24].

**Table 1.** Mean ±SD values of viability percentage of SPION-C595 with MCF-7 after incubation times, among six different concentrations.

| Concentration (µg Fe/ml) | Incubation time | | | | Total | P-value |
|---|---|---|---|---|---|---|
| | 2 [a] h | 8 hrs | 24 hrs | 48 hrs | | |
| 1.56 [b] | 106. 85 (6.07) [c] | 82. 41 (2.90) | 154.42 (7.89) | 83.24 (1.15) | 196.73 (30.85) | < 0.001 |
| 3.12 | 99.59 (2.54) | 90.20 (3.28) | 139.57 (17.35) | 77.07 (1.55) | 101.61 (25.54) | < 0.001 |
| 6.25 | 99.02 (11.81) | 97.75 (0.96) | 132.89 (12.45) | 74.70 (0.99) | 101.09 (22.89) | < 0.001 |
| 12.5 | 109.14 (3.62) | 106.00 (2.77) | 141.07 (11.04) | 77.75 (4.37) | 108.49 (24.05) | < 0.001 |
| 25 | 107.04 (10.39) | 109.05 (5.40) | 159.59 (8.71) | 84.84 (3.75) | 115.13 (29.30) | < 0.001 |
| 50 | 111.71 (5.30) | 122.43 (1.00) | 167.82 (6.06) | 68.24 (0.31) | 117.55 (37.15) | < 0.001 |
| Total | 105.56 (7.90) | 101.31(13.68) | 149.22 (15.77) | 77.64 (6.03) | 108.43 (28.38) | |
| P-value | NS | < 0.001 | 0.017 | < 0.001 | < 0.001 | |

[a] *Incubation times: after 2 hrs, 8 hrs, 24 hrs, and 48 hrs.*
[b] *Concentrations: 1.56, 3.12, 6.25, 12.5, 25, and 50 (µg Fe/ml).*
[c]*Arithmetic means [24].*
*NS: Not significant at 5% level.*

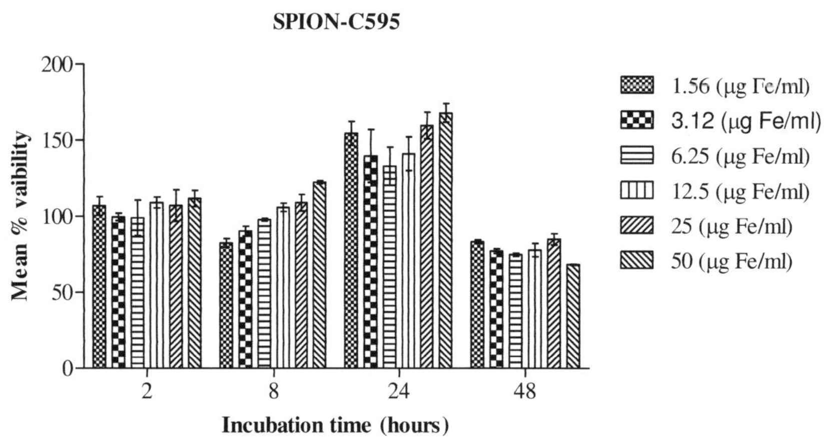


**Figure 1.** Cell Viability of MCF-7 cells after different incubation times (2-48 hrs) at different concentrations of SPION-C595 (6.25-100 (µg Fe/ml)).

### *3.2. Determination of efficacy of nanoprobe*

The z-average hydrodynamic diameter of SPION-C595 was obtained 87.4±0.7 nm with a polydispersity index of 0.3 after conjugation reaction, which was still an ideal size for application of transfection. The nanoparticle size before conjugation was measured at 51.3± 0.07 nm. The 30 nm increasing in size of nanoparticle revealed that the conjugation occurred [26, 27].

The Prussian blue technique was based on the conversion of ferrocyanide to the insoluble crystal of Prussian blue in the presence of $Fe^{3+}$ under acidic conditions which resulted in the formation of a blue pigment called Prussian blue. The cell was first treated with dilute hydrochloric acid to release ferric ions from the binding proteins. Figure 2 shows the interaction and attachment of nanoprobe to the cancer cells. Attachment to the cytoplasm of the MCF-7 cells due to over expression of MUC1 on MCF-7 cells is shown in blue colour.

The cellular uptake of the SPION-C595 nanoprobe was verified on cancerous (MCF-7) and normal (EA.hy926) cells. Additionally, AAS results showed a 9.95±0.09 ppm iron/cell cellular uptake by MCF-7 cells in comparison to normal cells at higher doses of nanoprobe (Table 2). The quantity of significant ($P<0.001$) SPION-C595 uptake by MCF-7 confirmed the selectivity of the nanoprobe and its capability for breast cancer detection. Moreover, the results strongly suggested there was no significant cellular uptake by the normal cells in comparison to the breast cancer cells after the administration of a higher SPION-C595 dose.

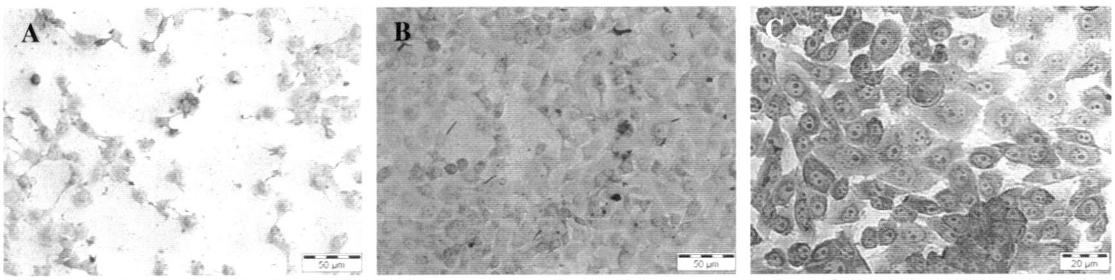


**Figure 2.** Presence of Fe in MCF-7 cells treated with varying concentration of SPION-C595 nanoprobe. **A** is cells treated with 100 µg Fe/ml of the nanoprobe at 10X magnification. **B** is cells treated with 25 µg Fe/ml of the nanoprobe at 20X magnification. **C** is the negative control untreated MCF-7 cells at 10 X magnification. *Note: Fe ions are stained blue, the nucleus of the cells is reddish and the cytoplasm in pink.

**Table 2.** Summary of iron content (Mean ±SD) of EA.hy926 and MCF-7 at different concentration of SPION-C595 after 6 hrs incubation. [a.] Mean value±SD [b.] According to two-independent samples nonparametric Mann-Whithney U test.[*.] Significant at level of 0.05.

| Cell line | Concentrations (µg Fe/ml) of iron content of SPION-C595 (ppm iron/cell) | | | | |
|---|---|---|---|---|---|
| | Control | 25 | 50 | 100 | 200 |
| EA.hy926 | 0.428±0.01[a] | 0.456±0.03 | 0.505±0.01 | 0.633±0.01 | 0.835±0.02 |
| MCF-7 | 0.030±0.01 | 0.372±0.01 | 0.479±0.01 | 0.954±0.11 | 9.950±0.09 |
| P-value [b] | **< 0.001***  | **< 0.001*** | **< 0.001*** | **< 0.001*** | **< 0.001*** |

### 3.3. *MR in vitro imaging*

Magnetic resonance contrast agents are effective in enhancing MR images by decreasing the $T_1$ and $T_2$ relaxation times through spin-interaction between electron spins of the metal-coating of contrast and the water protons (hydrogen) in tissue. All contrast agents produce changes in magnetic susceptibility by enhancing the local magnetic field. Efficacy of enhancing $T_1$ and $T_2$ weighted images depends on: (a) the time of interaction between the nuclear and paramagnetic substance, (b) the paramagnetic concentration which is directly proportion to the increasing the relaxation time at the lower concentration doses, and (c) the distance between proton nucleus and electronic field. Paramagnetic materials have unpaired electron in the outer orbital shell. By applying magnetic field, the unpaired electrons give rise to magnetic dipoles, and therefore, paramagnetic ions induce large fluctuating magnetic field experienced by nearby protons. If the frequency of fluctuation has a component close to the Larmor frequency, the significant enhancement of relaxation times will happen. Two mechanism of relaxations are inner-sphere and outer-sphere relaxation. The inner-sphere relaxation occurs between the ions and the water molecules of the tissue, unstable covalent bonds between ions and the water molecules cause chemical interaction of the water molecules and other water molecules in the surrounding environment, and thus enhances the relaxation time. The reduction of $T_1$ will be more obvious depending on the number of bonding. In other word, paramagnetic substance cause greaten transition between spins states. However, the outer-sphere relaxation occurs when the paramagnetic substances greaten the local field which the water molecules are experienced. There is no chemical interaction in this mechanism. The diffusion of the molecules is causing the influence of protons of tissue into the complex. The large heterogeneity of magnetic field around the nanoparticle through which water molecule diffuse, since diffusion induced dephasing of the proton magnetic moment resulting in $T_2$ reduction [28, 29]. In regards to this study, the results demonstrated that only at the 200 µg Fe/ml the signal intensity of the $T_1$ spin-lattice relaxation time decreased by 33% (Figure 3. **a** and **b**) [29]. However, the $T_1$ values of the rest of the concentrations were the same as the controls. Also, the percentage of signal intensity reduction ranged from 76% to 16% for the highest to the lowest concentrations. The $T_2$ relaxation time of the SPION-C595s treated cells decreased from 101.1 to 24.81 $ms^{-1}$ in comparison to the $T_2$ relaxation time of the untreated MCF-7 cells at iron concentrations ranging from 25 µg Fe/ml to 200 µg Fe/ml (Figure 3. **c** and **d**) [24, 29] .

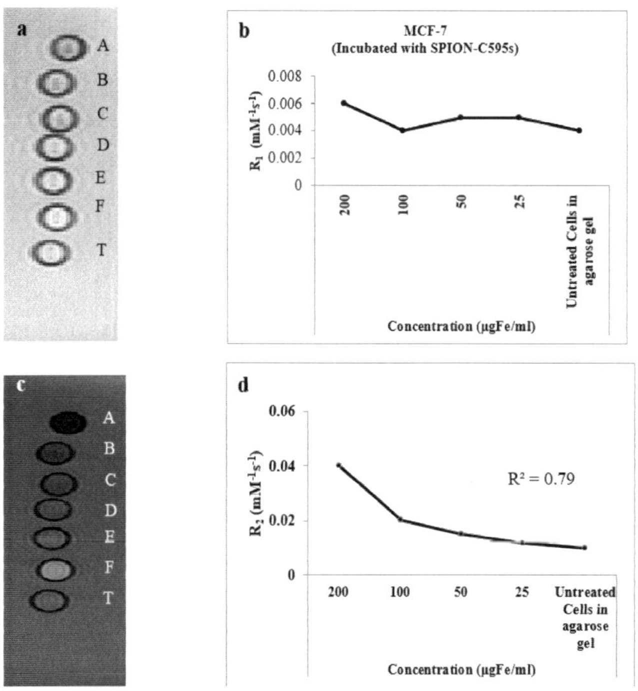


**Figure 3.** The *in vitro* $T_1$-weighted image of MCF-7 cells (**a**), The graph of $R_1$ versus different concentration of SPION-C595 after 6 h incubation with MCF-7 cells (**b**), The *in vitro* $T_2$-weighted image of MCF-7 cells (**c**) and the graph of $R_2$ versus different concentration of SPION-C595 after 6 h incubation with MCF-7 cells (**d**).

## 4. Conclusion

Early detection of breast cancer is crucial for better diagnosis and chance of survival. For early cancer detection, more accurate, sensitive, and selective disease characterisation is essential when utilising MR imaging techniques. The SPION-C595s exhibited good biocompability and safe profile with no significant toxicity was observed which proves good potential clinical applications. A linear dependence was observed between the relaxation rate constants and the iron content of each sample. Thus, the MR *in vitro* images of the breast cancer cells demonstrated that the nanoprobe had potential $T_2$ MR imaging contrast agent for early breast cancer detection.

## 5. References

[1] Nath, S. and P. Mukherjee, MUC1: a multifaceted oncoprotein with a key role in cancer progression. Trends in molecular medicine, 2014. 20(6): p. 332-342.
[2] Kaur, S., et al., Altered expression of transmembrane mucins, MUC1 and MUC4, in bladder cancer: pathological implications in diagnosis. PloS one, 2014. 9(3): p. e92742.
[3] Kufe, D.W., MUC1-C oncoprotein as a target in breast cancer: activation of signalingpathways and therapeutic approaches. Oncogene, 2013. 32(9): p. 1073-1081.
[4] Shahbazi-Gahrouei, D. and M. Abdolahi, Superparamagnetic iron oxide-C595: Potential MR imaging contrast agents for ovarian cancer detection. Journal of Medical Physics, 2013. **38**(4): p. 198-204.
[5] Genitsch, V., et al., MUC1 is upregulated in advanced prostate cancer and is an independent prognostic factor. Prostate cancer and prostatic diseases, 2016.
[6] Wurz, G.T., et al., Antitumor effects of L-BLP25 Antigen-Specific tumor immunotherapy in a novel human MUC1 transgenic lung cancer mouse model. Journal of translational medicine, 2013. 11(1): p. 1.
[7] Kasimanickam, R., V. Kasimanickam, and J. Kastelic, Mucin 1 and cytokines mRNA in endometrium of dairy cows with postpartum uterine disease or repeat breeding. Theriogenology, 2014. 81(7): p. 952-958. e2.
[8] Gnemmi, V., et al., MUC1 drives epithelial–mesenchymal transition in renal carcinoma through Wnt/β-catenin pathway and interaction with SNAIL promoter. Cancer letters, 2014. 346(2): p. 225-236.
[9] Ricardo, S., et al., Detection of glyco-mucin profiles improves specificity of MUC16 and MUC1 biomarkers in ovarian serous tumours. Molecular oncology, 2015. 9(2): p. 503-512.
[10] Beatty, P.L. and O.J. Finn, Preventing cancer by targeting abnormally expressed self-antigens: MUC1 vaccines for prevention of epithelial adenocarcinomas. Annals of the New York Academy of Sciences, 2013. 1284(1): p. 52-56.
[11] Regimbald, L.H., et al., The breast mucin MUC1 as a novel adhesion ligand for endothelial intercellular adhesion molecule 1 in breast cancer. Cancer research, 1996. 56(18): p. 4244-4249.
[12] Xing, P.-X., et al., Synthetic peptides reactive with anti-human milk fat globule membrane monoclonal antibodies. Cancer research, 1990. 50(1): p. 89-96.
[13] Singh, R. and D. Bandyopadhyay, MUC1: a target molecule for cancer therapy. Cancer biology & therapy, 2007. 6(4): p. 481-486.
[14] Devine, P.L. and I.F. McKenzie, Mucins: structure, function, and associations with malignancy. Bioessays, 1992. 14(9): p. 619-625.
[15] Möller, H., et al., NMR-based determination of the binding epitope and conformational analysis of MUC-1 glycopeptides and peptides bound to the breast cancer-selective monoclonal antibody SM3. European Journal of Biochemistry, 2002. 269(5): p. 1444-1455.
[16] Croce, M.V., et al., Patterns of MUC1 tissue expression defined by an anti-MUC1 cytoplasmic tail monoclonal antibody in breast cancer. Journal of Histochemistry & Cytochemistry, 2003. 51(6): p. 781-788.
[17] Weissleder, R., Molecular imaging in cancer. Science, 2006. 312(5777): p. 1168-1171.
[18] Shi, X., et al., Dendrimer-Functionalised Shell-crosslinked Iron Oxide Nanoparticles for In-Vivo Magnetic Resonance Imaging of Tumors. Advanced Materials, 2008. 20(9): p. 1671-1678.
[19] Eck, W., et al., PEGylated gold nanoparticles conjugated to monoclonal F19 antibodies as targeted labeling agents for human pancreatic carcinoma tissue. Acs Nano, 2008. 2(11): p. 2263-2272.
[20] Hui, J.Z., et al., Facile Method for the Site-Specific, Covalent Attachment of Full-Length IgG onto Nanoparticles. Small, 2014. 10(16): p. 3354-3363.
[21] Moradi Khaniabadi, P., et al., Magnetic iron oxide nanoparticles as T2 MR imaging contrast agent for detection of breast cancer (MCF-7) cell. Avicenna Journal of Medical Biotechnology, 2017. 9(4). In Press

[22] Shahbazi-Gahrouei, D. and M. Abdolah, A novel method for quantitative analysis of anti-MUC1 expressing ovarian cancer cell surface based on magnetic cell separation. Journal of Medical Sciences, 2012. 12(8): p. 256.
[23] Shahbazi-Gahrouei, D. and M. Abdolahi, Detection of MUC1-expressing ovarian cancer by C595 monoclonal antibody-conjugated SPIONs using MR imaging. The Scientific World Journal, 2013.
[24] DeNardo, S.J., et al., Development of tumor targeting bioprobes (111In-chimeric L6 monoclonal antibody nanoparticles) for alternating magnetic field cancer therapy. Clinical Cancer Research, 2005. 11(19): p. 7087s-7092s.
[25] Murphy, R., et al., Size and structure of antigen-antibody complexes. Electron microscopy and light scattering studies. Biophysical journal, 1988. 54(1): p. 45.
[26] Khosroshahi, M.E., et al., Evaluation of cell viability and T2 relaxivity of fluorescein conjugated SPION-PAMAM third generation nanodendrimers for bioimaging. Materials Science and Engineering: C, 2016. 62: p. 544-552.
[27] Szpak, A., et al., T1–T2 Dual-modal MRI contrast agents based on superparamagnetic iron oxide nanoparticles with surface attached gadolinium complexes. Journal of Nanoparticle Research, 2014. 16(11): p. 1-11.
[28] Shahbazi-Gahrouei, D., M. Roufeh, and M.B. Tavakoli, Gadolinium-diethylenetriaminepenta-acetic acid conjugated with monoclonal antibody C595 as new magnetic resonance imaging contrast agents for breast cancer (MCF-7) detection. Iranian Biomedical Journal, 2006. 10(4): p. 209-213.
[29] Lundquist, C.M., et al. Porous silicon-superparamagnetic iron oxide complexes and impact on negative contrast enhancement in MRI. in ASME 2013 2nd Global Congress on NanoEngineering for Medicine and Biology. 2013. American Society of Mechanical Engineers.

IOP Conf. Series: Journal of Physics: Conf. Series **851** (2017) 012015 doi:10.1088/1742-6596/851/1/012015

# Optimisation of pulsed and pseudo-continuous arterial spin labeling MRI techniques: A phantom study

**Norain Yusoff[1], Jihan Zukhi[1], Awatif Rusli[2], Rafidah Zainon[1]**

[1]Oncological and Radiological Sciences Cluster, Advanced Medical and Dental Institute, Universiti Sains Malaysia, 13200 Bertam, Kepala Batas, Pulau Pinang, Malaysia.
[2]Hospital Seberang Jaya, 13700 Prai, Pulau Pinang, Malaysia, 13700 Prai, Pulau Pinang, Malaysia.

E-mail: rafidahzainon@usm.my

**Abstract**. Arterial Spin Labeling (ASL) MRI is a non-invasive technique using a freely diffusible intrinsic tracer. The main objective of this study is to evaluate two different techniques of ASL MRI; pulsed ASL (PASL) and pseudo-continuous ASL (PCASL) in obtaining the best signal by manipulating the different imaging parameters. We used a fabricated Perspex flow phantom that is magnetically susceptible. The phantom has a straight tube that mimics carotid artery in adult patients and a U-shaped tube with 75% stenosis. We used a mixture of 60:40 distilled water and glycerol respectively as a substitute to blood. The fabricated phantom was scanned with 1.5T and 3T MRI Scanner using PCASL technique and PASL respectively. Two main imaging parameters were studied which were the field of view (FOV) and slice thickness (ST) to obtain the signal-to-noise ratio (SNR) of the region of interests. The 1.5 T PCASL technique gave SNR values of (13, 22, 30.1) for ST (5, 7, 9 mm) and FOV 240. When higher FOV = 320 was selected, the SNR values were (26.8, 15, 37) for different ST (5, 7, 9 mm). The 3.0 T PASL technique gave the SNR values of (9, 9.3, 11) for ST (5, 7, 9mm) and FOV 240mm. In the higher FOV = 320 mm, we obtained SNR values of (15.2, 17.5, 37.2) for ST (5, 7, 9 mm). As a conclusion, the images quality which can be measured by SNR value is affected by types of ASL and also different parameters.

## 1. Introduction

Many studies data has shown that ASL gave stable noise characteristics over the entire frequency spectrum, which makes it really suitable in studying low-frequency events in brain function [1]. ASL technique has the limitation of having long radiofrequency (RF) pulse, low signal-to-noise ratio (SNR) and difficulty in the planning process. The ASL technique has two main subcategories called pulsed ASL (PASL) and continuous (CASL). CASL has higher signal to noise ratio (SNR) than PASL, but less available on commercial MRI scanners due to the requirement of long and continuous radiofrequency (RF) transmission. The latest technique in ASL is known as pseudo-continuous ASL (PCASL). It divides the long continuous radiofrequency (RF) pulse in CASL into multiple separates short pulses but eventually will have same labeling effects as CASL. As a result, better SNR is achieved compared to CASL and can be convenient to be used in most hardware setup of MRI scanners [2]. This study is to

review two different techniques of ASL MR imaging with the fabricated phantom using PCASL and PASL techniques both at different magnetic strengths of 1.5 T and 3 T respectively.
The non-invasive technique of Arterial Spin Labeling (ASL) technique in imaging will provide as an alternative to the needs of injectable tracers without compromising the results [3]. This method has several advantages as compared with other perfusion techniques such as positron emission tomography (PET), single-photon emission computed tomography (SPECT), CT perfusion and dynamic susceptibility contrast (DSC) magnetic resonance imaging (MRI). All of these techniques need exogenous tracer administration, but in the ASL method no exogenous contrast injection is needed prior the scanning session. The tracer for ASL is the magnetically labelled water [4]. In ASL MRI, the hydrogen protons from the arteries in neck region are inverted and after a while allowing the labeled protons reaching brain tissue, images will be captured. This is called as labeled image. Control images also will be captured to get perfusion weighted images when subtracting them with labeled images [5].

## 2. Materials and Methodology

### *2.1. Phantom design*

The body of the phantom was made from Perspex because it is solid, stable, non-hazardous, and easy to handle. Perspex also has low mass magnetic susceptibility of 0.50 $m^2$ /kg which nearly reflects the tissue composition of tissue. The phantom is compact and portable that can be well suited for reassembly in multi-center research setting. This phantom has a set of straight tube with the diameter of 0.8 cm mimicking adult carotid artery and with u-tube that has 75 % stenosis. The phantom mimics diameter of the carotid arteries because it is capable of reproducibly generating the requisite flow rates and patterns.

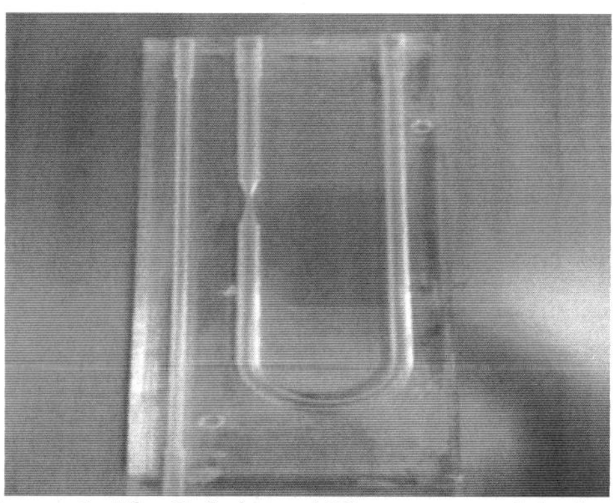

**Figure 1**. The Fabricated Phantom used during scan.

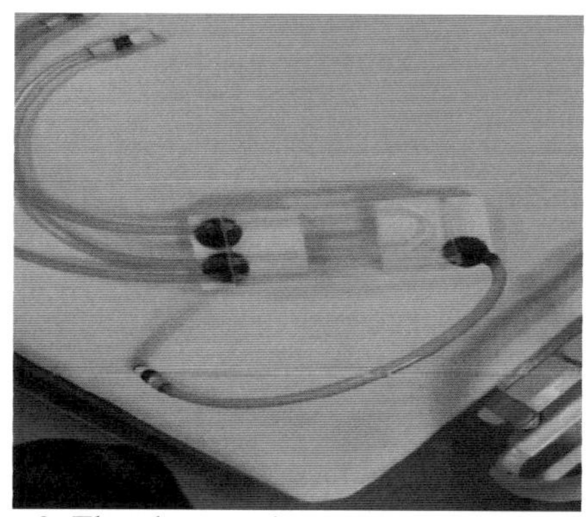

**Figure 2**. The phantom is permanently secured with water catcher basement.

### *2.2. Phantom set-up.*

In this study, a fabricated flow phantom was scanned with different strength of MRI scanners. A combination of glycerol and distilled water 40% and 60% respectively were used to mimic blood [6]. This water–glycerol mixture possesses a similar density (ρ) and viscosity (μ) as human blood (ρ=1053 kg/m3, μ=3 mPa.s) [7]. Water pump is placed in the container to pump the blood mimic into the phantom and circulate it during the experiment.

The water pump is switched on to ensure that there was no bubble in the tubes and phantom during circulation of blood mimic. The phantom and tubes were checked to ensure there was no leakage. The velocity of blood mimic was measured to make sure it is close to reference with 75 cm/s for adult and 97 cm/s for paediatric[8]. Velocity was measured by calculating how long the mixture travels in certain distance of tube.

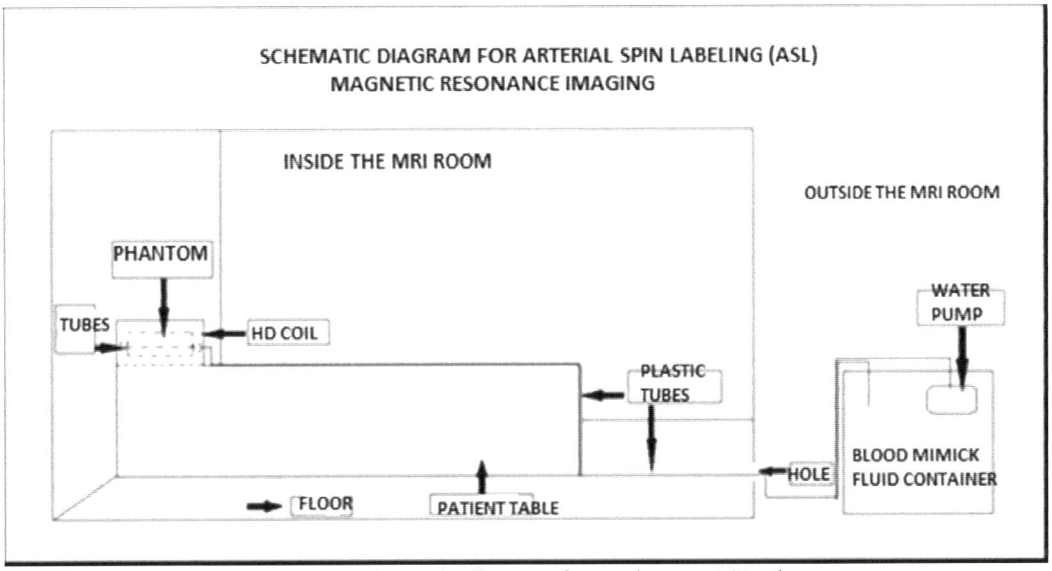


**Figure 3**. The schematic diagram for setting up ASL phantom.

### *2.3. Imaging Protocol and Hardware*

We used two different magnetic field strengths of ASL MRI that are available in our university campus in AMDI and HUSM. We want to gather data of new advanced technique in ASL and by testing different parameters using the new fabricated phantom.

i. PCASL :

All data and imaging sequences were carried out at Advanced Medicine and Dentistry Institute (AMDI), Universiti Sains Malaysia. The phantoms were scanned using 1.5 Tesla magnetic field MRI scanner (GE Signa HDxt) with 16 channel array head coil.

ii. PASL :

The image acquisition of the phantoms was performed by using Philips Achieva 3.0 T X- series MR imaging in Radiology Department of Hospital Universiti Sains Malaysia.

As a result of low signal in ASL protocol, agarose gel blocks were used to provide the background signal (figure 4). This will help in increasing the SNR and also the image quality [9]. Agarose gel slabs (35 g agar, 80 mL glycerol per liter of water) were prepared prior to scan. They will cut according to phantom size and put above and under the phantom.

**Figure 4**. The agarose gel slabs placed on the top and bottom of the fabricated phantom.

*2.4. Parameters*
The parameters that are manipulated in this study are field of view (FOV) and slice thickness (ST). We used two different FOV values which were 240 mm and 320 mm. Slice thicknesses were gradually increased from 5 mm, 7 mm and 9 mm. Matrix size is kept constant at intermediate value which is 256 x 256 throughout the study. Meanwhile, the Bandwidth is kept constant at value of 62.5 kHz.

*2.5. Data Analysis*
We used imageJ software for the analysis of SNR. ImageJ is a freely downloadable image analysis software developed at the National Institute of Health (NIH) to assist in clinical and scientific image analyses [10]. Relevant MRI slices were chosen and downloaded as JPEG images. The images were retrieved with window explorer and opened in imageJ by dragging them to the imageJ main window. The SNR is measured by drawing a circle at region of interest (ROI) in each slice of image. The ROI is the highest signal region obtained during scanning. The selected slice of image with clearest signal was picked out individually for different parameters used. The darkest region of image is the noise background. Same size of circle is also drawn in this region.

Mathematically, the SNR is the quotient of the signal intensity measured in a region of interest (ROI) and the standard deviation of the signal intensity in a region outside the anatomy or object being imaged (i.e. a region from which no tissue signal is obtained) as shown in equation 1 (Gudbjartsson & Patz, 1995). Signal-to-noise ratios (SNR) were calculated according to equation 1.

$$SNR = \frac{Signal\ of\ Region\ of\ Interest\ (ROI) - Signal\ of\ background}{Standard\ deviation\ of\ background} \tag{1}$$

## 3. Results and Discussions

We found that in 1.5 T PCASL technique (figure 5), with the value of FOV=240 and by increasing ST parameters from 5 mm, 7 mm, and 9 mm, the values of SNR were also increasing from 13, 22, to 30.1. Meanwhile, when FOV=320, the SNR value decreased at ST=7 mm (SNR= 26.8, 15, 37). For 3.0T PASL technique (figure 6), the SNR values using FOV=240 gradually increased with ST. The values for SNR obtained were 9, 9.3, and 11. By using higher FOV=320, the SNR values also increased with higher ST. The SNR values were 15.2, 17.5, and 37.2.The images quality and SNR values are affected by several parameters such as field of view (FOV) and slice thickness. In this study, matrix size is kept constant at medium value (256x256). Generally, SNR increase with higher FOV, slice thickness and low matrix size. When FOV increase, more area will receive the signal. But thinner slices are associated with more noise thus lower the SNR value. Increase the matrix size will decrease the signal because there are fewer photons per photon so the signal is less. This study protocols are intended to evaluate two different ASL techniques using the fabricated flow phantom. The images acquired by 1.5 T and 3 T MR imaging scanner of two different ASL MRI techniques are satisfactory. There are still images that are appeared not so clear.

IOP Conf. Series: Journal of Physics: Conf. Series **851** (2017) 012015 doi:10.1088/1742-6596/851/1/012015

But, we still able to captured images by testing several parameters and measure the mean SNR values in the region of interests (ROIs). Adjusting the parameters can also reduce the artifacts in the images [11]. Some of the images captured by ASL technique have flow artifacts. It is known that flow effects in MRI produce a range of artifacts due to flowing blood flow. Blood flow artifact is considered as a special sub-group of motion artifacts. It can be reduced with flow compensation and pre-saturation process [12]. In this study, the artifacts most probably are caused by turbulence flow of liquid inside the phantom. Correctly identified common imaging artifacts will help precise interpretation. We also suspected that the phantom might has some debris of metal dust which is left during its cutting process. The presence of foreign body produce susceptibility effects that produce artifactual dark signal which can be misinterpreted as region of low signal [13]. Our findings in the present study show the capability of noninvasive ASL to enable optimal image acquisition measured by the SNR. With the use of 1.5 T MR imaging scanner, PCASL has showed that it also can give reasonable amount of SNR in the region of interest. While in higher strength of scanner using 3 T PASL technique, the SNR also has relatively same value when highest FOV and ST are used. The application of PCASL has better balance between tagging efficiency and SNR, as well as reducing Magnetization Transfer effects and Radio Frequency power deposition [14]. This experiment shows that the best parameters value to be set for the phantom was FOV: 320 x 320 mm and slice thickness: 9 mm regardless of the different ASL techniques. In this study, the various labeling parameters are tested one by one to enable the best images to be captured. As proposed by previous studies, higher magnetic field MRI scanners (> 1.5 Tesla) able to give longer T1 and increase in SNR. Longer T1 value can increase the amount of labeled protons thus it will affect how quickly the net magnetisation vector (NMV) recovers to its ground state in the direction of external magnetic field (B0) [15]. This can significantly enhance image quality. The consensus from the Perfusion Study Group of the International Society for Magnetic Resonance in Medicine (ISMRM) 2012 has suggested that the optimal label duration is greatly influenced by the relaxation time of the label (T1) and the repetition time (TR).

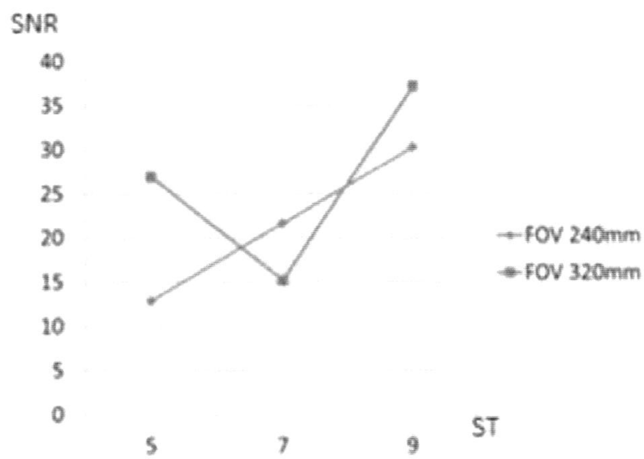


**Figure 5**. SNR against slice thickness for FOV 240mm and FOV 320mm using PCASL.

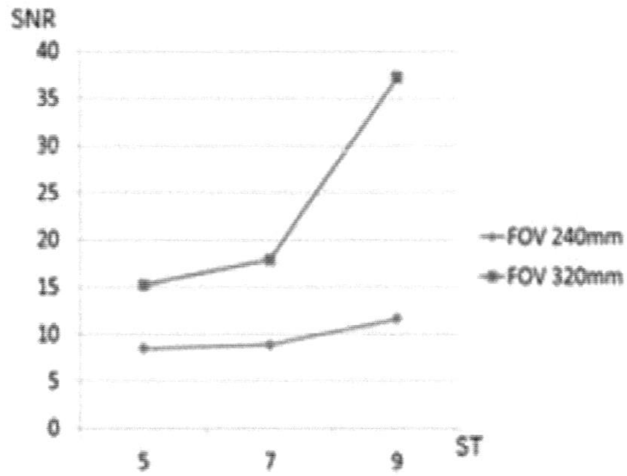


**Figure 6**. SNR against slice thickness for FOV 240mm and FOV 320mm using PASL.

## 4. Conclusion

SNR is important to be quantified in each image. It will show how much the true signal is gained at the ROI versus how much background noise a particular image has. Improvement of the SNR of ASL measurements is crucial so that it could take a more central role in Magnetic Resonance Imaging (MRI) studies. Compared with pulsed ASL, pulsed continuous ASL is more advantageous as we can obtain good SNR values even with lower magnetic strength field. Vendors are currently taking the advantages of promoting the PCASL technique as it requires less hardware settings, yet can produce desirable SNR values. More studies using ASL are actively conducted and researchers are encouraged to share the results so that this technique can be utilized widely in clinical platform. It can bring a tremendous benefit for patients using this non-invasive method MR imaging and preferable as it also can be used in continuous monitoring of patients such as in pharmacological and chronic renal disease.

## 5. References

[1] Jiongjiong Wang, G.K.A., Daniel Y. Kimberg, Anne C. Roc,Lin Li, and John A. Detre, Recent data suggest that ASL contrast shows stable noise characteristics over the entire frequency spectrum, which makes it suitable for studying low-frequency events in brain function. Magnetic Resonance in Medicine, 2003. 49: p. 796-802.
[2] Sung-Hong Park, D.J.J.W., Timothy Q. Duong, Balanced steady state free precession for arterial spin labeling MRI: Initial experience for blood flow mapping in human brain, retina, and kidney. Magnetic Resonance Imaging, 2013. 31: p. 1044-1050.
[3] Tracy R. Melzer, et al., Arterial spin labelling reveals an abnormal cerebral perfusion pattern in Parkinson's disease. Brain A Journal of Neurology, 2011. 134: p. 845-855.
[4] Noll, Douglas C, and D Ph. "A Primer on MRI and Functional MRI." 0: 1–14.
[5] Jill B. De Vis, et al., Regional changes in brain perfusion during brain maturationmeasured non-invasively with Arterial Spin Labeling MRI in neonates. European Journal of Radiology, 2013. 82: p. 538-543.
[6] Majid Y. Yousif, D.W.H., Tamie L. Poepping., A blood-mimicking fluid for particle image velocimetry with silicone vascular models.
[7] Emma M.L. Chung, James P. Hague, Marie-Anne Chanrion, Kumar V. Ramnarine, Emmanuel

Katsogridakis, David H. Evans., Embolus Trajectory Through a Physical Replica of the Major Cerebral Arteries. Stroke, 2010. 41: p. 647-652.
[8] Monica S Vavilala, M.S.K., Saipin L Muangman, Pilar Suz, Irene Rozet and Arthur M Lam, Gender Differences in Cerebral Blood Flow Velocity and Autoregulation between the Anterior and Posterior Circulations in Healthy Children. Pediatric Research, 2005. 58: p. 574-578.
[9] Villemaire, L., Brain Phantoms for Ultra High Field MRI, 2010, Department of Medical Biophysics, University of Western Ontario.
[10] Ls, Jaba, V Shanthi, and D J Singh. 2011. "Estimation of Hippocampus Volume from MRI Using ImageJ for Alzheimer ' S Diagnosis." 1(1): 15–20.
[11] Bdef, Katarzyna Krupa, and Monika Bekiesińska-figatowska Abde. 2015. "Artifacts in Magnetic Resonance Imaging." : 93–106.
[12] MRI Artifacts, 2015: Magnetic Resonance-Technology Information Portal.
[13] Petcharunpaisan, Sasitorn, Joana Ramalho, and Mauricio Castillo. 2010. "Arterial Spin Labeling in Neuroimaging." World journal of radiology 2(10): 384–98.
[14] Pollock, Jeffrey M et al. 2010. 17 Arterial Spin Labeled MRI Perfusion Imaging: Clinical Applications.
[15] Marincek, D.W.V.D.K.B., How Does MRI Work?, ed. H. Dr. Ute Heilmann. 2006, Germany: Springer. 171.

**Acknowledgment**
The authors would like to thank the financial support from Universiti Sains Malaysia (USM) through the Short Term Grant Scheme.

IOP Conf. Series: Journal of Physics: Conf. Series **851** (2017) 012016 doi:10.1088/1742-6596/851/1/012016

# Cortical activation during power grip task with pneumatic pressure gauge: an fMRI study

**M Mohamad[1], N H Mardan [2], S S Ismail[1]**

[1]Diagnostic Imaging & Radiotherapy Programme, Faculty of Health Sciences, Universiti Kebangsaan Malaysia, Kuala Lumpur,50300, Malaysia;
[2]Department of Cognitive Science, Universiti Malaysia Sarawak, Kota Samarahan, 94300, Malaysia

E-mail: mazlyfarina@ukm.edu.my

**Abstract**. Aging is associated with a decline in cognitive and motor function. But, the relationships with motor performance are less well understood. In this study, functional magnetic resonance imaging (fMRI) was used to assess cortical activation in older adults. This study employed power grip task that utilised block paradigm consisted of alternate 30s rest and active. A visual cue was used to pace the hand grip movement that clenched a cylindrical rubber bulb connected with pressure pneumatic gauge that measure the pressure (Psi). The objective of this study is determined the brain areas activated during motor task and the correlation between percentage signal change of each motor area (BA 4 and 6) and hand grip pressure. Result showed there was a significant difference in mean percentage signal change in BA 4 and BA 6 in both hemispheres and negative correlation obtained in BA 4 and BA 6. These results indicate that a reduced ability in the motor networks contribute to age-related decline in motor performance.

## 1. Introduction

Decline in cognitive and motor function of older adults is pointed as the reasons for a range of neurodegenerative and neurochemical changes with advancing age [1, 2]. Functional magnetic resonance (fMRI) is one of the approaches that has been used to map age-related changes [4, 5]. Previous studies depicted that the regions of motor area activation on studying the movements of normal right and left fingers via finger tapping stimulation were shown in the primary motor cortex [6, 7, 8]. In a published article of Cortical Activity in Precision-Versus Power-Grip Tasks which utilised auditory cues, power-grip task in right handed 5 young and healthy subjects were associated with stronger activity in the contralateral primary motor cortex (M1), primary sensory cortex (SI) and parietal operculum (PO) than in the precision-grip task [9]. However in this current study, power-grip task was employed which utilised visual cues instead of auditory cues in 14 older adult male and female subjects to identify brain areas activated in older adults while subjects were performing power-grip task.

One of the methods to map functioning human brain is blood oxygenation level dependent (BOLD) which relies on the magnetic properties of blood to measure the changes in local hemodynamic related to neuronal activity [10]. Hence, in finding the activation in motor regions in this fMRI experiment, BOLD contrast was utilised in this study because of its non-invasive nature and it proved to identify patterns of activation across the whole brain [11].

Previous research [12] showed BOLD activation increased in relation to force amplitude in the primary motor/somatosensory cortex (M1/S1) in precision grip task in healthy adults. A study of age-dependent changes in the neural correlation with force modulation, involving subjects from age of 25 to 70 showed positive correlation between BOLD signal and grip force in contralateral primary motor cortex (Brodmann area (BA 4), superior cingulated sulcus (BA 6) and primary visual motor (BA 7) [5]. However, diminished correlation was found in primary motor cortex and superior cingulated sulcus in older adults. Due to diminish correlation seen in older adults thus this study focused on a relationship percentage of signal change in motor cortex and pressure (Psi) applied during power grip task. The aim of this study is therefore, by using simple motor task via visual cue, to investigate the differences of brain activation (BA 4 and BA 6) in motor regions in both hemisphere and the correlation between percentage signal change of each motor area (BA 4 and 6) and hand grip pressure. Therefore this study hypothesised that brain activations in left hemisphere is higher compare to right hemisphere in both regions and there will be positive correlation between hand grip pressure and percentage signal change.

## 2. Method

### *2.1. Subjects*
Fourteen healthy subjects (40-55 years old, mean 44.4 years old, S.D. 4.4 years) comprising of 5 male and 9 female participated in the study and gave their informed consent to a protocol approved by Ethic Committee of The Universiti Kebangsaan Malaysia (No: MN052). All subjects were right-handed as judged by Edinburgh Handedness Inventory [13].

### *2.2. Tasks*
The subjects performed a power grip task with the right hand in the presence of visual cues by applying maximum grip to a bulb connected to pneumatic pressure gauge which measured pressure applied by subjects. Prior to scanning, but whilst lying in the scanner, subjects were asked to grip the bulb which connected to pneumatic pressure gauge using maximum force (1-2 Psi). The measurement was recorded for each subject. During the tasks, the subjects rested comfortably in a supine position in the MR scanner. The extended right arm was oriented parallel to the trunk in a relaxed position. Visual cues which employed block-design paradigm consists of 12 alternate active and 12 rest blocks, 30 sec each was used to pace hand movement. Ten measurements were taken for each block with total of 250 measurements noted including 10 baseline measurements. The measurements of the force were monitored during the scanning session.

BOLD fMRI was performed using an optimised gradient-echo EPI protocol using parallel-imaging (GRAPPA) acquisition/reconstruction on a Siemens Trio 3T. Acquisition parameters were: TR/TE=3000/60 ms, 64 × 64 acquisition matrix, 192 mm × 100 mm FOV, 21 slices of 4.5 mm covering the entire brain with a tilted axial orientation.

### *2.3. Image Preprocessing and Data Analysis*
Images were prepared for analysis using statistical parameter mapping (SPM8) (http://www.fil.ion.ucl.ac.uk). For regions of interest analysis (ROI), two regions were selected which were BA4 and BA6. Using Marsbar, a sphere size of 6 mm was utilised to obtained ROI in both BA4 and BA6 in right and left hemisphere.

## 3. Results

### *3.1. Brain areas activated*
Result from this study which showed number of voxels activated in Broadmann Areas were described in Table 1 ($p<0.001$) uncorrected [14] in descending order. However in this study, regions of interest were only BA 4 and BA 6.

**Table 1.** Number of voxels activated in power grip task.

| Broadmann Area | Anatomical location | Voxels activated |
|---|---|---|
| 6 | Precentral gyrus | 2379 |
| 40 | Supramarginal gyrus | 893 |
| 3 | Postcentral gyrus | 377 |
| 4 | Precentral gyrus | 369 |
| 2 | Postcentral gyrus | 237 |
| 7 | Precuneus | 168 |
| 8 | Middle frontal gyrus | 168 |
| 1 | Postcentral gyrus | 46 |

### *3.2. Percentage of signal change in BA 4 & BA 6*
From the ROIs analysis using Marsbar, four ROIs were obtained which were left BA 4 (-34, -20, 55), right BA 4 (4 -38 72), left BA 6 (-4, 0, 52) and right BA 6 (12, 2, 60) (refer Fig 1). Result showed a significant different ($p<0.001$) between mean signal change for right and left hemisphere in BA 4 and BA 6. Mean percentage signal change in BA 4 right (0.353), BA 4 left (1.969), BA6 right (0.67) and BA 6 left hemisphere (1.062).

### *3.3. Grip pressure and percentage signal change*
Pearson correlation of both areas in both hemispheres showed weak negative correlation. BA4R correlation coefficient was -0.249 while -0.304 for BA4L with pressure. BA6 correlation coefficient was -0.286 and -0.299 for right and left hemispheres respectively.

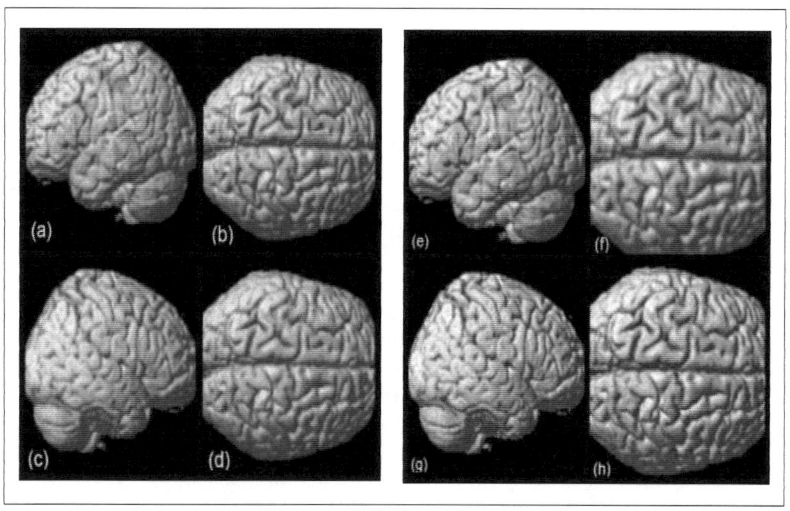


**Figure 1.** (a) & (b) Activation shown in left BA4. (c) & (d) Activation shown in right BA4.(e) & (f) Activation shown in left BA6. (g) & (h) Activation shown in right BA6.

## 4. Discussion

### *4.1. Brain areas activated (BA 4 & BA 6)*

Previous study reported activation in BA 4 in power grip task [9] but in this study, the results demonstrated a highest activation in BA6 which aligned with findings [15] that described more activation was seen in BA6 rather than BA4 with advancing age due to possibility of older adults forming the task given more like precision grip thus accounting for the trends towards BA 6 since it was reported that there was a age-related decline in accuracy or speed at making fine movements for older adults [16].

### *4.2. Percentage of signal change in BA 4 & BA 6*

In several PET and fMRI experiments a preference for the left hemisphere has been observed in, both in primary motor or sensorimotor regions and in medial and lateral premotor regions. Activations have been reported to be primarily contralateral during right hand movement but notably less strongly lateralised during left-hand movement [17]. Overall, there left hemisphere in both BA 4 and BA 6 showed higher mean percentage signal change in keeping with studies suggesting contralateral (left) primary motor cortex and SMA were strongly in right-handed subjects.

### *4.3. Grip pressure and percentage signal change*

In this study there was a weak negative correlation in BA6L (-0.289) than that in BA6R (-0.299) between grip pressure and percentage signal change. The negative correlation results obtained with older adults may due to the efficiency of the whole brain diminished with increasing age [5, 18] and cortical regions known to contribute to corticospinal tract are less able to increase output-related activity with advancing age. However, in this study only mean value of pressure was measured. Therefor further studies should be done to investigate the relationship between grip pressure and percentage signal change in older adults.

## 5. Conclusion

Percentage of signal change showed significant difference between both BA4 and BA6 in right and left hemisphere. Weak negative correlation in BA4 and BA6 may due to aging effects. This findings will help understanding the functions in both region and neuroplasticity process to reduce disability in neurological diseases such as stroke.

## 6. Limitations

The measurement of the grip pressure was collected as a mean value and may vary through out the scanning session.

## 7. References

[1] Hackel ME, Wolfe G A, Bang S M and Canfield J S 1992 Changes in hand function in the aging adult as determined by the Jebsen Test of Hand Function *Physical Therapy* **72** 373-377

[2] Kolb B, Forgie M, Gibb R, Gorny G and Rowntree S 1998 Age, experience and the changing brain *Neuroscience & Biobehavioral Reviews* **22** 143-159

[3] Meunier D, Achard S, Morcom A and Bullmore E 2009 Age-related changes in modular organization of human brain functional networks *Neuroimage* **44** 715-723

[4] Park C, Boudrias M, Rossiter H & Ward N S 2012 Age-related changes in the topological architecture of the brain during hand grip *Neurobiology of Aging* **33** 833.e827-833.e837.

[5] Wang L, Li Y, Metzak P, He Y ad Woodward T S 2010 Age-related changes in topological patterns of large-scale brain functional networks during memory encoding and recognition *Neuroimage* **50** 862–872

[6] Yusoff A N, Hashim M H, Ayob M M and Kassim I 2006. Pengaktifan otak akibat gerakan jari

bagi subjek dominan tangan kanan dan kiri *Jurnal Sains Kesihatan Malaysia* **4(2)** 63-83
[7] Yusoff A N, Mohamad M and Hamid K A 2011 Pemerolehan, analisis dan interpretasi data fMRI: kehubungan efektif dalam korteks auditori primer manusia.Sains Malaysiana **40(6)** 665-678
[8] Yusoff A N, Mohamad, M, Hamid K, Hamid, A I A, Manan H A and Hashim M H 2010 Activation characteristics of the primary motor (M1) and supplementary motor (SMA) areas during robust unilateral finger tapping task *Jurnal Sains Kesihatan Malaysia* **8(2)** 43-49
[9] Ehrsson H H, Fagergren A, Jonsson T, Westling G, Johansson R S and Forssberg H 2000 Cortical Activity in Precision-Versus Power-Grip Tasks: An fMRI Study. *J Neurophysiology* 83(1) 528-536
[10] Ogawa S., Tank, D W, Menon R, Ellermann J M, Kim S G , Merkle H and Ugurbil K 1992 Intrinsic signal changes accompanying sensory stimulation: functional brain mapping with magnetic resonance imaging. *Proceedings of the National Academy of Sciences of the United States of America* **89(13)** 5951–5955
[11] Kwong K K, Belliveau J W, Chesler D A, Goldberg I E, Weisskoff R M, Poncelet B P, Kennedy D N, Hoppel B E, Cohen M S and Turner R 1992 Dynamic magnetic resonance imaging of human brain activity during primary sensory stimulation *Proceedings of the National Academy of Sciences of the United States of America* **89(12)** 5675-5679
[12] Planetta P J, McFarland N R, Okun M S and Vaillancourt D E 2014 MRI reveals brain abnormalities in drug-naive Parkinson's Disease *Exercise And Sport Sciences Reviews* **42(1)** 12-22
[13] Oldfield R C 1971 The assessment and analysis of handedness: the Edinburgh inventory *Neuropsychologia* **9(1)** 97–113.
[14] Langan J, Peltier S J, Bo J, Fling B W, Welsh R C and Seidler R D 2010. Functional implications of age differences in motor system connectivity *Frontiers In Systems Neuroscience* **4**
[15] Ward N S, Swayne O B C and Newton J M 2008 Age-dependent changes in the neural correlates of force modulation: An fMRI study *Neurobiology of Aging* **29** 1434-1446 -17
[16] Spirduso W W, Francis K L and MacRae P G 2005 Physical Dimensions of Aging Human Kinetics (2 ed., pp. 1076) Champaign

[17] Kawashima R, Yamada K, Kinomura S, Yamaguchi T, Matsui H, Yoshioka S and Fukuda H 1993 Regional cerebral blood flow changes of cortical motor areas and prefrontal areas in humans related to ipsilateral and contralateral hand movement *Brain Resources* **623** 33-40
[18] Zhu W, Wen W, He Y, Xia A, Anstey K J and Sachdev P 2010 Changing topological patterns in normal aging using large-scale structural networks *Neurobiology of aging* **33(5)** 899-913

**Acknowledgement**
We thank from Department of Radiology, Universiti Kebangsaan Malaysia Medical Centre, for the assistance in fMRI scans. This work is supported by the Research University Grant GGPM-2012-107.

IOP Conf. Series: Journal of Physics: Conf. Series **851** (2017) 012017 doi:10.1088/1742-6596/851/1/012017

# Lu-177 DOTATATE dosimetry for neuroendocrine tumor: single center experience

**Said MA[1], Masud MA[1], Zaini MZ[1], Salleh RA[1], BN Lee[1], Zainon R[2]**

[1]Nuclear Medicine department, Institut Kanser Negara, Presint 7, 62250, Putrajaya, Malaysia
[2]Advanced Medical and Dental Institute, Universiti Sains Malaysia, Bertam, 13200 Kepala Batas, Pulau Pinang, Malaysia

E-mail: aminhpj@gmail.com

**Abstract.** Lu-177 labelled with DOTATE is widely acceptable to treat Neuroendocrine Tumor (NET) disease and it better improvement of quality of patients' life since few years ago. However, the radionuclide toxicity becomes the main limitation of the (NET) treatment. Therefore, we performed a pilot study aimed to estimate radiation absorbed doses to dose-limiting organs to develop a systemic therapy with Lu-177 in NET patients. In this study, five set of planar whole body images was acquired every 0.5 hour, 4 hours, 24 hours, 48 hours and 72 hours after Lu-177 administrations. The planar image acquisition was done using Philip Brightview X with Medium Energy General Purpose Collimator (MEGP) collimator. All patients' images in Conjugate View (CV) format were transferred into PMOD 3.7 Software for Region of Interest (ROI) analysis. The ROI were drawn at selected organs such as kidneys, liver, spleen and bladder. This study found that the mean absorbed dose for kidneys 0.62 ± 0.26 Gy/GBq, liver 0.63 ± 0.28 Gy/GBq, spleen 0.83 ± 0.73 Gy/GBq and bladder 0.14 ± 0.07 Gy/GBq. The radionuclide kinetic for the whole body 99.7 ± 0.1 percent at 0.5 hours, 79.5 ± 10.7 percent at 4 hours, 56.6 ± 10.3 percent at 24 hours, 43.2 ± 7.9 percent at 48 hours and 37.1 ± 9.0 percent at 72 hours. This study verifies that this planar quantitative method able to determine organ at risk and the result line with other published data.

## 1. Introduction

Lu-177 label with DOTATOC is widely acceptable to treat Neuroendocrine Tumor (NET) disease and showing patients increase of quality of life since few years ago due to its properties able to emitted Beta particle and Gamma rays simultaneously. This radionuclide also good for dosimetry assessment [1-4]. In current clinical practice, Lu-177 DOTATATE treatment is normally infused between 5.55 to 7.4 GBq per cycle with time interval between per cycle about 6–16 weeks apart two to six number of cycle together with kidney protection agents [4-5]. The radionuclide toxicity to the kidneys become the main limitation of the (NET) treatment. The role of radionuclide dosimetry is to assess of optimum radionuclide therapy to administered to the patients and assess the organs at risk condition. The dose-limiting organs for NET therapy are the kidney less than 23-28 Gy[1-3]. In order to avoid toxicity the amount of radiation dose given to these dose-limiting organs has to be estimated. Recently, the dosimetry technique used in nuclear medicine was proven to assess the absorbed dose for the critical organ such as kidneys [2, 3, 5]. In general the dose estimates by using MIRD scheme, with the basic equation (1)

$$D=\tilde{A}S \tag{1}$$

Where Ã is the integral activity in the organ and S is a dose conversion factor of the radionuclide and the target [6]. The value of S should be corrected for the actual organ volume and mass of the organ. However, there are limited data regarding the radiation absorbed dose given to the normal tissues for systemic therapy with Lu-177 for this region. Therefore, we aimed a pilot study to estimate radiation absorbed doses to dose-limiting organs in order to develop a systemic therapy with Lu-177 with suitable NET patients.

## 2. Methods

In this study, five set of planar whole body images from patients was acquire every 0.5 hour, 4 hour, 24 hour, 48 hour and 72 hour post Lu-177 administrations as shown at Figure 1. The planar image acquisition was done by using Philip Brightview X with MEGP collimator and window energy was set at 208 keV (20%) and 111 keV (20%) at 12 cm/min [2], [7]. The (Ant-Post) planar images were convert into conjugate view (CV) by using Philip Jetpack 2.0 software for every each patient's scanning acquisitions hours. All Patients images in Conjugate View (CV) format were transfer into PMOD 3.7 Software for Region of Interest (ROI) analysis. The ROI were drawn at selected organs such as Kidneys, liver, spleen and bladder. The ROI's were draw starting 24 hour, 0.5 hour, 4 hour, 48 hour and 72 hour. All the ROI's data was transfer in MS Excel spreadsheet. The time activity curve (TAC) is obtained by using fractionation factor (f-factor) for every ROI's count for every organ. The resident time for every each organ obtained from Area under curve (AUC) graph fitting obtained from bi or three exponential fit by using MS EXCEL worksheet. The estimation of Absorbed Dose (S-Value) to the organ were based resident time from each organ. The actual kidney weight was obtained from Radiotherapy Planning System, MONACO and patient's body weight is compulsory for accurate result by using OLINDA-EXM 1.1 software.

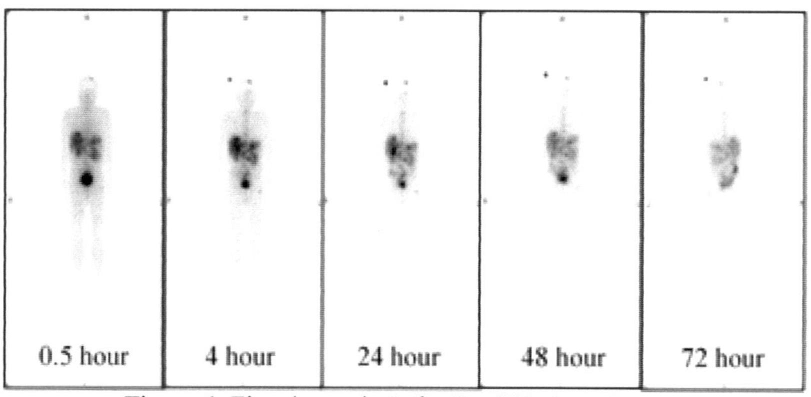


**Figure 1**. Five data point's for Lu-177 planar images

## 3. Results

The results of the organ dosimetry S-Value absorbed doses for kidneys, spleen liver and bladder of all six patients are presented in Figure 1. Mean absorbed dose for kidneys is 0.64 ± 0.41 Gy/GBq, liver 0.76 ± 0.32 Gy/GBq, spleen 1.23 ± 0.59 Gy/GBq and bladder 0.17 ± 0.11 Gy/GBq. The kinetics of Lu-177 DOTATE at patient's whole body are also shown in Graph 2. The IA kinetic for the whole body is 99.7 ± 0.1 percent at 0.5 hours, 79.5 ± 10.7 percent at 4 hours, 56.6 ± 10.3 percent at 24 hours, 43.2 ± 7.9 percent at 48 hours and 37.1 ± 9.0 percent at 72 hours.

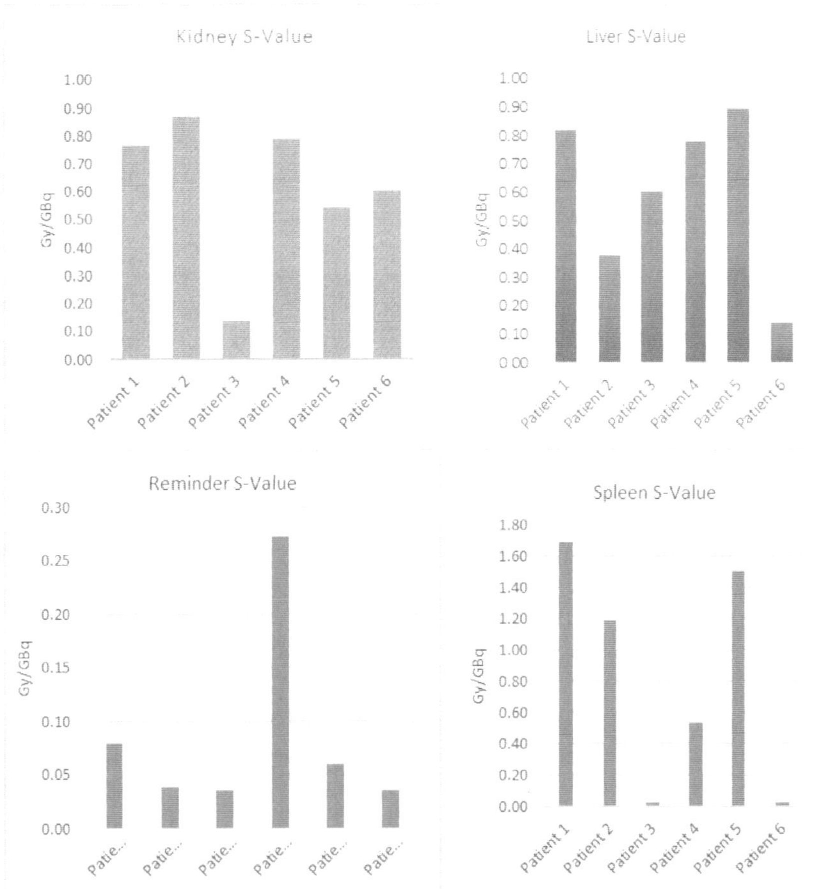


**Figure 2.** S Value Lu-177 for kidney, liver, spleen and reminder

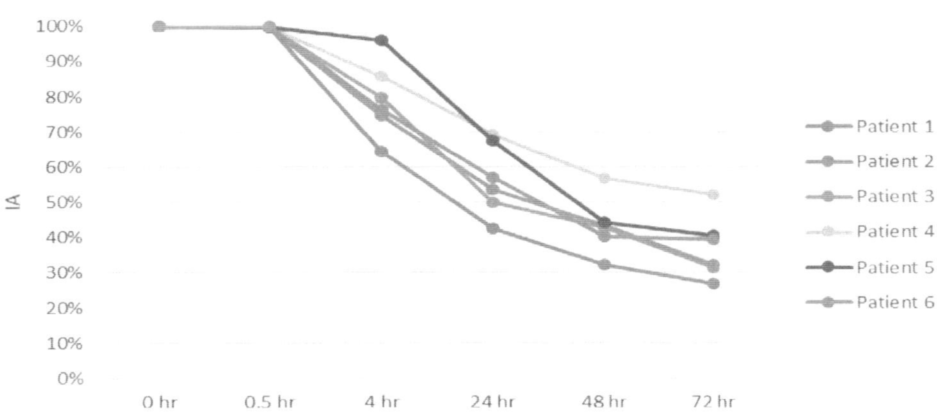


**Figure 3.** The whole body Lu-177 DOTATATE kinetic.

## 4. Discussions
In Lu-177 DOTATATE therapy, patient specific dose calculation is important technique to determine organ absorbed dose due to different pharmacokinetic for every each patients and organ weight. The general equations for absorbed dose (D) is defined in Equation (2):

$$D = d\varepsilon / dm \quad (2)$$

Where D is the mean energy imparted by ionising radiation to matter of mass *dm*. The units of absorbed dose are typically J/kg and refer as Gray (Gy) as SI unit. In radionuclide therapy with specific peptides receptor such as 177-Lu DOTATATE, kidneys and bone marrow highly sensitive with radionuclide toxicity [4-10]. Agreed with previous works, the absorbed radiation dose to the kidneys varies widely between patients treated with 177Lu-DOTATATE [2], [4], [7], [11-15]. Our current result for kidney range between 0.15 to 1.27 Gy/GBq, Liver 0.34 to 1.28 Gy/GBq, spleen 0.55 to 2.14 Gy/GBq and reminder 0.07 to 0.14 Gy/GBq. The maximum limit absorbed dose to the kidneys in our center is 27 Gy and in average it's would be reached with a total cumulative administered dose up to 46 GBq. Agreed with [7], bone marrow dosimetry in previous studies had shown that the absorbed dose to the bone marrow is usually low by others researchers. The difference in calculated absorbed doses to the kidneys depending on the dosimetry method used and the administration of amino acids has proven to decrease the doses to the kidney effectively and improved the toxicity profile of the treatment [13]. Although the calculation method at our institution is different with previous works [3], [5-6], but the dosimetry concept still similar. In our works, the actual organ weight measurement were done in radiotherapy planning software will make the absorbed dose to the critical organ more realistic approach. Our works are not covering the planar lesion dose study to the tumour due to the accuracy issues [2].

Current measured absorbed doses for kidneys are in good agreement with absorbed doses measured by other published works [13-15]. The residence time of Lu-177 in the whole body in our works is 0.5 to 5.6 days and longer compared with other published data (range 0.89 – 3.0 days) [10]. In this work, all the patients result show the absorbed dose to the critical organ are below the maximum limit.

## 5. Conclusions

The patient-specific dosimetry procedure were introduced here helps optimise the planning of administered radionuclide therapy. The patients' self-assessed quality of life increases significantly after treatment with 177Lu-DOTATOC. This procedure is highly recommend to the patients would go for the radionuclide therapy and also to optimise the treatment.

## 6. References

[1] B. L. R. Kam, J. J. M. Teunissen, E. P. Krenning, W. W. de Herder, S. Khan, E. I. van Vliet, and D. J. Kwekkeboom, "Lutetium-labelled peptides for therapy of neuroendocrine tumours.," Eur. J. Nucl. Med. Mol. Imaging, vol. 39 Suppl 1, pp. S103–12, Feb. 2012.

[2] F. Guerriero, M. E. Ferrari, F. Botta, F. Fioroni, E. Grassi, A. Versari, A. Sarnelli, M. Pacilio, E. Amato, L. Strigari, L. Bodei, G. Paganelli, M. Iori, G. Pedroli, and M. Cremonesi, "Kidney dosimetry in 177Lu and 90Y peptide receptor radionuclide therapy: Influence of image timing, time-activity integration method, and risk factors," *Biomed Res. Int.*, vol. 2013, 2013.

[3] G. K. Lammers, J. Paul Esser, P. C. M. Pasker, M. E. Sanson-van Praag, and J. M. H. de Klerk, "Can I-124 PET/CT Predict Pathological Uptake of Therapeutic Dosages of Radioiodine (I-131) in Differentiated Thyroid Carcinoma?," *Adv. Mol. Imaging*, vol. 04, no. 03, pp. 27–34, Jul. 2014.

[4] J. Svensson, J. Söderström, T. Magnander, E. Wikberg, B. Wängberg, and P. Bernhardt, "An image based method for bone marrow dosimetry in 177Lu-DOTATATE therapy," *Eur. J. Nucl. Med. Mol. Imaging*, vol. 42, no. 1, p. S9, 2015.

[5] M. Sandström, U. Garske-Román, D. Granberg, S. Johansson, C. Widström, B. Eriksson, A. Sundin, H. Lundqvist, and M. Lubberink, "Individualized dosimetry of kidney and bone marrow in patients undergoing 177Lu-DOTA-octreotate treatment.," *J. Nucl. Med.*, vol. 54, no. 1, pp. 33–41, Jan. 2013.

[6] J. Svensson, J. Söderström, T. Magnander, E. Wikberg, B. Wängberg, and P. Bernhardt, "An image based method for bone marrow dosimetry in 177Lu-DOTATATE therapy," *Eur. J. Nucl. Med. Mol. Imaging*, vol. 42, no. 1, p. S9, 2015.

[7] J. Svensson, G. Berg, B. Wängberg, M. Larsson, E. Forssell-Aronsson, and P. Bernhardt, "Renal function affects absorbed dose to the kidneys and haematological toxicity during $^{177}$Lu-DOTATATE treatment.," *Eur. J. Nucl. Med. Mol. Imaging*, vol. 42, no. 6, pp. 947–55, May 2015.

[8] M. E., "Individual variation in renal dosimetry of lu-177 octreotate," *Internal Medicine Journal*, vol. 43. p. 8, 2013.

[9] L. Bodei, J. Mueller-Brand, R. P. Baum, M. E. Pavel, D. Hörsch, M. S. O'Dorisio, T. M. O'Dorisio, T. M. O'Dorisiol, J. R. Howe, M. Cremonesi, D. J. Kwekkeboom, and J. J. Zaknun, "The joint IAEA, EANM, and SNMMI practical guidance on peptide receptor radionuclide therapy (PRRNT) in neuroendocrine tumours.," *Eur. J. Nucl. Med. Mol. Imaging*, vol. 40, no. 5, pp. 800–16, May 2013.

[10] B. Brans, L. Bodei, F. Giammarile, O. Linden, M. Luster, W. J. G. Oyen, and J. Tennvall, "Clinical radionuclide therapy dosimetry: the quest for the 'Holy Gray'.," *Eur. J. Nucl. Med. Mol. Imaging*, vol. 34, no. 5, pp. 772–86, May 2007.

[11] A. Delker, H. Ilhan, C. Zach, J. Brosch, F. J. Gildehaus, S. Lehner, P. Bartenstein, and G. B??ning, "The Influence of Early Measurements Onto the Estimated Kidney Dose in [$^{177}$Lu][DOTA$^{0}$,Tyr$^{3}$]Octreotate Peptide Receptor Radiotherapy of Neuroendocrine Tumors," *Mol. Imaging Biol.*, vol. 17, no. 5, pp. 726–734, 2015.

[12] C. Schuchardt, H. R. Kulkarni, V. Prasad, C. Zachert, D. Müller, and R. P. Baum, "The Bad Berka dose protocol: comparative results of dosimetry in peptide receptor radionuclide therapy using (177)Lu-DOTATATE, (177)Lu-DOTANOC, and (177)Lu-DOTATOC.," *Recent results cancer Res. Fortschritte der Krebsforsch. Progrès dans les Rech. sur le cancer*, vol. 194, pp. 519–36, Jan. 2013.

[13] J. Heikkonen, H. Mäenpää, and M. Tenhunen, "Methodological effects in kidney dosimetry in Lu-177 radionuclide therap," *Eur. J. Nucl. Med. Mol. Imaging*, vol. 40, p. S265, 2013.

[14] M. Sandström, E. Ilan, A. Karlberg, S. Johansson, N. Freedman, and U. Garske-Román, "Method dependence, observer variability and kidney volumes in radiation dosimetry of (177)Lu-DOTATATE therapy in patients with neuroendocrine tumours.," *EJNMMI Phys.*, vol. 2, no. 1, p. 24, 2015.

[15] C. Schuchardt, H. R. Kulkarni, S. Wiessalla, M. Shahinfar, S. Shafaei, S. Kropf, H. J. Wester, and R. P. Baum, "Dosimetry using Lu-177 dotatate, Lu-177 HA-dotatate and lU-177 dotatoc: Comparative results in patients undergoing peptide receptor radionuclide therapy (PRRT)," *Eur. J. Nucl. Med. Mol. Imaging*, vol. 41, pp. S224–S225, 2014.

**Acknowledgement**
We would like to thank the Director General of Health Malaysia for his permission to publish this article.

# Determination of effective atomic numbers from mass attenuation coefficients of tissue-equivalent materials in the energy range 60 keV- 1.33 MeV

**Noorfatin Aida B. Amin[1], J Zukhi[1] , N. A. Kabir[2], R. Zainon[1]**

[1] Oncological and Radiological Sciences Cluster, Advanced Medical and Dental Institute, Universiti Sains Malaysia, 13200 Kepala Batas, Pulau Pinang, Malaysia;
[2] School of Physics, Universiti Sains Malaysia, 11800 Pulau Pinang, Malaysia

E-mail: rafidahzainon@usm.my

**Abstract.** The main aim of this study was to establish a cost-effective tissue-equivalent material for phantom fabrication. Effective atomic numbers ($Z_{eff}$) and effective electron densities ($N_{eff}$) were calculated based on mass attenuation coefficient values. The linear and mass attenuation coefficients of two samples of paraffin wax and NaCl compositions were measured using Si detector for NaI (Tl) detector of 1.5" resources. Radioactive source was placed in front of detector and the sample was placed between the source and the photomultiplier tube (PMT) of the detector. The real time was set for 6000 seconds. The photopeak, full width at half maximum (FWHM) and net area of photopeak were measured using Meastro software. The attenuation coefficient values obtained from this study were used to calculate $Z_{eff}$ and $N_{eff}$ of paraffin wax and NaCl compositions. The measured results were compared with the theoretical values from XCOM and ICRU Report 44. The relative percentage difference of mass attenuation coefficients between experimental and human tissue for both paraffin wax and NaCl mixture are below 5%, whereas the relative percentage difference of $Z_{eff}$ and $N_{eff}$ are above 5%. The measured values of $Z_{eff}$ and $N_{eff}$ of paraffin wax and NaCl help us to establish the optimal mixtures to fabricate a cost-effective tissue-equivalent material.

## 1. Introduction

Studies on tissue equivalent materials for anthropomorphic physical phantom fabrication are important for radiation dosimetry, calibration and medical imaging [1]. These tissue equivalent materials are being constructed from several chemicals [2] that exactly represent the anatomical shape of human organs [3-4] and have similar interaction of ionizing radiation to human body. In order to be accepted as a tissue-equivalent material, the absorbed and the scattered radiations should be as close as possible to those experienced by the irradiated tissue under similar conditions [5]. To understand the physical properties of tissue equivalent materials, a study of parameters such as mass attenuation coefficient ($\mu_m$), linear attenuation coefficient ($\mu_l$), total atomic cross-section ($\sigma_t$), electronic cross-section ($\sigma_e$), effective atomic number ($Z_{eff}$), electron density ($N_{eff}$), mean free-path ($\lambda$) are important [6].

Mass attenuation coefficient is defined as the measure of a material's ability to absorb or scatter electromagnetic radiation in any form (e.g., x-rays, gamma ray, etc.) per unit of mass. The mass attenuation coefficient is then used to derive other photon interaction parameters. Linear attenuation coefficient ($\mu_l$) describes the fraction of a beam of x-rays or $\gamma$- rays that is absorbed or scattered per unit thickness of the absorber. The scattering and absorption of gamma radiations are related to the density and atomic number of each element. For composite material, it is related to effective density and effective atomic number. The knowledge of mass attenuation coefficients is prime importance in the determination of effective atomic number [6]. A single number of atomic number therefore, cannot represent the atomic number uniquely across the entire energy range, since the partial interaction cross-sections depend on different element numbers [7].

The $Z_{eff}$ of the materials is one of the most important parameter representing radiation interaction, and this parameter has widely been used in medical imaging, radiotherapy and nuclear medicine. In this study, the mass attenuation coefficients and other photon interaction parameters of composite materials, paraffin wax and NaCl samples were determined at different gamma energies. The measured values were compared with the values obtained using semi empirical relations based on mixture rule, and values from XCOM and ICRU-44 report. Other photon interaction parameters were derived from the mass attenuation coefficient.

## 2. Experimental Method

### *2.1. Sample preparation*
Two mixtures of paraffin wax and NaCl were prepared to fabricate soft tissue equivalent materials and bone tissue equivalent material. The paraffin wax was melted at 60 °C. Then, the NaCl was added quickly into paraffin wax. The mixtures were then poured into fabricated mould with dimension of 3 cm x 3 cm x 0.5 cm. Two difference samples were prepared to represent soft tissue equivalent material of thyroid and bone tissue equivalent material of spinal cord. Both samples have difference ratio of NaCl added into paraffin wax.

### *2.2. Measurement of linear attenuation coefficients of paraffin and NaCl compositions*
The linear attenuation coefficients of soft-tissue equivalent and bone tissue equivalent samples were measured for energies between 88 keV and 1.33 MeV from exponential attenuation law. These energy range was chosen due to this range are used in medical imaging, radiotherapy and nuclear medicine department. The measurement was performed using a NaI (Tl) detector for gamma spectrometry. The detector was shielded with a lead housing to reduce the scattered radiation coming directly from the source or from the background. The attenuation measurements were measured using multichannel analyser (MCA). The samples were placed between detector and radiation source. The real time which is the total time it takes for a detector to count (maybe several seconds longer than live-time) was set for 6000 seconds. The photopeak, full width at half maximum (FWHM) and net area of photopeak were measured using Meastro software. The radioactive sources used to gained differences in gamma energies were Am-241 (87.83KeV), Na-22 (1274.53KeV), Cs-137 (661.66KeV), Mn-54 (834.85KeV) and Co-60 (1173.24 KeV and 1332.5 KeV). The linear attenuation coefficients of both samples were determined from the XCOM software. XCOM software is a web database used in calculating the photon cross sections and total attenuation coefficients for element, compound or mixture ($Z \leq 100$) at energies from 1 keV to 100 GeV.

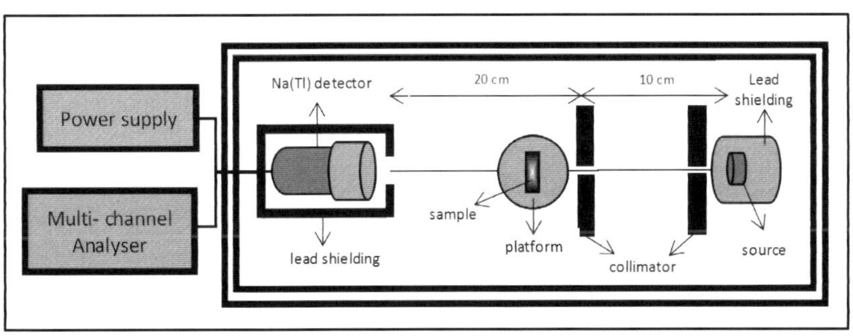

**Figure 1.** The experimental set up of MCA and Na(Tl) detector to measure mass attenuation coefficient of samples

*2.3. Comparison between measured mass attenuation coefficients, XCOM software and ICRU-44 report*

Two comparisons were carried out in this study to verify the soft-tissue and bone tissue equivalent materials. The relative percentage difference of mass attenuation coefficients of soft-tissue and bone tissue equivalent samples between experimental values and XCOM software values were determined. Next, the relative percentage difference of mass attenuation coefficients between samples and ICRU-44 were evaluated. The soft tissue and bone tissue equivalent samples were compared with ICRU-44 soft tissue and ICRU-44 bone tissue respectively for the energies of the radionuclide used.

*2.4. Calculating mass attenuation coefficient, effective atomic number and effective electron density of paraffin wax and NaCl compositions.*

Determination of mass attenuation coefficients of the samples were performed at difference energies based on Equation (1).

$$I = I_o \exp(-\mu_m t) \tag{1}$$

where $I_0$ is unattenuated photon intensity, $I$ is attenuated photon intensity, $\mu m$ = μ/ρ ($cm^2/g$) is the mass attenuation coefficient and $t$ ($g/cm^2$) is sample mass thickness (the mass per unit area)

For any chemical compound or mixture of elements, the total mass attenuation coefficient μis given by mixture rule shown in Equation (2):

$$\mu_m = \sum_i w_i (\mu_m)_i \tag{2}$$

where $w_i$ is the weight fraction, $(\mu_m)$i is the mass attenuation coefficient of ith element.

For a material composed of multi elements the fraction by weight is given by Equation (3):

$$w_i = \frac{n_i A_i}{\sum_i n_i A_i} \tag{3}$$

where Ai is the atomic weight of the ith element and ni is the number of formula units.

From $\mu m$ values measured, the total atomic cross-section (σt) for materials can be generated using the following relation shown in Equation (4):

$$\sigma_t = \frac{\mu_m N}{N_A} \tag{4}$$

where $N = \sum_i n_i A_i$ is atomic mass of materials and $NA$ is the Avogadro's number.

Total electronic cross-section (σe) for the element is expressed by the following equation (5):

$$\sigma_e = \frac{1}{N_A} \sum \frac{f_i N_i}{Z_i} (\mu_m)_i = \frac{\sigma_t}{Z_{eff}} \tag{5}$$

where fi denotes the fractional abundance of the element i with respect to the number of atoms such that

f1+f2+f3+f4+........fi =1

Zi is the atomic number of ith element

The effective atomic number ($Z_{eff}$) of the material can be calculated using equation (6)

$$Z_{eff} = \frac{\sigma_t}{\sigma_e} \quad (6)$$

The effective electron number or electron density (*Neff*) (number of electrons per unit mass) can be calculated using the following Equation (7):

$$N_{eff} = \frac{N_A}{N} Z_{eff} \sum n_i = \frac{\mu_m}{\sigma_e} \quad (7)$$

The photon mean free path ($\lambda$) which is the average distance between two successive interactions, is calculated by Equation (8):

$$\lambda = \frac{\int_0^\infty x \exp(-\mu x) dz}{\int_0^\infty \exp(-\mu x) dx} = \frac{1}{\mu_1} \quad (8)$$

where $\mu_l$ is linear attenuation coefficient and x is the absorber thickness.

## 3. Results and discussion

The relative percentage difference between the mass attenuation coefficients of soft-tissue and bone tissue equivalent materials were evaluated between experimental values and XCOM software. The results are presented in Table 1. For both samples, at energy 88 keV, the relative percentage difference are above 5%. The results also shown below 5% for energies between 662 keV and 1.3 MeV for both samples. Therefore, the use of soft tissue and bone tissue equivalent materials may be considered as a substitute material for soft tissue and bone tissue for radiotherapy and nuclear medicine uses.

**Table 1.** Absolute percentage difference of mass attenuation coefficients (cm$^2$/g) of soft tissue equivalent and bone tissue equivalent samples

| $E_\gamma (KeV)$ | **88** | **662** | **835** | ***1173*** | **1275** | **1333** |
|---|---|---|---|---|---|---|
| $\mu_{STE}(cm^2 g^{-1})$ | 0.1860 | 0.0860 | 0.0750 | 0.0630 | 0.0610 | 0.0590 |
| $\Delta_l(E_\gamma)_{STE}(\%)$ | 7.2 | 3.49 | 1.33 | 4.76 | 0.21 | 1.69 |
| $\Delta_l(E_\gamma)_a(\%)$ | 4.49 | 3.24 | 1.35 | 5.12 | 0 | 1.72 |
| $\mu_{BTE}(cm^2 g^{-1})$ | 0.195 | 0.079 | 0.070 | 0.062 | 0.057 | 0.055 |
| $\Delta_l(E_\gamma)_{BTE}(\%)$ | 5.4 | 0.38 | 0.43 | 0.32 | 0.88 | 0.9 |
| $\Delta_l(E_\gamma)_b(\%)$ | 1.52 | 0.38 | 0.43 | 0.32 | 0.35 | 0.36 |

Δl (Eɣ)soft tissue equivalent(%); Absolute percentage difference between mass attenuation coefficients of experimental and stimulation
Δl (Eɣ)a (%); Absolute percentage difference between mass attenuation coefficients of XCOM value and ICRU-44

The relative percentage difference of mass attenuation coefficient between measured values and ICRU-44 report was determined for the gamma energies of common radionuclide used. Differences below 5 % are observed in Table 1 for soft tissue equivalent sample. It shows that soft tissue equivalent has a good relationship in respect to soft tissue ICRU-44, except in the gamma energy of 1173 KeV which is 5.12%. For bone tissue equivalent material and ICRU-44, the differences below 0.5% were obtained of energies range 662 KeV and 1.3 MeV and 1.52% for energy 88 keV. It is clear that mass attenuation coefficient depends on photon energy and chemical content. The mass attenuation coefficient of a material decreases due to probability of absorption reduces with increasing incident photon energies which results in the increase in the transmission of photons through it. The total experimental uncertainty of mass attenuation coefficient values depend on the uncertainties of peak area evaluation, mass thickness measurements, experimental system, counting statistics, and efficiency errors.

The other parameters such as linear attenuation coefficient ($\mu_l$), total atomic cross-section ($\sigma_t$), electronic cross-section ($\sigma_e$), effective atomic number ($Z_{eff}$), electron density ($N_{eff}$), mean free-path ($\lambda$) were calculated using mass attenuation coefficient for both samples at different photon energies. The results were displayed in Tables 2 and 3. The photon mean free path ($\lambda$) for soft-tissue and bone tissue equivalent materials increases with the photon energy (Figure 2). This is due to decrease in the probability of interaction of photons in the material with the increase in energy. At low energies, total atomic cross-section ($\sigma_t$) and electronic cross-section ($\sigma_e$) decreases when energies increases (Figures 3 and 4). However, only at low energies, $\sigma_t$ and $\sigma_e$ the dependence of photon energy is dominant. The $Z_{eff}$ and the $N_{eff}$ remain constant and are found to be independent of photon energy for a compound except for energy 622 KeV which is slightly higher. Both effective atomic number and electron density have close relations to each other and thus have the same qualitative energy dependence.

The agreement between the calculated values from semi empirical relations, XCOM and experimental results is good. In a real practice, the uniformity of NaCl in the paraffin wax were varied as the mixture of these two elements are inhomogeneous, that a change in chemical surroundings and distortion. This factor may affect the mass attenuation coefficient value gained from experiment. However these factors are not considered when the values of mass attenuation coefficient were computed by XCOM and empirical relations. Hence the experimental values for the mass attenuation coefficients of soft tissue equivalent and bone tissue equivalent samples differ from the calculated values. This difference in value between experimental and calculated (XCOM value) might arise in the values of $\mu_m$ obtained from experiment and calculation due to experimental set up and its counting efficiency errors.

**Table 2.** $\mu_l$, $\mu_m$, $\sigma_t$ , $\lambda$ ,$\sigma_e$, $Z_{eff}$ and $N_{eff}$ values for soft tissue equivalent sample at difference gamma energies.

| Energy(KeV) | 87.8 | 622.0 | 834.9 | 1173.2 | 1274.5 | 1332.5 |
|---|---|---|---|---|---|---|
| $\mu_l$ (cm$^{-1}$) | 0.202 | 0.093 | 0.081 | 0.068 | 0.066 | 0.064 |
| $\mu_m$(cm$^2$/g) | 0.186 | 0.086 | 0.075 | 0.063 | 0.061 | 0.059 |
| $\lambda$(cm) | 4.95 | 10.75 | 12.35 | 14.71 | 15.15 | 15.63 |
| $\sigma_t$ (10x$^{23}$) barn/atoms | 4.05 | 1.96 | 1.63 | 1.37 | 1.33 | 1.28 |
| $\sigma_e$ (10x$^{25}$) barn/atoms | 4.14 | 1.82 | 1.67 | 1.40 | 1.36 | 1.31 |
| $Z_{eff}$ (electrons) | 9.78 | 10.72 | 9.78 | 9.78 | 9.78 | 9.78 |
| $N_{eff}$ (10x$^{23}$) electrons/z | 4.49 | 4.72 | 4.49 | 4.49 | 4.49 | 4.49 |

**Table 3.** $\mu_l$, $\mu_m$, $\sigma_t$ , $\lambda$ ,$\sigma_e$, $Z_{eff}$ and $N_{eff}$ values for bone tissue equivalent sample at difference gamma energies.

| Energy(KeV) | 87.8 | 622.0 | 834.9 | 1173.2 | 1274.5 | 1332.5 |
|---|---|---|---|---|---|---|
| $\mu_l$ (cm$^{-1}$) | 0.276 | 0.111 | 0.099 | 0.088 | 0.081 | 0.078 |
| $\mu_m$(cm$^2$/g) | 0.195 | 0.079 | 0.07 | 0.062 | 0.057 | 0.055 |
| $\lambda$(cm) | 3.62 | 9.01 | 10.10 | 11.26 | 12.35 | 12.83 |
| $\sigma_t$ (10x$^{23}$) barn/atoms | 5.56 | 2.17 | 1.99 | 1.77 | 1.63 | 1.58 |
| $\sigma_e$ (10x$^{25}$) barn/atoms | 3.25 | 1.26 | 1.16 | 1.04 | 0.97 | 0.92 |
| $Z_{eff}$ (electrons) | 17.10 | 17.25 | 17.10 | 17.10 | 17.10 | 17.10 |
| $N_{eff}$ (10x$^{23}$) electrons/z | 6.0 | 6.25 | 6.0 | 6.0 | 6.0 | 6.0 |

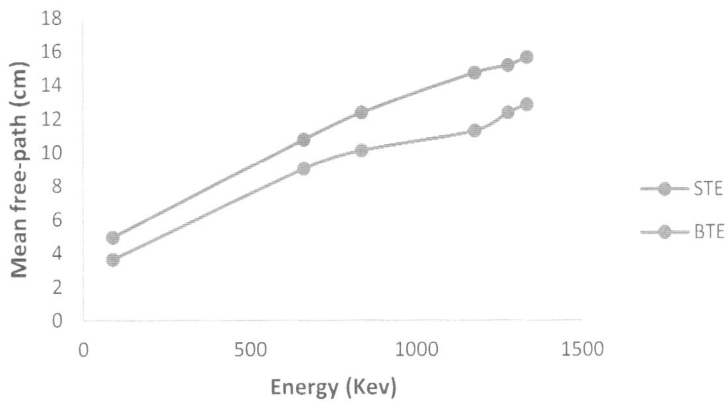


**Figure 2**. Mean free-path against photon energy

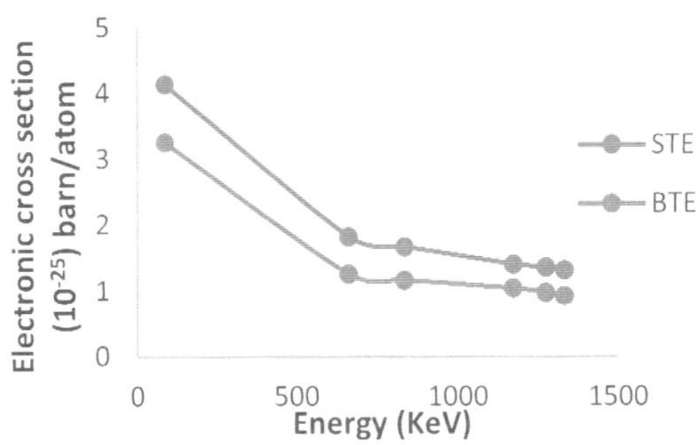


**Figure 3.** Electronic cross section against photon energy

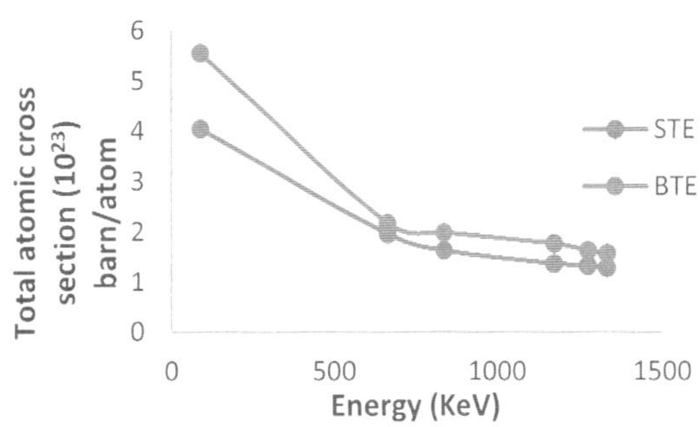


**Figure 4.** Total atomic cross section against photon energy

## 4. Conclusion

The above experimental study were carried out to determine the mass attenuation coefficients and other related photon interaction parameters for soft tissue equivalent and bone tissue equivalent samples. In the interaction of photon with matter, the $\mu_m$ values are dependent on the physical and chemical composition of the elements in the sample.The $\mu_m$ values of both samples decrease with increase in photon energy. Both soft tissue equivalent and bone tissue equivalent samples have mass attenuation coefficient values close to human tissue, with percentage difference below 5%. The $\sigma_t$ and $\sigma_e$ decreases as the photon energy increases but the influence dominant at low energies. It can be concluded that the mass attenuation coefficients can be used to generate and determine the Zeff and Neff and other photon interaction parameters for soft tissue equivalent of thyroid and bone tissue equivalent of spinal cord. The $Z_{eff}$ and $N_{eff}$ are independence on photon energy. However, the $Z_{eff}$ and $N_{eff}$ calculated are above 5% of its percentage difference when compare to $Z_{eff}$ and $N_{eff}$ human tissue. All photon interaction parameters of soft tissue equivalent and bone tissue equivalent obtained in this study will enhance the understanding on how the mass attenuation coefficients change with variation of the atomic and electronic number for difference amount of NaCl added into paraffin wax. This study also aimed to establish soft tissue equivalent and bone tissue equivalent as tissue equivalent materials for phantom fabrication.

## 5. Reference

[1] SM Naderi, S Sina, M Karimipoorfard, F Lotfalizadeh, M Entezarmahdi, H Moradi and R Faghihi. 2015. *Radiation Protection Dosimetry* ,pp. 1–6.

[2] H Hasanzadeh, A Abedelahi. 2011. *Journal pf Pramedical Sciences* **2**(4): 2008-4978.

[3] White, D. R. 1993. *Radiation Protection Dosimetry* **49** (1/3), 359-369.
[4] Cerqueira, R. A. D. and Maia, A. F. 2014. *Radiat. Phys. Chem* **95**, 174–176.
[5] L.S. Del Lama, L.D.H. Soares et al. 2014. *Nuclear Instruments and Methods in Physics Research* **784** (2015) 597-601.
[6] A.S. Madhusudhan Rao, A Getachew. 2015. *International Journal of Advance Research In Science And Engineering* **4**(01), 2319-8354.
[7] Singh, N., Singh, K.J., Singh, K. and Singh, H., 2006. *Radiat. Meas.* **41**, pp.84-88.
[8] International Commission on Radiation Units and Measurements ICRU. 1992. *ICRU Report* **48,** 96–101.
[9] M Kurudirek. 2014. *Applied Radiation and Isotopes* **94** (2014), 1-7.
[10] H Yücel, E Güllüoğlu, Ş Çubukçu and Yiğit Ali Ü. 2016. *Journal of Physical Science*, **27** (1), 111–128
[11] Chaudhari L. M and Raje D. V. 2013. *Journal of Chemical, Biological and Physical Sciences* **3** (2) 1504-1510
[12] A. Hermosilla, G. Dı´az London˜o, M. Garcı´a, F.Ruı´z, P. Andrade and A. Pe´rez. 2014. *Radiation Protection Dosimetry* **162** (4), 508–514
[13] Parisa Akhlaghi, Hashem Miri Hakimabad, Laleh Rafat Motavalli. 2015. *Radiation Physics and Chemistry* **112** (2015), 169–176.
[14] C C Ferreira , R E Ximenes, C A B Garcia, J W Vieira, A F Maia. 2010. *Journal of Physics: Conference Series* **249** (2010) 012029

**Acknowledgements**
The authors would like to acknowledge the financial support from Ministry of Higher Education through Fundamental Research Grant Scheme (FRGS).

IOP Conf. Series: Journal of Physics: Conf. Series **851** (2017) 012019 doi:10.1088/1742-6596/851/1/012019

# High dose rate $^{192}$Ir source calibration: A single institution experience

**R Abdullah[1], N H Abdullah[1], M Mohamed[2, 3], N R N Idris[2, 3], A L Yusoff[2, 3], S C Chen[1], and A Zakaria[1]**

[1]School of Health Sciences, Health Campus USM, Universiti Sains Malaysia, Kelantan, Malaysia
[2]Department of Nuclear Medicine, Radiotherapy and Oncology, School of Medical Sciences, Health Campus USM, Universiti Sains Malaysia, Kelantan, Malaysia
[3]Department of Nuclear Medicine, Radiotherapy and Oncology, Hospital Universiti Sains Malaysia, Kelantan, Malaysia.

E-mail: reduan@usm.my

**Abstract.** Measurement of source strength of new high dose rate (HDR) $^{192}$Ir supplied by the manufacturer is part of quality assurance recommended by Radiation Safety Section, Ministry of Health of Malaysia. The source strength is determined in reference air kerma rate (RAKR). The purpose of this study was to evaluate RAKR measurement of $^{192}$Ir using well-type ionisation chamber with RAKR stated in the certificate provided by the manufacturer. A retrospective study on 19 MicroSelectron HDR $^{192}$Ir Classic from 2001 to 2009 and 12 MicroSelectron HDR $^{192}$Ir V2 sources from 2009 to 2016 supplied by manufacturer were compared. From the study, the agreement between measured RAKR and RAKR stated in the certificate by manufacturer for all 32 sources supplied were within ±2.5%. As a conclusion, a threshold level of ±2.5% can be used as suitable indicator to spot problems of the brachytherapy system in Department of Nuclear Medicine Radiotherapy and Oncology, Hospital USM.

## 1. Introduction

HDR brachytherapy is an adjuvant treatment to the external beam radiotherapy (EBRT) in treating many types of cancer such as cervical cancer, endometrial cancer, breast cancer, prostate cancer, and bone sarcoma. Nowadays, $^{192}$Ir source has become the most commonly used high dose rate (HDR) brachytherapy source. It has high specific activity; thus, a very small volume is needed to deliver a very high dose. Smaller volume of $^{192}$Ir source can easily travel through curvature anatomy of human body for example through intracavity, intraluminary, and interstitial. The only disadvantage of $^{192}$Ir is that it has a short half -life of 73.83 days, requiring relatively frequent source replacement after 4 to 5 months to maintain short treatment times [1]. Thus, Department of Nuclear Medicine, Radiotherapy and Oncology, Hospital USM needs to spend about USD 30,000 to USD 50,000 per year to purchase the $^{192}$Ir sources and this cost is increasing year by year.

Our department needs to get approval from the Radiation Safety Section of Ministry of Health to purchase the new source. After getting the approval, the department can make an order from a local vendor with a license to do sales operation to bring in the source from the manufacturer's country. The license is endorsed by the Atomic Energy Licensing Board as local authorised agency that controls any activities involved with radiation. After the source arrived at the national airport, the local vendor must assign a transportation company with a license to conduct transportation operation relating to radioactive source to transport the source to the department. The same company will transport back the old source to the manufacturer's country. At the department, the trained engineer from the local vendor company will carry out the exchange process by retracting the old source into the Dual Channel Transport Container (Nucletron, Mallinckrodt Medical B.V., Netherland) and transfer the new source from the container into the brachytherapy treatment unit. Each new $^{192}$Ir source delivered for exchange comes with a certificate which the manufacturer states its strength. The source strength is determined in reference air kerma rate (RAKR). All the procedures are supervised by our physicist based on a guideline prescribed by national authority, followed to ensure the safety of public, worker, and environment, and also security of radioactive material. Currently, there are 13 centres using the same brand of brachytherapy treatment unit manufactured by Mallinckrodt Medical (Netherlands) and $^{192}$Ir source in this country.

Independent verification as well as exact measurement of source strength is quite significant to assure the quality of brachytherapy treatment [2]. As part of quality assurance programme for brachytherapy system in our department, the RAKR of a new source delivered by the vendor needs to be verified. Well-type ionisation chambers are preferred for routine calibration of HDR $^{192}$Ir source because of its ease, precision in use, fastness and reproducibility of source positioning [1-3].The well-type ionisation chambers have been specifically designed for calibration of HDR sources. The advantages of using these chambers are that they can minimise uncertainties from wall scattering effect, room humidity, and transit dose effect. The source was programmed to enter the chamber and dwell at the maximum response dwell position of the chamber. This point was determined by varying in small steps of the source position in the central axis from the top to the bottom defining a small plateau and selecting most representative values [4].

Measured RAKR values were calculated and compared with values specified by manufacturer. The American Association of Physicists in Medicine (AAPM) report No. 40, gave an option of using either the manufacturer's stated RAKR or the user's measured activity if they were within a ±3% agreement [5]. Nevertheless, the latest AAPM report No. 56 recommended that a calibration is performed by qualified medical physicist as a part of quality assurance programme and that the user calibration be used as the basis for treatment planning and prescription [5-6].

## 2. Material and methods

The retrospective study was carried out on 19 Microselectron HDR $^{192}$Ir Classic sources from 2001 to 2009 and 13 MicroSelectron HDR $^{192}$Ir V2 sources from 2009 to 2016. The study was conducted to assess the accuracy of HDR $^{192}$Ir supplied by manufacturer to our department.

### *2.1. Reference Air Kerma Rate (RAKR) measurement of MicroSelectron HDR $^{192}$Ir Classic Source*

A MicroSelectron HDR Classic brachytherapy machine (Nucletron, Mallinckrodt Medical B.V., Netherland) was installed at our department on 3rd of August 1999. The first MicroSelectron $^{192}$Ir HDR Classic source (Nucletron, Mallinckrodt Medical B.V., Netherland) was installed into the machine on 5th May 1999. The source was encapsulated by stainless steel, AISI 316L. This miniature source has dimensions of 1.5 mm diameter and 5.0 mm length and source pellet dimensions of 0.6 mm diameter and 3.5 mm length. The first clinical case of cervical cancer patient was done on 22nd July 1999.

For the first five $^{192}$Ir sources, RAKR stated on the certificate were used to be entered into Plato treatment planning system (TPS) version 14.1 (Nucleteron, Netherland) and into treatment console system (TCS). In July 2001, the department received the sixth MicroSelectron $^{192}$Ir HDR Classic source and our physicist performed RAKR measurement. The HDR 1000 Plus well-type ionisation chamber (Standard Imaging, USA), calibrated at the Accredited Dosimetry Calibration Laboratory of the University of Winconsin (ADCL-UW) was used to measure RAKR. The well-type chamber was connected to the Max-4000 electrometer (Standard Imaging, USA) outside the treatment room through a connection cable. Blunt side of the blue endobronchial catheter was inserted through the top of well-type ionisation chamber while open end of the catheter was connected through the gynaecological transferring tube which was then connected to the first indexer of treatment unit. The source was programmed to be at the most sensitive position of the well chamber. Production of ion pair (negative and positive charges) occurred when photon was interacted with argon gas molecule inside the well chamber. Next, the charges were collected and recorded. Temperature and pressure of the treatment room were recorded after the measurement procedure. RAKR value was calculated using Equation (1). After the measured RAKR was determined, the value needs to be compared using Equation (2)

$$\text{RAKR} = A \times N_k \times K_{TP} \times A_{ion} \tag{1}$$

$$D\,(\%) = ((M - C) / C) \times 100\% \tag{2}$$

RAKR is the reference air kerma or air kerma source strength of $^{192}$Ir, A is the current, $N_k$ is air kerma strength calibration factor of well-type ionisation chamber in $mGy^{-1}h^{-1}$ at 1 $m^2$, $K_{TP}$ is temperature and pressure correction coefficient, $A_{ion}$ is charge collection efficiency correction at the time of measurement, D is percentage deviation in %, M is measured RAKR in $mGyA^{-1}h^{-1}$ at 1 $m^2$ and C is certificate-stated RAKR in $mGyA^{-1}h^{-1}$ at 1 $m^2$ (the value needs to re-calculated to consider the decay or source strength from the time source manufactured to the time of measurement done)

### *2.2. Reference Air Kerma Rate (RAKR) measurement of MicroSelectron HDR $^{192}$Ir V2 Source*

In May 2009, a MicroSelectron HDR V3 brachytherapy machine (Nucletron, Mallinckrodt Medical B.V., Netherland) was installed to replace the MicroSelectron HDR Classic treatment unit (Nucletron, Mallinckrodt Medical B.V., Netherland). In the last 10 years of using the old treatment unit, 282 patients were treated.

For the new treatment unit, the new source needs to suit with the new machine. The first MicroSelectron $^{192}$Ir HDR V2 source (Nucletron, Mallinckrodt Medical B.V., Netherland) was installed into the machine on 18th May 2009. Smaller source was encapsulated compared to the older source with dimensions of 0.9 mm diameter and 4.5 mm length and source pellet dimensions of 0.6 mm diameter and 3.5 mm length. The new treatment unit came with new set of quality assurance tools including a well-type ionisation chamber (HDR Chamber Type 3304) and an electrometer (Unidose E). These tools are manufactured by PTW, Germany. The procedures in measuring RAKR for the new source were slightly different compared to the older treatment unit. HDR Chamber Type 3304 is specially designed for brachytherapy calibration and measuring the RAKR of HDR sources. One end of the gynaecological transfer tube was connected to source holder that was kept inside the well of the chamber and the other end of the transfer tube connected to the first indexer of treatment unit. Unidose E electrometer was switched to the charge mode and high range was selected. From our experience with the system, the most sensitive dwell position was at position 22 with 138.8 cm, which was measured earlier using the source simulator ruler. The source was programmed to dwell in this position for 600 s. The charge was collected at a 60 s interval and two readings were taken each for + 400 V, - 400 V, + 200 V and - 200 V. At this maximum response of dwell position, the chamber charge is maximised and uncertain in the RAKR determination, due to minimised positional variations [1]. Temperature and pressure of the brachytherapy treatment room were recorded after measurement was done. The RAKR value stated on the certificate was corrected backwards for initial source decay and was compared with measured values. All these data were recorded and was entered into the Excel spread sheet provided in the department based on Equations 1 and 2.

## 3. Results and Discussion

As part of quality assurance (QA) based on AAPM recommendations, the source calibration or RAKR measurements were carried out every time the HDR $^{192}$Ir source was exchanged. Based on this retrospective study, Table 1 shows that the percentage difference between measured RAKR and

IOP Conf. Series: Journal of Physics: Conf. Series **851** (2017) 012019 doi:10.1088/1742-6596/851/1/012019

RAKR stated in the certificate provided by manufacturer of MicroSelectron HDR $^{192}$Ir Classic source ranges from - 2.50% to 1.60%. Among 19 measured RAKR of the HDR source, it can be observed that 9 sources have a deviation more than 1% and all other 10 sources have a deviation within ±1%.Meanwhile for MicroSelectron HDR $^{192}$Ir V2 source as tabulated in Table 2, the percentage difference between measured RAKR and RAKR stated in the certificate provided by manufacturer's values ranges from - 1.77% to 1.69%. Only 3 sources out of 12 sources that have been measured of RAKR have deviations more than 1%. From the study, all deviations were within the tolerance limits stated by national authority and international recommendations, which is ±3%.

**Table 1.** Measured RAKR, RAKR stated in the certificate, and respective percentage differences of MicroSelectron HDR $^{192}$Ir Classic source.

| No. | Source no. | Measurement date | Stated RAKR (corrected for decay) $mGym^2h^{-1}$ | Measured RAKR using well type ionisation chamber $mGym^2h^{-1}$ | Percentage difference (%) |
|---|---|---|---|---|---|
| 1 | D35Q-568 | * | * | * | * |
| 2 | D35S-154 | * | * | * | * |
| 3 | D35S-772 | * | * | * | * |
| 4 | D35T-446 | * | * | * | * |
| 5 | D35U-365 | * | * | * | * |
| 6 | D35V-003 | 7/19/2001 | 43.352 | 43.786 | 1.00 |
| 7 | D35V-736 | 11/27/2001 | 43.030 | 42.518 | - 1.19 |
| 8 | D35W-992 | 7/17/2001 | 46.043 | 45.440 | - 1.31 |
| 9 | D35Y-270 | 2/24/2003 | 40.210 | 39.981 | - 0.57 |
| 10 | D35Z-024 | 6/28/2003 | 38.620 | 39.238 | 1.60 |
| 11 | D35Z-307 | 9/7/2003 | 43.779 | 43.061 | - 1.64 |
| 12 | D35A0149 | 6/30/2004 | 45.497 | 44.360 | - 2.50 |
| 13 | D35A1495 | 3/3/2004 | 40.211 | 39.660 | - 1.37 |
| 14 | D35A1720 | 9/26/2004 | 42.937 | 42.559 | - 0.88 |
| 15 | D35A1951 | 1/10/2005 | 36.722 | 37.053 | 0.90 |
| 16 | D35A2995 | 6/28/2005 | 44.477 | 43.350 | - 2.31 |
| 17 | D35A3585 | 12/28/2005 | 38.580 | 38.900 | 0.83 |
| 18 | D35A4298 | 7/31/2006 | 43.532 | 43.793 | 0.60 |
| 19 | D35A4853 | 12/26/2006 | 40.340 | 40.090 | - 0.62 |
| 20 | D35A5510 | 8/28/2007 | 30.663 | 30.326 | - 1.10 |
| 21 | D35A5680 | 1/1/2008 | 39.110 | 39.094 | - 0.04 |
| 22 | D35A6231 | 6/10/2008 | 44.797 | 44.389 | - 0.91 |
| 23 | D35A6823 | 11/30/2008 | 46.366 | 45.930 | - 0.94 |
| 24 | D35A7141 | 3/22/2009 | 36.215 | 36.070 | - 0.40 |

*RAKR was not measured; RAKR stated on certificate was used

The associated uncertainty with a source certificate calibration is ±5% for $^{192}$Ir source based on measurement done for initial RAKR and the uncertainty in well-type chamber calibration recommended by national and international bodies is ±3%. The well-type chamber uncertainty may be smaller than the source certificate but if an error were to occur with the departmental ionisation chamber calibration, the extent of the error would be far greater than error on an individual source calibration. This is due to the length of time that an ionisation chamber is used in a departmental before re-calibration, with ionisation chamber calibrations to be performed every 2 years [7]. In Malaysia, only the Nuclear Agency provides the service, or another option is to send it to other calibration centres in Europe to re-calibrate the chamber, which will be very costly.

**Table 2.** Measured RAKR, RAKR stated in the certificate, and respective percentage differences of MicroSelectron HDR $^{192}$Ir V2 source.

| No. | Source no. | Measurement date | Stated RAKR (corrected for decay) $mGym^2h^{-1}$ | Measured RAKR using well type ionisation chamber $mGym^2h^{-1}$ | Percentage difference (%) |
|---|---|---|---|---|---|
| 1 | D36B9249 | 7/20/2009 | 25.78 | 25.92 | 0.55 |
| 2 | D36C2132 | 2/4/2010 | 38.46 | 38.20 | - 0.68 |
| 3 | D36C4659 | * | * | * | * |
| 4 | D36C6751 | 2/14/2011 | 48.68 | 48.35 | - 0.68 |
| 5 | D36C9646 | 10/10/2011 | 42.61 | 42.40 | - 0.50 |
| 6 | D36E1350 | 3/22/2012 | 42.40 | 42.40 | 0.01 |
| 7 | D36E4036 | 10/15/2012 | 37.73 | 37.65 | - 0.22 |
| 8 | D36E7071 | 5/8/2013 | 41.44 | 41.62 | 0.43 |
| 9 | D36F0151 | 1/28/2014 | 37.30 | 37.93 | 1.69 |
| 10 | D36P1156 | 6/18/2014 | 42.65 | 42.44 | - 0.50 |
| 11 | D36F4122 | 1/7/2015 | 42.77 | 42.40 | -0.86 |
| 12 | D36F6233 | 4/26/2016 | 3.510 | 3.448 | - 1.77 |
| 13 | D36G0541 | 7/14/2016 | 39.65 | 39.16 | - 1.23 |

*RAKR was not measured; RAKR stated on certificate was used

Measuring the RAKR using well-type ionisation chamber is easier, faster and straight forward. If an institution determines the ratio of RAKR with manufacturer's values and it differs more than 3%, that institution should review the data for calculations and measurements [8]. Nevertheless, the consideration due to positional of well-type chamber uncertainty and source transit effect requires attention during the measurement. In order to minimise the positional uncertainty, the chamber should be placed at the same position for every exchange source procedures. Moreover, based on the contribution of scattered radiation from room's wall, the chamber should be placed at the centre between two walls and above the floor (at least 1 m away or height from any wall or floor). Moreover, the chamber should be kept energised at the corner of the treatment room and should maintain overnight a day before the measurement was carried to attain thermal equilibrium and electronic stabilisation. Besides that, instead of using charge mode, the current mode was recommended to avoid extraneous current generated in chamber during source transit [9].

## 4. Conclusion

Radiation Safety Section under Ministry of Health as national authority stated that RAKR measurement should be carried out every time the source exchanges. The tolerance deviation between measured RAKR and RAKR stated in the certificate is ±3% [10]. From this retrospective study, the percentage difference between measured and stated RAKR values of 32 sources of brachytherapy $^{192}$Ir sources is within limit ±3%. As a conclusion, the percentage difference of RAKR ≤2.5% can be achieved between measured and certificate stated values. The ±2.5% of deviation is suitable to be a threshold level practiced as department standard procedure, as an indicator to spot any problems in calibrating HDR $^{192}$Ir source. It has been a standard operation procedure in our department to evaluate the HDR $^{192}$Ir source calibration supplied by the vendor prior to clinical use. Based on department experiences, either manufacturer's stated RAKR or measured RAKR may be used in our TPS and TCS since the deviations are less than ±3%. In order to improve the brachytherapy treatment accuracy, it is strongly recommended that physicists should carry out RAKR measurement at the time of source replacement.

## 5. References

[1] Nagappan B, Kumar Y, P. Patel N, Dhull A K and Kaushal V 2016 Brachytherapy source calibration, reviews and consistency of $^{192}$ Ir high-dose rate afterloading sources supplied over the period of 10 years: A retrospective analysis *Radiat. Prot. Environ.* **38** 144-50

[2] Gadhi M A, Buzdar S A, Afzal M, Fatmi S, Akhtar M S and Nizamani A H 2013 Calibration of iridium-192 source by ionization chamber for high dose rate brachytherapy *Inter. J. Radiat. Res.* **11** 189-93

[3] Nor Faranaz S N A, Ahmad Z, Reduan A and Norha A H 2012 comparison QA methods of brachytherapy using well ionization chamber and in air method *Malays. J. Fund. App. Sci* **8** 246-52

[4] De Almeida C E, Pereira A J, Marechal M H, Pereura G, Cruz J C, Ferraz J C, Giordani A J, Khalil C M, Martins R H, Menuggussi G, Moreira D, Rocha J R and Pinto M A 1999 Intercomparison of calibration procedures for $^{192}$Ir HDR sources in Brazil *Phys. Med. Biol.* **44** N31-N38

[5] Austerlitz C, Wolfe M, Campos D and Sibata C H 2012 Consistency of vendor-specified activity values for $^{192}$Ir brachytherapy sources *Medical Dosimetry* **37** 67-70

[6] Nath R, Anderson L L, Olch A J, Stitt J A and Williamson J F 1997 Code of practice for brachytherapy physics: Report of the AAPM Radiation Therapy Committee Task Group No. 56. American Association of Physicist in Medicine *Med. Phys.* **24** 1557-98

[7] Demsey C 2011 Lessons learned from a HDR brachytherapy well ionisation chamber calibration error *Australas. Phys. Eng. Sci. Med.* **34** 529-33

[8] Lee J H, Su S H, Hsieh M T, Chen I J Liang J A and Hsu S M 2011 Air kerma standard and measurement comparison on source strength determination for high-dose rate $^{192}$ Ir brachytherapy in Taiwan *Radiation Protection Dosimetry* **146** 100-02

[9] International Atomic Energy Agency 1999 *Calibration of Brachytherapy sources, IAEA-TECDOC-1079. Guidelines on standardized procedures at Secondary Standards Dosimetry Laboratories and Hospitals* Vienna

[10] Ministry of Health 2013 *Manual Implementation of Quality Assurance Program (QAP) in Radiotherapy Services* Kuala Lumpur

## Acknowledgments

This study was supported financially by the Short-term Grant (304/PPSK/61313048) of Universiti Sains Malaysia. This work was made possible with the cooperation of the Nuclear Medicine, Radiotherapy and Oncology Department, School of Medical Sciences, Hospital Universiti Sains Malaysia and School of Health Sciences, Universiti Sains Malaysia (USM).

IOP Conf. Series: Journal of Physics: Conf. Series **851** (2017) 012020 doi:10.1088/1742-6596/851/1/012020

# Radiotherapy dose verification on a customised head and neck perspex phantom

**K Y Eng[1], S Kandaiya[1], N Z Yahaya[2]**

[1] School of Physics, Universiti Sains Malaysia, Penang, Malaysia.
[2] School of Distance Education, Universiti Sains Malaysia, Penang, Malaysia

E-mail: eky7411@hotmail.com

**Abstract**. IMRT dose planned for head and neck radiotherapy was verified using a customised acrylic head-and-neck phantom. The dosimeters used were calibrated Gafchromic EBT2 film and metal–oxide–semiconductor-field-effect-transistor (MOSFET). Target volumes (TV) and organs-at-risk (OAR) which were previously contoured by an oncologist on selected nasopharynx (NPC) patients were transferred to this phantom by an image fusion procedure. Three radiotherapy plans were done: Plan1 with 7-fields intensity-modulated radiotherapy (IMRT) of prescribed dose 70 Gy using 33 fractions; Plan2 with 7-fields IMRT plan at 70 Gy and 35 fractions; and Plan3 which was a mid-plane-dose (MPD) plan of 66 Gy at 33 fractions. The dose maps were first verified using MapCheck2 by SNC-PatientTM software. The passing rates from gamma analysis were 97.7% (Plan1), 93.1% (Plan2) and 100% (Plan3). Percentage difference between Treatment Planning System (TPS) calculated dose and MOSFET measured dose was comparatively higher than those from EBT2. Calculated dose and EBT2 measured doses showed differences of within the range of ±3% for TV and <±10% for OARs. However MOSFET had differences of within the range of ±6% for TV and within the range of ±10% for OARs between measured and planned doses. An overdose treatment may occur as TPS calculated doses were lower than the measured doses in these plans. This may be due to the effects of leaf leakage, leaf scatter and photon backscatter into the measuring tools (Pawlicki *et al.*, 1999 and Ma *et al.*, 2000). More IMRT plans have to be studied to validate this conclusion. However, the dose measurements were still within the 10% tolerance (AAPM Task Group 119). In conclusion, both GafchromicEBT2 film and MOSFET are suitable for IMRT radiotherapy dosimetry.

## 1. Introduction

*In vivo* dosimetry plays an important role in quality assurance (QA) in radiotherapy especially in advanced radiotherapy techniques like IMRT. Conventional dosimeter such as ionisation chamber (IC), semiconductor detector and thermoluminescence dosimeter (TLD) often fail to meet the requirement to provide high spatial resolution of dose distributions and absolute dose determination at point of interest. Gafchromic EBT2 film provides high spatial resolution of dose distribution which is necessary for IMRT QC whereas MOSFET's characteristic like easy to use, small size, light weight, instant readout and reducing time for processing. Therefore, MOSFET and Gafchromic EBT2 film were used as clinical dosimeters in this study. The main objective of this study is to verify the exposed radiation dose of TV and OAR for NPC radiotherapy, in conventional MPD and IMRT plan. Dose verification can detect any differences between planned and delivered doses; it is performed on a customised acrylic (Perspex) head and neck phantom using Gafchromic EBT2 film and MOSFET.

### *1.1. Simulation*
In the very beginning for nasopharynx (NPC) radiotherapy, the oncologist will mark the desired treatment area on the patient's skin by estimation from anatomical structure. With improvement in technology, NPC radiotherapy went from conventional simulation using a simulator named two-dimension (2D) treatment planning to Computed Tomography (CT) simulation. The latter was named three-dimension (3D) treatment planning and now even four-dimension (4D) treatment planning [1].

In radiotherapy planning, it is crucial to have a comfortable positioning for patient for high reproducibility. An immobilisation tool, for example a beam directional shell (BDS), is commonly used in NPC radiotherapy to ensure high reproducibility of treatment.

### *1.2. Radiotherapy Planning*
Radiotherapy planning is the process of determining the best method of treating a TV using high energy radiation. The main purpose of radiotherapy planning is to ensure TV receives a uniform and adequate radiation dose whilst the healthy tissues and OARs are protected.

Lead block or multi-leaf collimators (MLCs) will enhance the radiotherapy plan by shielding the OARs risk and healthy tissues from irradiation. Manual calculation of dosage or complex computerised planning will be the next step of radiotherapy planning.

MPD plan is a parallel opposed field beam configuration [1] used in 2D treatment planning. This field arrangement is suited for treatment of midline structures. Since the human head is symmetrical and nasopharyngeal tumor is assumed to be located along the midline of the clivus bone, NPC can be treated with lateral MPD plan. IMRT is an advanced conformal radiotherapy that utilizes computer optimised inverse treatment planning and computer controlled treatment delivery [2]. It is currently the most favoured choice for head and neck cancer radiotherapy, including NPC. This radiotherapy involves multiple beam angles and segments which have varied or modulated dose intensity. It is able to produce a customised radiation dose intensity to maximize the TV dose while minimizing the dose to adjacent OAR and healthy tissues.

### *1.3. Dose Verification and Radiotherapy Delivery*
After treatment planning, the next procedure is dose verification prior to radiotherapy delivery. Since IMRT has high complexity dose patterns and techniques, quality control (QC) should be done for both equipment operation and delivery of treatment to individual patients [3]. In Cancer Centre Pantai Hospital Penang, for equipment operation, QC was done using MapCheck2 to check the dose map whereas for patient, portal film or electronic portal imaging (EPID) was taken before the start button was pressed. Measurements on phantom where the human anatomy is reproduced are often recommended in addition to in-vivo measurements [4]. Therefore, a series of tests using *in-vivo* measurement tools like Gafchromic EBT2 Film and MOSFET were performed on the phantom.

Radiotherapy is delivered once the verification in dose and patient positioning passed the requirement. In normal cases, NPC radiotherapy is delivered over a total of 33 - 37 fractions with five fractions per week.

## 2. Methodology

### *2.1. Fabrication of Phantom*
A set of CT scans for a NPC case which can represent an average patient size was selected from the XiO TPS, Cancer Center Pantai Hospital Penang. The head and neck contours were printed out first in paper at 1 cm thickness in transverse plane from the level vertex of head to chest. The papers were traced onto a 1 cm Acrylic (Perspex) slices according to the slices sequences. Acrylic slabs were cut using electronic hand cutter and polished to make shape smooth. The cut acrylic slices were then stacked together using adhesive double side tape to ensure no air gaps remained in between the slabs. A customised head and neck phantom is formed when all the slices were stacked together (Figure 1).

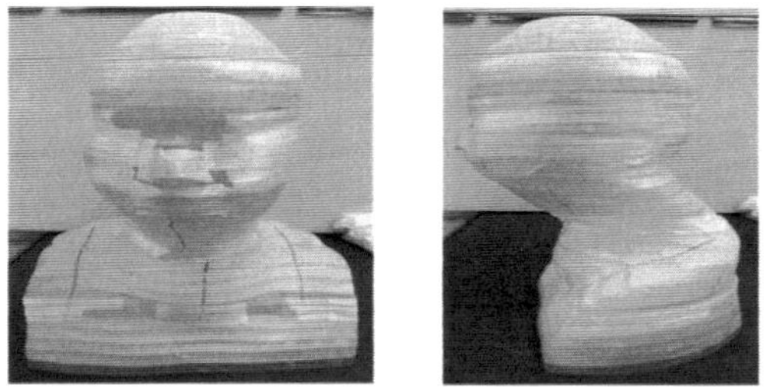

**Figure 1.** Customised Perspex head and neck phantom using 1 cm slice of acrylic (Perspex). Both the front and side views are shown.

### *2.2. Calibration of dosimeters*

The dosimeters were irradiated with 6 MV photons at source to surface distance (SSD) of 100 cm and field size used was 10 cm x 10 cm. The dosimeters were placed at 5 cm depth with 10 cm slab of solid water phantom to provide sufficient backscattering. Various parameters such as reproducibility, fading, field size dependence and angular dependence were studied.

### *2.3. CT simulation*

The phantom was set to immobilisation aids like head rest and sponge wedges, and then a BDS was custom-made for this acrylic head and neck phantom. Three lead markers which were visible in CT images were placed on the BDS as setup reference points. The phantom was scanned using Toshiba Aquillion 64 slices CT scanner then reconstructed to 2 mm slices from vertex of head to chest level (Figure 2). The CT images were exported to CMS XiO TPS through Digitising Imaging and Communication in Medicine (DICOM) system.

### *2.4. Treatment Planning*

According to the International Commission on Radiation Units and Measurements 62 (ICRU 62) [6], the oncologist will contour the target volume with the desired treatment dose. The medical physicist will design a customised radiotherapy planning with appropriate radiation field arrangements.

In this study, the dose verification of three radiotherapy plans planned by XiO TPS; MPD plan and two different seven fields IMRT plans were done on a customised acrylic head and neck phantom.

TV and OARs (brainstem, left eye, right eye, left parotid, right parotid and oral cavity) which were previously contoured by oncologist and medical physicist on the selected NPC patient was transferred to this phantom by image fusion procedure. Two 7 field IMRT plans were planned at gantry angle 0°, 51°, 102°, 153°, 207°, 258° and 309°. The differences of these two IMRT plans were the prescribed dose and fractions. A simple MPD plan using two lateral opposed fields onto the PTV area and with MLCs to shield the OARs and healthy tissues was planned. As a summary, Plan 1 is 7-field IMRT plan with prescribed dose of 70 Gy in 33 fractions; Plan 2 is 7-field IMRT plan with 70 Gy in 35 fractions; and Plan 3 is mid-plane dose plan with 66 Gy in 33 fractions.

### *2.5. Treatment Delivery and Verification*

Then these three plans were optimised, the doses were verified using MapCheck2 as per routine for dose QC at Cancer Center Pantai Hospital Penang. All the fields were set to gantry angle 0° and then exposed to the MapCheck2 to get a dose map. This measured dose map was then compared with the planned dose map by the SNC Patient™ software for all the three plans.

The phantom was properly aligned on the LINAC couch according to the three lead markers on the BDS with setup same as per CT simulation process (Figure 3) aided by electronic portal imaging device (EPID)[7]. Both the Gafchromic EBT2 film pieces and MOSFET were set on the points of interest (POI) for measurements (Figure 4). The three plans were delivered to the phantom and measurements were recorded. The minimum dose, maximum dose, mean dose and point doses for selected TV and OARs by TPS calculation were recorded. In order to validate the comparison in dose (cGy), the exposed Gafchromic EBT2 films were scanned and the pixel value converted to optical density then to dose (in cGy); MOSFET readings in millivolts (mV) were converted into dose (cGy).

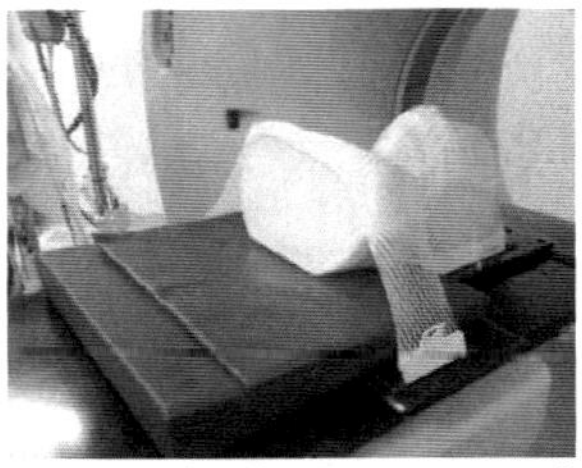

**Figure 2.** Customised Perspex phantom with immobilisation aids and BDS was scanning by CT scanner (Source: Picture was taken at CT simulation room, Cancer Center Pantai Hospital Penang).

**Figure 3.** Setup of Perspex phantom on the LINAC couch (Source: Picture taken at LINAC treatment room, Cancer Center Pantai Hospital Penang).

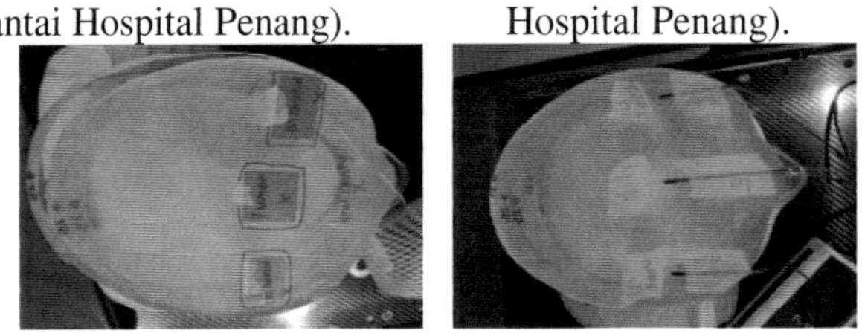

(a) (b)

**Figure 4.** Set up of Gafchromic EBT2 film and MOSFET onto the phantom's Perspex slice.

## 3. Results and Discussion

*3.1 Calibration of Dosimeters*

Accordingly to ISO methodology as described in IAEA TRS-398 dosimetry protocol, the uncertainties of Gafchromic EBT2 Film and MOSFET measurements were derived [5]and given in Table 1. The combined standard uncertainty, $u(x)$, from source of uncertainties of $ua(x)$, $ub(x)$, $uc(x)$ was calculated by equation 1.

$$u(x) = \sqrt{(ua(x))^2 + (ub(x))^2 + (uc(x))^2 + \cdots etc.} \quad (1)$$

The dosimeters Gafchromic EBT2 and MOSFET were first characterized. Gafchromic EBT2 had a 5.05% combined standard uncertainty which were from uncertainties in reproducibility, film darkening effect, field size and angular dependence. MOSFET had 2.94% combined uncertainty from uncertainties like reproducibility, fading effect, field size and angular dependence.

**Table 1.** Summary of uncertainties and calculated combined standard uncertainty for Gafchromic EBT Film and MOSFET.

| | Uncertainty (%) | |
|---|---|---|
| | **Gafchromic EBT2 Film** | **MOSFET** |
| Reproducibility | 0.25 | 1.36 |
| Film Darkening Effet (EBT2 Film) / Fading effect (MOSFET) | 0.02 | 0.64 |
| Field Size Dependence | 4.73 | 1.54 |
| Angular Dependence | 1.76 | 2.00 |
| **Combined standard uncertainty:** | **5.05** | **2.94** |

*3.2 Treatment Delivery and Verification*

The dose calculated by TPS was first verified by comparing the calculated dose map with the measured dose map from MapCheck2 at DTA of 3 mm and dose difference of 3%. The results are shown in Table 2.

**Table 2.** Comparison of selected points between TPS calculated doses with Gafchromic EBT2 Film measurements and MOSFET measurements for the three plans.

| Variables | Tool | Plan 1 | | | | | Plan 2 | | | | Plan 3 | | | |
|---|---|---|---|---|---|---|---|---|---|---|---|---|---|---|
| | | n | Median (IQR) | $X^2$ statistic | p-value | % Difference with TPS | Median (IQR) | $X^2$ statistic | p-value | % Difference with TPS | Median (IQR) | $X^2$ statistic | p-value | % Difference with TPS |
| Tumor Volume | TPS | 5 | 173.69 (0.17) | | | - | 148.52 (0.42) | | | - | 199.48 (0.95) | | | - |
| | MOSFET | 5 | 184.22 (1.54) | 12.500 | 0.002 | -5.86 | 153.21 (2.08) | 12.500 | 0.002 | -3.33 | 197.03 (2.02) | 12.020 | 0.002 | 1.19 |
| | EBT 2 | 5 | 175.33 (0.98) | | | -1.04 | 149.92 (1.01) | | | -1.13 | 203.67 (2.41) | | | -2.00 |
| Brainstem | TPS | 5 | 153.42 (0.10) | | | - | 150.88 (0.96) | | | - | 182.49 (1.39) | | | - |
| | MOSFET | 5 | 168.88 (1.07) | 12.545 | 0.002 | -9.98 | 165.12 (3.78) | 12.500 | 0.002 | -9.62 | 191.67 (2.03) | 12.500 | 0.002 | -5.02 |
| | EBT 2 | 5 | 157.56 (0.75) | | | -2.71 | 143.02 (1.40) | | | 4.85 | 180.04 (1.10) | | | 1.42 |
| Oral Cavity | TPS | 5 | 76.08 (0.13) | | | - | 80.32 (0.53) | | | - | 16.89 (0.64) | | | - |
| | MOSFET | 5 | 82.43 (0.99) | 12.522 | 0.002 | -8.28 | 87.98 (1.03) | 12.020 | 0.002 | -9.87 | 18.14 (0.18) | 10.659 | 0.005 | -7.86 |
| | EBT 2 | 5 | 79.48 (1.42) | | | -4.41 | 82.03 (1.42) | | | -1.77 | 17.90 (0.56) | | | -5.50 |
| Right Parotid | TPS | 5 | 106.98 (0.45) | | | - | 71.43 (0.93) | | | - | 195.08 (1.19) | | | - |
| | MOSFET | 5 | 113.04 (0.91) | 12.020 | 0.002 | -5.82 | 73.09 (0.05) | 12.613 | 0.002 | -2.45 | 204.61 (2.33) | 12.522 | 0.002 | -4.76 |
| | EBT 2 | 5 | 111.23 (1.09) | | | -4.09 | 66.99 (1.75) | | | 5.12 | 184.89 (2.05) | | | 5.54 |
| Left Parotid | TPS | 5 | 85.68 (0.52) | | | - | 74.18 (0.22) | | | - | 183.27 (2.16) | | | - |
| | MOSFET | 5 | 89.51 (1.14) | 12.500 | 0.002 | -4.47 | 79.65 (1.36) | 12.522 | 0.002 | -6.94 | 197.23 (1.57) | 12.020 | 0.002 | -7.09 |
| | EBT 2 | 5 | 93.27 (1.56) | | | -9.10 | 69.98 (1.59) | | | 6.25 | 187.90 (2.74) | | | -2.06 |
| Right Eye | TPS | 5 | 37.15 (0.13) | | | - | 22.81 (0.23) | | | - | 183.99 (1.03) | | | - |
| | MOSFET | 5 | 38.44 (1.30) | 9.081 | 0.011 | -3.48 | 23.53 (0.11) | 10.693 | 0.005 | -3.28 | 200.67 (1.78) | 10.820 | 0.004 | -9.39 |
| | EBT 2 | 5 | 39.02 (1.31) | | | -5.31 | 24.14 (0.97) | | | -6.36 | 187.43 (2.91) | | | -1.60 |
| Left Eye | TPS | 5 | 31.8 (0.17) | | | - | 15.68 (0.13) | | | - | 185.04 (1.81) | | | - |
| | MOSFET | 5 | 34.11 (0.53) | 9.500 | 0.009 | -7.64 | 16.79 (0.10) | 12.500 | 0.002 | -6.90 | 190.94 (1.98) | 9.500 | 0.009 | -3.57 |
| | EBT 2 | 5 | 32.03 (1.36) | | | -0.84 | 17.18 (0.55) | | | -9.68 | 191.02 (2.37) | | | -3.35 |

The measured dose maps from all the three radiotherapy plans were in the 10% tolerance as accepted in the protocol of Cancer Center Pantai Hospital Penang. The passing rates from gamma analysis were 97.7 %, 93.1 % and 100 % for Plan 1, Plan 2 and Plan 3 respectively. This shows that the calculated dose by TPS and measured map were in agreement. Hence, the calculated dose from TPS can be used to compare the measured dose by the dosimetry tools.

By Kruskal-Wallis test and Mann-Whitney test (the post hoc analysis), the medians of three dosimeters (TPS, MOSFET, EBT2 film) are significantly different ($p<0.05$). On the whole a percentage difference of less than ± 3 % were found between TPS calculated dose and the Gafchromic EBT2 film measured dose distribution for TV. Higher percentage differences for OARs of less than ± 10 % were found. Plan 1 and Plan 2 are IMRT plans; IMRT plans have high complexity and sharp dose gradients between TV and spared healthy tissues. The percentage difference is high for some measured POI, but it is still within the range between the minimum and maximum dose. However, Gafchromic EBT2 film showed a lower percentage difference of less than ± 6 % in Plan 3 which is the MPD plan. MPD plan has a more uniform dose in the region of treatment, where the dose has a nice square dose contour or a slightly hour glass shape of isodose lines [1]. Since the dose distribution for MPD plan is comparatively uniform, the measured doses agreed with the calculated dose by TPS.

On the other hand, the percentage difference between the calculated dose by TPS and the measured dose by MOSFET was comparatively higher than that of Gafchromic EBT2 film. A highest difference of 21.78 cGy occurred in brainstem in Plan 2. This may be due to MOSFET measuring only at a point whereas Gafchromic EBT2 film measured over a small area and average dose is over this small area.

In conclusion the TPS calculated dose and the MOSFET measured dose showed a percentage difference of less than ± 6 % for TV and ± 10 % for the OARs. Brainstem is the organ which is just next to TV. Although this is in a high dose gradient, the brainstem is under the tolerance dose. Hence brainstem showed the highest percentage differences of 9.98 % and 9.62 % in both IMRT plans.

The point of measurement was difficult to define as the point was often located in a steep gradient dose in IMRT treatment [8]. Therefore, the uncertainty in positioning of the detector on the phantom would introduce the dose difference in the measurement. The misplacement error should be eliminated to ensure the measured dose occurs at the same location at each measurement [7]. Therefore, verification process using portal imaging is highly recommended to reduce misplacement error.

Nevertheless, the TPS calculated dose was normally lower than the measured dose by Gafchromic EBT2 film and MOSFET. This may be due to the effects of leaf leakage, leaf scatter and photon backscatter into the measuring tools [8-13].

## 4. Conclusion

An overdose treatment may occur to patient as the TPS calculated doses were lower than the measured doses either by Gafchromic EBT2 film or MOSFET for the IMRT plans. More IMRT plans have to be studied to validate this conclusion. However, the dose measurements were still within the 10% tolerance which is the acceptance protocol of Cancer Center Pantai Hospital Penang (AAPM Task Group 119) [14]. In conclusion, both Gafchromic EBT2 film and MOSFET are suitable for use in radiotherapy dosimetry.

## 5. References

[1] Barrett, A., et al. (2009). Practical Radiotherapy Planning Fourth Edition, CRC Press.
[2] Chuang, C. F., et al. (2002). "Investigation of the use of MOSFET for clinical IMRT dosimetric verification." Medical physics **29**(6): 1109-1115.
[3] Williams, P. C. (2014). "IMRT: delivery techniques and quality assurance." The British journal of radiology.
[4] Gwi A Cho *et al* 2014. In vivo and phantom measurement versus eclipse TPS prediction of near surface dose for SBRT treatments. *J. Phys.: Conf. Ser.* **489** 012008
[5] Pedro, A., et al. (2006). IAEA TRS-398, Absorbed Dose Determination in External Beam Radiotherapy: An International Code of Practice for Dosimetry based on Standards of Absorbed Dose to Water. V.12 169.
[6] ICRU Report 62 – Prescibing, Recording and Reporting Photon Beam Therapy (Supplement to ICRU Report 50), 1999 (ICRU: Bethesda, MD) ISBN 0-913394-61-0
[7] Walter C et al (2007). "Phantom and in-vivo measurements od dose exposure by image-guided radiotherapy (IGRT): MV portal images vs. kV portal images vs. cone beam CT." Radiotherapy Oncology 2007 Dec; 85 (3): 418-23.
[8] Marcié, S., et al. (2005). "In vivo measurements with MOSFET detectors in oropharynx and nasopharynx intensity-modulated radiation therapy." International Journal of Radiation Oncology* Biology* Physics **61**(5): 1603-1606.
[9] C.-M Ma et al. (1999). " Clinical implement of a Monte Carlo treatment planning system." Medical Physics 26, 2133 (1999)
[10] C-M Ma et al. (2000). " Monte Carlo verification of IMRT dose distribution from a commercial treatment planning optimization system.
[11] Wang X, Spirou S, LoSasso T, Stein J, Chui C and Mohan R 1996 Dosimetric verification of intensity modulated fields *Med. Phys.* **23** 317–28
[12] Webb S 1997, *The Physics of Conformal Radiotherapy: Advances in Technology* (Bristol: Institute of Physics Publishing)
[13] Chui C S, LoSasso T and Spirou S 1994 Dose calculations for photon beams with intensity modulation generated by dynamic jaw or multileaf collimators *Med. Phys.* **21** 1237–43
[14] Ezzell et al. (2009), IMRT commissioning: Multiple institution planning and dosimetry comparisons, a report from AAPM Task Group 119. Medical Physics 36(11) 5369-5373

## Acknowledgement

Authors acknowledge the financial support for this research from the Department of Higher Education, Ministry of Education Malaysia through Fundamental Research Grant Scheme, FRGS (203.PJJAUH.6711489), Project USM - Islamic Development Bank and Universiti Sains Malaysia, providing the Short Term Grant (304.PJJAUH.6313152)

IOP Conf. Series: Journal of Physics: Conf. Series **851** (2017) 012021 doi:10.1088/1742-6596/851/1/012021

# Homogeneous and inhomogeneous material effect in gamma index evaluation of IMRT technique based on fan beam and Cone Beam CT patient images

**W E Wibowo[1,2], M Waliyyulhaq[2], and S A Pawiro[2,*]**

[1] Department of Radiotherapy, National General Hospital Dr. Cipto Mangunkusumo, DKI Jakarta, 10430, Indonesia
[2] Department of Physics, Faculty of Mathematics and Natural Sciences, Universitas Indonesia, Kampus UI, Depok, 16424, Indonesia

E-mail: supriyanto.p@sci.ui.ac.id

**Abstract:** Patient-specific Quality Assurance (QA) technique in lung case Intensity-Modulated Radiation Therapy (IMRT) is traditionally limited to homogeneous material, although the fact that the planning is carried out with inhomogeneous material present. Moreover, the chest area has many of inhomogeneous material, such as lung, soft tissue, and bone, which inhomogeneous material requires special attention to avoid inaccuracies in dose calculation in the Treatment Planning System (TPS). Recent preliminary studies shown that the role of Cone Beam CT (CBCT) can be used not only to position the patient at the time prior to irradiation but also to serve as planning modality. Our study presented the influence of a homogeneous and inhomogeneous materials using Fan Beam CT and Cone Beam CT modalities in IMRT technique on the Gamma Index (GI) value. We used a variation of the segment and Calculation Grid Resolution (CGR). The results showed the deviation of averaged GI value to be between CGR 0.2 cm and 0.4 cm with homogeneous material ranging from -0.44% to 1.46%. For inhomogeneous material, the value was range from -1.74% to 0.98%. In performing patient-specific IMRT QA techniques for lung cancer, homogeneous material can be implemented in evaluating the gamma index.

## 1. Introduction

Modern Linear Accelerator (LINAC) is typically equipped with Multi-Leaf Collimator (MLC), which shapes the beam to conform the target. The conformal radiotherapy will increase the target dose and reduce the dose to surrounding normal tissues [1]. In planning techniques using MLC, an error can occur if either (a) calculated leaf sequences not generated accurately on the fluence used by TPS for dose calculation; (b) information from treatment planning system is not transferred perfectly to the accelerator; or (c) radiotherapy accelerator is not functioning correctly, both mechanically and dosimetrically [2]. Unlike conventional radiotherapy, IMRT uses a number of sub-fields to modulate the intensity of the field on the target. This difference is to deliver high conformity radiotherapy to the target with a gradient of high doses to organs at risk [3].

In addition to the number of sub-fields in IMRT technique, determination of the Calculation Grid Resolution (CGR) in the dose calculation in the TPS also determines the accuracy of the dose calculation [4-5]. Therefore, it is necessary to revisit the problem on discretizing the impact of dose distributions due to a sub-field. In addition, dose accuracy is also required in terms of dose fluence in such planning, being evaluated using a program analysis based on the Gamma Index with criteria % dose difference - DD (difference doses) and distance to agreement (DTA) [6].

The traditional method of quality assurance of IMRT techniques in determining the value of Gamma Index (GI), may yield only limited information about the accuracy of dose distributions of actual patients, since it only accommodates homogeneous material and does not incorporate the effect of the presence of the inhomogeneous material, i.e. air spaces, bone structure and soft tissue [7]. Moreover, for radiotherapy on lung cancer, Kappas et. al described that inhomogeneous material with high range of density difference did present in the volume of interest. The study presented the presence of the inhomogeneous material, such as the lungs, would correct the dose at the water for more than 30%, implying the need of taking inhomogeneous material into consideration for dose calculation [8].

The commercial phantom for the quality assurance verification for IMRT setup verification employs a phantom with homogeneous materials, particularly phantoms with 2D-array capabilities during whose usage the direction of gantry is only on one direction, i.e. perpendicular with field detector. Although GI measurements tool which runs by 3-dimensional concept have been commercially available [9-10], the development of these dosimeters in general is still limited in homogeneous material when a beam passes through the phantom dosimeter [11]. This, due to the presence of the inhomogeneous material, may decrease GI value and accuracy of TPS dose calculations for inhomogeneity, with respect to the dose calculation algorithm [12-13].

In general, CBCT imaging with flat panel detector is primarily used for planning verification but Sriram [14] proposed CBCT to be used for replanning and dosimetric verification. The result of that the percentage dose difference based on between Fan Beam CT (FBCT) and kV-CBCT did not exceed 5%. On the other hand, the calibration of CT numbers has been performed using CBCT for the pelvic region planning also does not exceed 5% [15].

The pelvic region is homogenous area compared to the chest area which represented inhomogeneous material.The importance of the accuracy of dosing in IMRT technique in terms of dose distributions, as well as the location and the material. We carried out a study on the effect of homogeneous and inhomogeneous material on evaluation of Gamma Index in IMRT techniques (Intensity Modulated Radiotherapy) based on the image of the patient's Fan Beam and Cone Beam CT.

## 2. Materials and methods

### *2.1. Materials*

The study was performed at Radiotherapy Department, RSUPN Cipto Mangunkusumo, Jakarta. This study was performed with LINAC Elekta Synergy S which equipped with 6 MV photon energy and CBCT, slab phantom (RW3), homogeneous phantom as in Figure 1(a), inhomogeneous phantom as in Figure 1(b) used cork, acrylic, and resin; cork is equivalent with lungs density and acrylic or resin is equivalent to bone density. All of this material were manufactured from Bayer Materialscience Indonesia Factory. This experiment, we employed PTW 2D Array as detector. Matrixscan and Verisoft software was used to measure the planar dose and evaluate of gamma index, with Philips Pinnacle Planning System as planning software. Images acquired with FBCT modality have been acquired using GE Brightspeed, and images acquired using CBCT modality were acquired using XVI system (integrated with the LINAC).

### *2.2. Planning system*

This work used FBCT- and CBCT-acquired images as illustrated in Figure 2, for five male patients of 30-60 years old with right lungs cancer. Parameter for FBCT were used Chest protocol institution (kV 120; mA 10; and 2.5 mm slice thickness) and CBCT used Medium FOV (kV 120; mA 25, and ms 40). The data acquired from those modalities, planning was generated with IMRT technique. CBCT image from XVI was transferred to TPS by combining the target and OAR FBCT delineation. FBCT and CBCT re-planning was carried out using two of CT number calibration curves. CT number calibration curves are from FBCT and CBCT from a previous study [20], IMRT technique re-planning was performed by using seven coplanar fields [16]. This study used Philips Pinnacle Planning System with 0.2 cm and 0.4 cm calculation grid resolutions (CGR) and used 30, 50, and 70 segments for each planning. The CGR variation was used to observe its impact on dose distribution, especially in 0.4 cm [9] because the recommendation of maximum CGR for lungs cancer is 0.3 cm [4].

IOP Conf. Series: Journal of Physics: Conf. Series **851** (2017) 012021 doi:10.1088/1742-6596/851/1/012021

*2.3. Phantom setup for measurements*

The homogeneous and inhomogeneous materials phantoms were scanned using FBCT and CBCT. The setup for homogeneous slab phantom consist of (RW3, $\rho$=1.045 gr/cc) [17] from PTW with various depth of the detector which were ; 1.5 cm, 5 cm, 10 cm with 5 cm for back scatter. The inhomogeneous material simulated the chest area which was arranged as :

- Slab phantom with 3 cm thickness for upper and lower position [18].
- Resin material with 1 cm thickness ($\rho$=1.363 gr/cc) [19] and 0.4 cm thickness of acrylic ($\rho$=1.060 gr/cc) [20] as bone simulation set for upper and lower position.
- Cork material with 15 cm thickness ($\rho$=0.210 gr/cc) [20] as lung tissues simulation. The lungs and bones thickness was obtained from measurement data with FBCT modality for lung cancer patient in Radiotherapy RSCM in 2012 until March 2015. The PTW 2D Array detector was set in 1.5 cm, 5 cm, and 10 cm thickness of lung tissue.

The image of phantom was transferred to Philips Pinnacle Planning System to generate planar dose calculation. The isocenter was on 100 cm SAD with effective point of detector (0.5 cm from the surface of the PTW 2D Array detector) and on the center of the detector. All gantry directions were modified into $0^{\circ}$ angle on every planning and modality, so that its planar dose might be obtained.

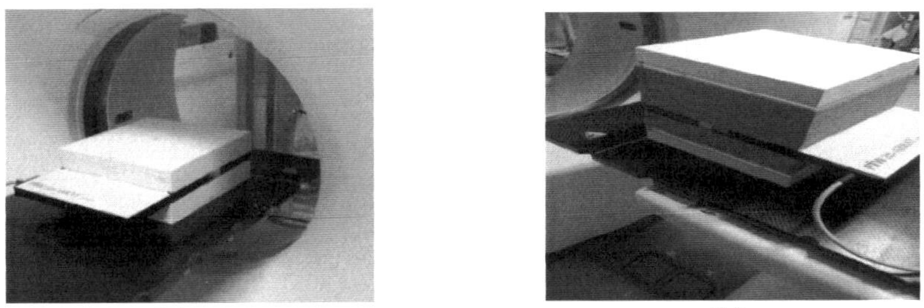

(a) (b)

**Figure 1.** Set up Phantom with : (a) Homogeneous and (b) Inhomogeneous phantom

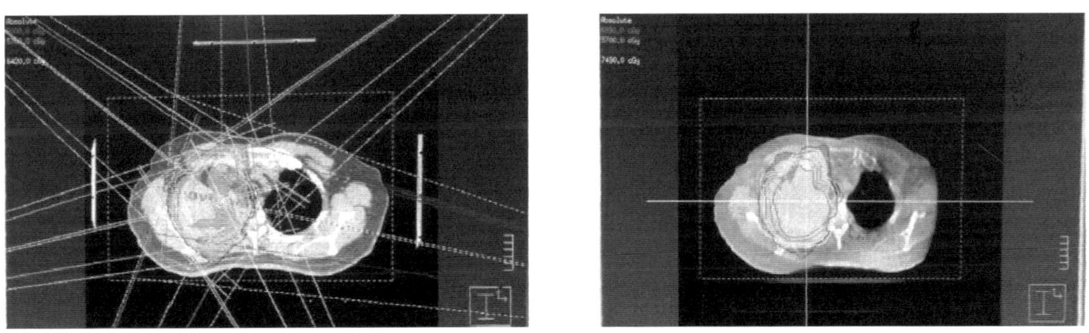

(a) (b)

**Figure 2.** IMRT Treatment Planning with: (a) FBCT- and (b) CBCT-acquired images

*2.4. Gamma index evaluation*

The setup on the measurement corresponded with the setup on the planning. Planar dose from measurement was compared with planar dose from TPS for all planning with DTA and DD criteria 2 mm/2% and 90% for the passing rate [21, 9, 22].

## 3. Results

*3.1. Gamma index evaluation of calculation grid resolution*

*3.1.1. Homogeneous material FBCT modality* Table 1(a) shows a deviation of GI value between the Calculation Grid Resolution (CGR) of 0.2 and 0.4 cm in images of homogeneous material acquired with FBCT modality. A CGR of 0.2 cm is used as a reference. The greatest deviation of GI value for IMRT of 70 segments at a depth of 1.5 cm has a positive sign (+), meaning that CGR of 0.4 cm is higher than 0.2 cm. The deviation with a negative sign (-) has a meaning that CGR of 0.2 cm is higher than 0.4 cm. In deviation sign (-) as seen in Table 1(a), 50 segments at a depth of 1.5 cm is the greatest. Deviation value (from a negative value to a positive value) is increase along with the addition of depth for 50 segments. Different phenomena are shown in Table 1(a) for 70 segments, where deviation decreases with increasing of depth. Table 1(a) shows deviation is 0% at a depth of 5 cm, which indicates the value of CGR 0.2 cm is identical to CGR 0.4 cm. At the depth of 10 cm, all variations of segment have positive sign (+). At a depth of 10 cm, the deviation decreases with the increase of the number of segments with the greatest deviation value in 30 segments is 0.47% and the least value in 70 segments is 0.04%.

*3.1.2 Homogeneous material CBCT modality* Table 1(b) shown a deviation of GI value between the CGR of 0.2 and 0.4 cm in images of homogeneous material acquired using CBCT modality. As seen in Table 1(b) at the overall depth and technique, the deviation is negative (-), meaning that the average value of GI on CGR 0.4 cm was lower than the average value of GI on CGR 0.2 cm. the greatest deviation in 30 segments at a depth of 5 cm with a value of -1.13% and the least deviation in 50 segments at the same depth with a value of -0.04% was described in Table 1(b). the least deviation is in 30 segments with a value of -0.44% and -0.29% at 1.5 cm and 10 cm depths . In both depths, the greatest deviation value was found at the same technique (see Table 1(b)) i.e in 50 segments with a value of -1.00% for a depth of 1.5 cm and – 1.10% for a depth of 10 cm. For a depth of 5 cm, the least deviation value of -0.04% on the 50 segment and the greatest deviation value of -1.13% for 30 segments.

*3.1.3. Inhomogeneous material FBCT modality* Table 2(a) shows a deviation with the same parameters as Table 1(b) but using images of inhomogeneous material acquired with FBCT modality. Table 2(a) shows overall deviation denoting that average value of GI on CGR 0.4 cm being lower than CGR 0.2 cm. The greatest deviation (Table 2(a)) at a depth of 5 cm IMRT technique 30 segments is -2.54% and the least deviation of GI value at 50 segment is -0.04% at a depth of 10 cm. When compared to all three variations of depth to the entire segment, the least deviation value is at a depth of 1.5 cm and the greatest deviation of GI value is at a depth of 5 cm.

For a depth of 1.5 cm, Table 2(a) shown that the deviationof GI value increases along with the increase of number of segments, with the least value at this depth being -0.11% and the greatest value was being -0.55%. Meanwhile at a depth of 5 cm, the deviation of GI value inversely proportional to the depth of 1.5 cm. With the trend at a depth of 5 cm, a deviation decreases with increasing the number of segments (see Table 2(a)). The highest and lowest deviation of GI value at this depth were - 2.54% and -1.42%. At a depth of 10 cm, as seen in Table 2(a), the least deviation of GI value was found at 50 segments, with a value of -0.04% and the greatest deviation value was 0.47%, found on 30 segments.

*3.1.4. Inhomogeneous material CBCT modality* For CBCT-acquired inhomogeneous material, a deviation of GI value between CGR is 0.2 cm and 0.4 cm as seen in Table 2(b). Almost all deviation values were negative, except at a depth of 5 cm segments 50 (close to 0%), and depth of 10 cm which was 0.98%. Table 1(b) also shown the greatest deviation of GI values in the sign (-) at a depth of 1.5 cm segment 70 with a value of -1.74% and the greatest deviation in sign (+) at a depth of 10 cm 50 segments with a value of 0.98%. For 50 segments, it was seen that deviation was increasingly moving towards (+) with increasing depth. Furthermore in Table 2(a), the greatest deviation value for a depth of 1.5 cm, 5 cm and 10 cm were at 70 segments with a deviation value - 1.74%, -0.99%, and -1.24%, respectively, and the least being in 50 segments for the three depths, with the value being -0.40%, 0.01%, 0.98%, at the depth of 1.5, 5, and 10 cm, respectively.

| Technique | Deviation (%) | | |
|---|---|---|---|
| | Depth | | |
| | 1.5 cm | 5 cm | 10 cm |
| 30 Seg | 0,66 | 0,00 | 0,47 |
| 50 Seg | -0,29 | 0,00 | 0,43 |
| 70 Seg | 1,46 | 0,00 | 0,04 |

(a)

| Technique | Deviation (%) | | |
|---|---|---|---|
| | Depth | | |
| | 1.5 cm | 5 cm | 10 cm |
| 30 Seg | -0,44 | -1,13 | -0,29 |
| 50 Seg | -1,00 | -0,04 | -1,10 |
| 70 Seg | -0,77 | -0,30 | -0,66 |

(b)

**Figure 3.** Deviation of average GI value between CGR 0.2 cm and 0.4 cm in homogeneous materials: (a) FBCT Modality and (b) CBCT Modality

| Technique | Deviation (%) | | |
|---|---|---|---|
| | Depth | | |
| | 1.5 cm | 5 cm | 10 cm |
| 30 Seg | -0.11 | -2.54 | 0,47 |
| 50 Seg | -0,26 | -1,70 | -0,04 |
| 70 Seg | -0,55 | -1,42 | -0,29 |

(a)

| Technique | Deviation (%) | | |
|---|---|---|---|
| | Depth | | |
| | 1.5 cm | 5 cm | 10 cm |
| 30 Seg | -0,79 | -0,38 | -1,07 |
| 50 Seg | -0,40 | 0,01 | 0,98 |
| 70 Seg | -1,74 | -0,99 | -1,24 |

(b)

**Figure 4.** Deviation of average GI value between CGR 0.2 cm and 0.4 cm in inhomogeneous materials: (a) FBCT Modality and (b) CBCT modality

## 4. Discussion

In general, the use of CGR in the evaluation of GI resulting in GI values being more varied in CGR 0.4 cm in comparison to those of 0.2 cm. This might be due to reduced accuracy in the calculation of the dose with the use of CGR being greater than 0.2 cm. Generally, the deviation of GI value using CGR 0.2 cm and 0.4 cm at 1.5 cm depth increased along with increasing number of segment, except for the inhomogeneous phantom for CBCT modality. However, there were no trend with number of segment at 5 and 10 cm depths. Similarly, there was no trend for the effect of GI at same number of segment with different depth between CGR 0.2 cm and 0.4 cm. However, the use of CGR 0.4 cm still suffice for use on the purpose of therapeutic planning. Additionally, added presence of irregularities, thickness, and volume may present more inaccuracy in terms of dose calculations using CGR 0.4 cm in comparison with the same situation using CGR 0.2 cm [23, 24, 4, 5, 25, 13]. This study, however, had not observed a significant difference in both existing CGR.

## 5. Conclusion

Patient-specific IMRT QA techniques for lung cancer can be implemented using homogeneous material in evaluating the gamma index and optimum depth at 5 cm using CGR 0.2 cm.

## 6. References

[1] Jen-Sai T & Wazer D E 1998 Dosimetric Verification of the Dynamic Intensity-Modulated Radiation Therapy of 92 Patients *Int. J. Radiation Oncology Biol. Phys.* **40**(5) 1213–1230

[2] Van Zijtveld M. et al. 2006 Dosimetric Pre-treatment Verification of IMRT using an EPID; Clinical Experience *Radiotherapy and Oncology* **81**(2) 168–175

[3] DeLuca P M & Seltzer S M 2008 *Journal of the ICRU: ICRU Report 83 Prescribing, Recording, and, Reporting Photon-Beam Intensity-Modulated Radiation Theraphy (IMRT)*
[4] Benedict S H et al. 2010 Stereotactic Body Radiation Therapy: The Report of AAPM Task Group 101 *Medical Physics* **37**(8) 4078–4101
[5] Li J S et al. 2010 Uncertainties in IMRT Dosimetry *Medical physics* **37** 2491–2500
[6] Donmez Kesen N & N N 2014 Pre-Treatment Dose Verification of IMRT Using Gafchromic EBT3 Film and 2DArray *Journal of Nuclear Medicine & Radiation Therapy* **05**(02) 3–8
[7] Chung K et al. 2013 Radiochromic Film Based Transit Dosimetry for Verification of Dose Delivery with Intensity Modulated Radiotherapy *Medical physics* **40**(2) 021725
[8] Kappas C & Rosenwald J -C. 1995 Quality Control of Inhomogeneity Correction Algorithms in Treatment Planning System. *Science*, **32**(3) 847–858
[9] Van Esch A, Basta K & Evrard M 2014 The Octavius1500 2D Ion Chamber Array and its Associated Phantoms: Dosimetric Characterization of a New Prototype *Medical physics* **41**(9) 091708
[10] Oldham M et al. 2008 An investigation of the Accuracy of an IMRT Dose Distribution Using Two-and Three-Dimensional Dosimetry Techniques *Medical physics* **35**(5) 2072–2080
[11] Low D A et al. 2011 Dosimetry Tools and Techniques for IMRT *Medical physics* **38**(3) 1313–1338
[12] Lye J et al. 2014 A 2D Ion Chamber Array Audit of Wedged and Asymmetric Fields in an Inhomogeneous Lung Phantom *Medical physics* **41**(10) 101712–1 –101712–11
[13] Rock L et al. 2004 *Tissue Inhomogeneity Corrections for Megavoltage Photon Beams*
[14] Sriram P 2011 A Study on Evaluation of kV-CBCT-Image-Based Treatment Planning Using Anthropormorphic Phantom *Journal of Medical and Biological Engineering* **31**(6) 429
[15] Kustriyadi H 2013 *Perbandingan Distribusi dosis Radioterapi antara Citra Volumetrik kV-Cone Beam dengan Fan Beam CT* Universtias Indonesia
[16] Kartutik et al. 2016 Comparison of Radiotherapy Dosimetry for 3D-CRT, IMRT, and SBRT Based on Electron Density Calibration. *Journal of Physics: Conference Series 694 (2016) 012017*
[17] Hill R F, Brown S & Baldock C, 2008 Evaluation of the Water Equivalence of Solid Phantoms Using Gamma Ray Transmission Measurements. *Radiation Measurements* **43**(7) 1258–1264
[18] Mckenzie E M et al. 2014 Toward Optimizing Patient---Specific IMRT QA Techniques in the Accurate Detection of Dosimetrically Acceptable and Unacceptable Patient Plans *Medical physics* **41** 121702–1 –121702– 15
[19] Heilemann G, Poppe B & Laub W 2013 On the Sensitivity of Common Gamma-Indeks Evaluation Methods to MLC Misalignments in Rapidarc Quality Assurance *Medical Physics*, 40(3), p.031702.
[20] Park J. et al. 2014. Optimal set of Grid Size and Angular Increment for Practical Dose Calculation Using the Dynamic Conformal Arc Technique: a Systematic Evaluation of the Dosimetric Effects in Lung stereotactic body radiation therapy *Radiation Oncology* **9**(5) 1–12
[21] ESTRO 2013 Oncology & Radiotherapy *Radiotherapy and oncology: journal of the European Society for Therapeutic Radiology and Oncology* **115**(9)
[22] Hsu S -H Et al. 2010 Dose Discrepancies in the Buildup Region and Their Impact on Dose Calculation for IMRT fields *Medical physics* **37**(5) 2043–53

**Acknowledgments**
Authors wish to thank the Head of Radiotherapy Departement, Cipto Mangunkusumo Hospital for the support to this work. This study was supported also by DRPM Universitas Indonesia through Hibah PITTA 2016 (contact number 2051/UN2.R12/HKP.05.00/2016)

IOP Conf. Series: Journal of Physics: Conf. Series **851** (2017) 012022 doi:10.1088/1742-6596/851/1/012022

# Dosimetric evaluation of IMRT plan for homogenous and inhomogeneous medium using AAPM TG-119 protocol

**L A N Fatimah[1], W E Wibowo[2] and S A Pawiro[1]**

[1]Department of Physics, Faculty of Mathematics and Natural Sciences, Universitas Indonesia, Depok, Indonesia
[2]Department of Radiotherapy, National General Hospital Dr. Cipto Mangunkusumo, Jakarta, Indonesia

E-mail: supriyanto.p@sci.ui.ac.id

**Abstract**. The American Association of Physicists in Medicine (AAPM) TG-119 protocol has been applied for dose verification in IMRT technique. However, some criteria in the protocol need to be verified for inhomogeneous medium and small volume targets. Hence, the purpose of this study was to verify the assessment criteria of dose verification in AAPM TG-119 for inhomogeneous medium and small volume targets. The work has been conducted by dose verification for homogeneous (phantom A) and inhomogeneous phantoms (phantom B and C) on two geometrical targets: C-shape and circular targets. The targets were simulated using 7 static dMLC IMRT fields at two different depths of 5 g/cm² and 10 g/cm². The dose optimisation and calculation were done by using Pinnacle³ for 6 MV photons beam. The planning objectives were set according to AAPM TG-119 parameters. The plan analysis was conducted by Conformity Index and Homogeneity Index . The point dose measurements were conducted with Exradin A16, Semiflex 0.125cc, and Gafchromic EBT3. The plan results show that CI for C-shape target is in the range of 0.710-0.999 at 10 g/cm² depth and 0.691-1.613 at 5 g/cm². In addition, HI for C-shape and circular were in the range of 6.3%-58.7% and 5.4%-87.1% for 10 g/cm² depth. The measurement results show that the dose measurement at inhomogeneous medium and small volume targets are much lower than the criteria in AAPM TG-119. In conclusion, the criteria in the AAPM TG-119 cannot be fully implemented for inhomogeneous medium and small volume targets.

## 1. Introduction

Modern radiotherapy techniques have been widely used as promising and more effective cancer treatment. These techniques, such as Intensity Modulated Radiation Therapy (IMRT), Image-Guided Radiation Therapy (IGRT), Stereotactic Body Radiation Therapy (SBRT) and Stereotactic Radiosurgery (SRS), have been developed and applied to achieve optimum doses to the target and spare healthy tissue around the targets [1][2]. Within those techniques, IMRT was the most been used for various diseases. In 2009, the AAPM has published Task Group 119 for verification protocol in IMRT which contains some geometric tests, such as multitargets, mock prostate, mock head and neck, and C-shape target. This TG-119 has suggested performing all the tests on point dose measurement, planar dose measurement, and planar dose field by field measurement [3]. The AAPM TG-119 protocol has been applied for dose verification in IMRT technique.

The procedures of dose verification in AAPM TG-119 have been carried out in a solid water phantom (PMMA), which is a homogeneous medium. In fact, human body consists of materials having different densities such as lungs, soft tissues and bones which may represent inhomogeneity condition. Hence, the procedures of dose verification in the protocol might not fully represent the dose verification for human body (patients) and might cause a dose perturbation due to the inhomogeneity issues [4]. Moreover, the dose perturbation due to inhomogeneity issues might be more complex due to the use of small field technique for treating small volume targets. The combination of the inhomogeneity problem and small volume target might cause a significant dose perturbation. The dose perturbation might be caused by some issues such as the lack of electronic equilibrium; this condition might occur because the field size is too small so that the energy which enters and leaves the volume target is not in the equilibrium state. Other issues are steep dose gradient, source occlusion and volume averaging effect, which occur when the range of the secondary electrons is too large compared to the chamber size, so that more electron will scatter laterally out of the field [1][5][6][7]. However, some criteria in the protocol need to be verified for inhomogeneous medium and small volume targets. For those reasons, this study has been done to verify the assessment criteria of dose verification in AAPM TG-119 for inhomogeneous medium and small volume targets.

## 2. Methodology

### 2.1. *Plan Optimisation*

This study has used the CT images which have been generated by GE Brightspeed CT simulator and Pinnacle3® treatment planning system (TPS) (Philips Radiation Oncology Systems, Fitchburg, WI) with an adaptive convolution algorithm. The dose measurement was performed using 6 MV photon beam-Synergy-S (Elekta AB, Stockholm, Sweden) accelerator. The measurement used homogeneous solid water (phantom A), inhomogeneous phantom that consist of lung, tissue, and bone (compact and spongy bone) equivalent materials with mass target that consists of tumour equivalent materials as the target in the lung (phantom B) and inhomogeneous phantom without mass target equivalent (phantom C). The inhomogeneous phantoms (phantoms B and C) were set by sandwich model. The plan and measurements also used C-shape and circular targets which have been adopted from AAPM TG-119 protocol. The treatment planning was generated by using IMRT technique with invers plan optimisation using seven gantry angles (14°, 34°, 138°, 180°, 229°, 327°, and 349°) and avoiding the lateral side of the phantoms using 200 cGy prescribed dose.

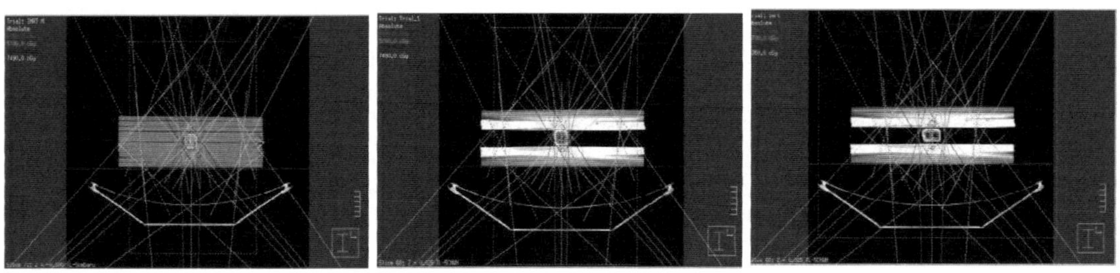

(a) (b) (c)

**Figure 1.** TPS plan for three type of phantom: **(a)** Homogeneous phantom (Phantom A); **(b)** Inhomogeneous phantom with tumour equivalent material (Phantom B); **(c)** Inhomogeneous phantom without tumour equivalent material (Phantom C).

### 2.2. *Point Dose Verification*

Point dose measurement was conducted using ionization chamber Exradin A16, Semiflex 0.125cc, and Gafchromic EBT3 in the isocentre, in 3 types of phantoms. The Semiflex, was only used in phantom A and phantom C since its size could not fit in the middle of mass target (phantom B). C-shape target and circular target volumes are 9 $cm^3$ and 8 $cm^3$ respectively with 0.2 cm PTV margins.

### 2.3. *Analysis of the Results*

The analysis was conducted based on the parameters of AAPM TG-119 such as Conformity Index (CI), Homogeneity Index (HI), dose goals, and measurement analysis based on the discrepancies between planning and measurement. Conformity index analysis was performed involving the ratio of volume that cover 95% dose ($V_{95\%}$) and volume of PTV ($V_{PTV}$)[8][9] using equation (1)

$$CI = \frac{V_{95\%}}{V_{PTV}} \tag{1}$$

Homogeneity index (equation (2)) analysis involves the ratio of the dose that covers 2% volume ($D_{2\%}$), dose covers 98% volume ($D_{98\%}$), and the dose cover 50% volume ($D_{50\%}$) of PTV. An HI of zero indicates that the absorbed-dose distribution is almost homogeneous[10].

$$HI = \frac{D_{2\%} - D_{98\%}}{D_{50\%}} \tag{2}$$

The measurement discrepancies follow equation (3)

$$discrepancy = \frac{Dmeasured - Dplan}{Dprescription} \tag{3}$$

The agreement criterion was determined according to International Commision on Radiation Unit and Measurements (ICRU) recommendation, which suggests the accuracy of ±5% in the delivery of an absorbed dose to a target volume [11].

## 3. Results and discussions

### *3.1. Conformity Index, Homogeneity Index, and dose goals*

The results of the CI and the HI in the small field IMRT technique are shown in Table 1 and 2. We know that IMRT technique can produce a good agreement in CI and HI [12]. In this research, the result of CI in phantom A has shown that it is in agreement with previous study Nainggolan[13] and Mynapati[14]. In contrast, the result in phantoms B and C has indicated slightly different CI and HI. This condition is caused by the inhomogeneity of the phantom material and significantly affected by the use of small field technique. For homogeneity index (HI), phantoms B and C have shown the higher value of HI while phantom A gives a lower value of HI, but it does not mean better from the previous study[13][14] since the implementation of small field technique in phantom A.

The result of treatment planning system plan (TPS plan) in 10 g/cm$^2$ and 5 g/cm$^2$ depth has shown the same patterns for phantoms B and C, which have high value for HI and a low value for CI. Consequently, inhomogeneity of the material and field size have affected the value of generated CI and HI, especially for the circular and C-shape targets. The patterns of CI and HI in the phantoms have been affected by the type of phantom materials and depth of the target position.

Table 3 until Table 5 have shown the result of the dose goals between the phantom plan and the criteria in AAPM TG-119. The results were analyzed for the value of $D_{95}$, $D_{10}$ for the target C-shape PTV, while the value of $D_5$ were calculated on the core areas in the isocenter.

**Table 1.** The Comparison of CI and HI in the depth of 10 g/cm² target position

| Comparison | Phantom A | | Phantom B | | Phantom C | | IMRT[13] | | IMRT[14] | |
|---|---|---|---|---|---|---|---|---|---|---|
| **Geometry** | C-shape | Circular | C-shape | Circular | C-shape | Circular | C-shape (hard) | Multi target | C-shape (hard) | Multi target |
| **Homogeneity Index (%)** | 6.3 | 5.4 | 58.7 | 77.4 | 57.1 | 87.1 | 15.4 | 4.2 | 11.02 | 4.56 |
| **Conformity Index** | 0.999 | 1.122 | 0.736 | 0.450 | 0.710 | 0.334 | 1.355 | 1.357 | 1.18 | 1.23 |
| **Amount of fields** | 7 | 7 | 7 | 7 | 7 | 7 | 9 | 7 | 9 | 7 |
| **Dose per fraction** | 200 | 200 | 200 | 200 | 200 | 200 | 200 | 200 | 200 | 200 |

**Table 2.** The Comparison Conformity Index and Homogeneity Index in 5 g/cm² depth

| Comparison | Phantom A | | Phantom B | | Phantom C | | IMRT[13] | | IMRT[14] | |
|---|---|---|---|---|---|---|---|---|---|---|
| **Geometry** | C-shape | Circular | C-shape | Circular | C-shape | Circular | C-shape (hard) | Multi target | C-shape (hard) | Multi target |
| **Homogeneity Index (%)** | 7.0 | 5.9 | 80.4 | 67.5 | 63.4 | 64.6 | 15.4 | 4.2 | 11.02 | 4.56 |
| **Conformity Index** | 1.613 | 1.165 | 0.691 | 0.847 | 0.693 | 0.729 | 1.355 | 1.357 | 1.18 | 1.23 |
| **Amout of fields** | 7 | 7 | 7 | 7 | 7 | 7 | 9 | 7 | 9 | 7 |
| **Dose per fraction** | 200 | 200 | 200 | 200 | 200 | 200 | 200 | 200 | 200 | 200 |

**Table 3.** The Dose goals comparing based on Phantom A plan

| | C-shape | | | Circular (multitarget) | |
|---|---|---|---|---|---|
| **Structure** | PTV | PTV | Core area (hard) | PTV | PTV |
| **parameters** | $D_{95}$ (cGy) | $D_{10}$ (cGy) | $D_5$ (cGy) | $D_{99}$ (cGy) | $D_{10}$(cGy) |
| **Dose goal (cGy)** | 5000 | <5500 | <1000 | >5000 | <5300 |
| **d=5 g/cm²** | 4858 (-2.8%) | 5121 (-6.9%) | 5069 (406.8%) | 5140 (2.8%) | 5723 (8.0%) |
| **d=10 g/cm²** | 4703 (-5.9%) | 6313 (14.8%) | 5955 (495.5%) | 4801 (-4.0%) | 5057 (-4.6%) |
| **AAPM TG-119** | 5010 | 5702 | 1630 | 4955 | 5455 |
| **IMRT [13]** | 5057 | 5596 | 1558 | 5023 | 5286 |
| **IMRT[14]** | 5000 | 5482 | 1585 | 5007 | 5358 |

**Table 4.** The Dose goals comparison based on Phantom B plan

| | C-shape | | | Circular (multitargets) | |
|---|---|---|---|---|---|
| **structure** | PTV | PTV | Core area (hard) | PTV | PTV |
| **parameters** | $D_{95}$(cGy) | $D_{10}$(cGy) | $D_5$(cGy) | $D_{99}$(cGy) | $D_{10}$(cGy) |
| **Dose goal (cGy)** | 5000 | <5500 | <1000 | >5000 | <5300 |
| **d=5 g/cm²** | 2504 (-49.9%) | 5357 (-2.6%) | 5245 (424.5%) | 1803 (-63.9%) | 5247 (-1.0%) |
| **d=10 g/cm²** | 4670 (-6.6%) | 5589 (1.6%) | 5619 (461.9%) | 886 (-82.2%) | 5132 (-3.2%) |
| **AAPM TG-119** | 5010 | 5702 | 1630 | 4955 | 5455 |
| **IMRT[13]** | 5057 | 5596 | 1558 | 5023 | 5286 |
| **IMRT[14]** | 5000 | 5482 | 1585 | 5007 | 5358 |

**Table 5.** The Dose goals comparing based on Phantom C plan

| | C-shape | | | Circular (multitarget) | |
|---|---|---|---|---|---|
| **structure** | PTV | PTV | Core area (hard) | PTV | PTV |
| **parameters** | $D_{95}$ (cGy) | $D_{10}$ (cGy) | $D_5$ (cGy) | $D_{99}$ (cGy) | $D_{10}$ (cGy) |
| **Dose goal (cGy)** | 5000 | <5500 | <1000 | >5000 | <5300 |
| **IMRT, d=5 g/cm²** | 3115 (-37.7%) | 5237 (-4.8%) | 5264 (426.4%) | 1578 (-68.5%) | 5218 (-1.6%) |
| **IMRT, d=10 g/cm²** | 3383 (-32.3%) | 5589 (1.6%) | 5619 (461.9%) | 1496 (-70.1%) | 5244 (-1.1%) |
| **AAPM TG-119** | 5010 | 5702 | 1630 | 4955 | 5455 |
| **IMRT [13]** | 5057 | 5596 | 1558 | 5023 | 5286 |
| **IMRT[14]** | 5000 | 5482 | 1585 | 5007 | 5358 |

Remarks :
$D_{95}$ : Dose which covers 95% of target volume (PTV)
$D_{10}$ : Dose which covers 10% of target volume (PTV)
$D_5$ : Dose which covers 5% of core areas (isocentre)
$D_{99}$ : Dose which covers 99% of target volume (PTV)

The Table 3 has shown that the dose goal of phantom A is close to the dose goal of TG-119 at about 6% and 15% for $D_{95}$ and $D_{10}$ respectively, but it has a big difference at about 495% compared to the dose goal from TG-119 for core area ($D_5$). Moreover, the result of analysis on the $D_{99}$ and $D_{10}$ for PTV of circular target has the discrepancy less than 5%. The evaluations of $D_{95}$, $D_{10}$ and $D_5$ in phantoms B and C give higher result than the dose goal in AAPM TG 119 up to 461.9%. Conversely, the achieved dose at $D_{95}$ and $D_{10}$ in phantom B is lower than dose goal in AAPM TG 119 at about to 49.9% and 37.7%.

### *3.2. Measurement discrepancies*
The discrepancy of the results in this study has been shown in Fig 2. The measurement on the inhomogeneous phantom with mass (phantom C) has presented that the discrepancy reached up to -17.7% by using Exradin A16 and -12.8% by using the Gafchromic EBT3 film.

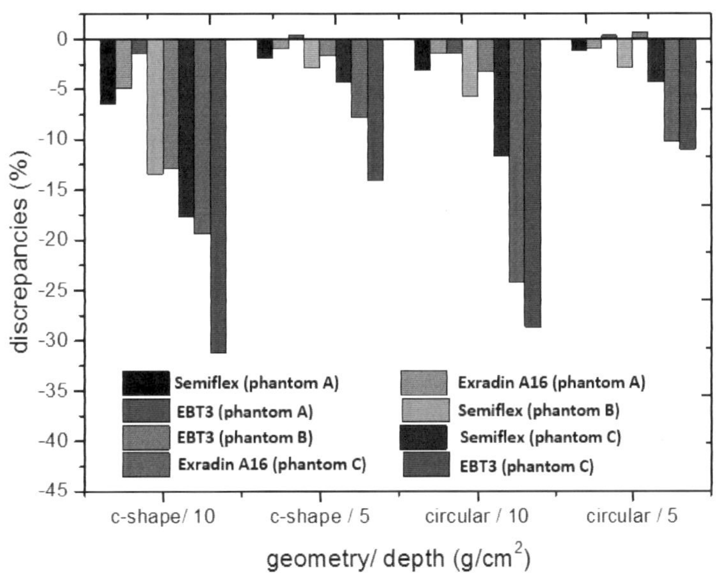


**Figure 2.** Discrepancies for all phantom measurement

However, the measurements in phantom C, have shown the different patterns that the dose reduction has reached -31.2% in EBT3 measurement. In addition, about the measurement in phantom C, the dose reduction is higher than the measurement in phantom B, because the location of measurement is in the low-density area [4][15]. The phenomena might be caused by the target mass attenuation which might involve in the measurements.

The highest dose reduction for phantom A, has occurred the C-shape target measurement. It has shown that the dose reduction of the Exradin A16 is lower than Semiflex measurement. This condition has occurred due to the smaller sensitive volume of Exradin A16 compared to Semiflex, so that the volume averaging effect can be minimised. On the other hand, EBT3 has produced the lowest dose reduction because of its highspatial resolution[15][16].

## 4. Conclusion
The result of dose measurement in inhomogeneous medium and small volume targets is much lower than the criteria in AAPM TG-119, and then criteria in the AAPM TG-119 could not be fully implemented for inhomogeneous medium and small volume targets.

## 5. References
[1] Das I J, Ding G X and Ahnesjö A 2007 *Medical physics.* **35**(1), 206-215
[2] Tanny S, Sperling N and Parsai E I 2015 *Medical Physics.* **42**, 5370-5376

[3] Ezzell G A, Burmeister J W, Dogan N, and LoSasso T J 2009 Med Phys. **36**(11), 5359-5373
[4] Li A X, Yu C, and Holmes T 2000 *Medical Physics.* **27**, 1011-1017
[5] Spang F J 2011 Monte Carlo study of the dosimetry of small-photon beams using CMOS active pixel sensors. *Department of Medical Physics and Bioengineering University College London, UCL*
[6] Alfonso R, Andreo P, Capote R, and Saiful Huq M 2008 *Medical physics.* **35**(11), 5179-5186
[7] Bouchard, H., Seuntjens, J., Duane, S., Kamio, Y., & Palmans, H. (2015). Detector dose response in megavoltage small photon beams. I. Theoretical concepts. *Medical Physics,* 42, 6033-6047.
[8] Vaezzadeh S, Allahverdi M 2012 *Iran. J. Radiat* **10** 145-150
[9] Ganesh. N, Drew G, Diana B, Ying L, Alonso G, Niko P, Sotirios S 2015 *Clinical Engineering and Radiation Oncology* **4** 104-112
[10] ICRU Report 83. (2010). *Prescribing, Recording and Reporting Photon Beam Intensity Modulated Radiation Therapy (IMRT).* The International Commission on radiation Units and Measurements
[11] International Atomic Energy Agency 2004 TRS 398-Absorbed Dose Determination in External Beam Radiotherapy: An International Code for Practice for Dosimetry based on Standards of Absorbed Dose to Water (Vienna: IAEA) p 13
[12] Kartutik K, Wibowo W E, Pawiro S A 2016 *Journal of Physics: Conference Series* **694** 012017
[13] Nainggolan A 2015 Uji akurasi perencanaan dan pemberian perlakuan volumetric modulated arc therapy (VMAT) menggunakan acuan AAPM-119. *Tesis Universitas Indonesia*
[14] Mynampati D K, Yaparpalvi R, Hong L, Kuo H-C, and Mah D 2012 *Journal of Applied Clinical Medical Physics* **13**(5)
[15] Bueno M, Carrasco P, Jornet N, Muñoz-Montplet C, and Duch M A. 2014 *Medical Physics* **41** 081710
[16] Fitriandini A, Wibowo W E, Pawiro S A 2016 *Journal of Physics: Conference Series* **694** 012006

**Acknowledgements**
The authors would like to express special thanks to the Department of Radiotherapy RS Cipto Mangunkusumo, also to the Laboratory of Medical Physics, University of Indonesia which provided equipment needed in this research. This study was supported by Hibah PITTA Universitas Indonesia 2016 (contract number 2051/UN2.R12/HKP.05.00/2016)

IOP Conf. Series: Journal of Physics: Conf. Series **851** (2017) 012023 doi:10.1088/1742-6596/851/1/012023

# Dosimetric characteristics of gafchromic EBT3 film on small field electron beam

**S Ulya[1], W E Wibowo[2], N Nuruddin[2], S A Pawiro[1]**

[1]Department of Physics, Faculty of Mathematics and Natural Sciences, Universitas Indonesia, Depok, West Java, 16424, Indonesia;
[2]Department of Radiotherapy, National General Hospital Cipto Mangunkusumo, Jakarta, 10430, Indonesia

E-mail: supriyanto.p@sci.ui.ac.id

**Abstract.** This study was aimed to determine the dosimetry characteristics of the EBT3 film on a small field electron beam. The experiments were conducted using Linac Synergy-Platform with 6, 8, 10, 12, and 15 MeV. EBT3 films were cut into 3.2×2.5 $cm^2$ in size and were irradiated with varied dose of 0 to 500 cGy on 1×1, 2×2, 3×3, 5×5, 8×8 and 10×10 $cm^2$ radiation field sizes. The irradiated film was scanned using Epson Perfection V700 flatbed scanner to evaluate the pixel value. Then, the calibration curve was generated from netOD and given a dose. The uncertainty and sensitivity of the EBT3 film were evaluated using in-house and Image-J software to check the consistency of the result. The results showed that the largest uncertainty of characteristics curve was 4.6% at 1×1 $cm^2$. Whereas the sensitivity value tends to be identical for radiation dose greater than 100 cGy for all energy, it depends on energy for radiation dose less than 100 cGy.

## 1. Introduction

Radiotherapy with small field radiation has been widely implemented to treat the cancer patient. However, there was a lack of electron lateral scatter equilibrium [1]. In order to obtain a successful radiotherapy, the accuracy of dosimetry has to be achieved during verification. Radiochromic film can be used to measure the accuracy of dosimetry with small field condition since it has equivalent density with tissue, high spatial resolution and also sensitivity [2].

Gafchromic EBT film was released in 2004 by International Specialty Product (ISP) and is continuously being developed to increase its performance. In 2009, Gafchromic EBT film was replaced by Gafchromic EBT2 film. The structure of Gafchromic EBT2 film has a combination of yellow marker dye in the active layer and synthesis polymer as material strap [3]–[6]. In 2012, Gafchromic EBT3 film was showed to be more suitable for the dose verification of radiotherapy than EBT2 film [7], [8]. A study on EBT3 film on electron beam and standard field size (10×10 $cm^2$) indicated that it was a good 2D dosimeter and had been meet the requirement for the dose distribution analyzer of the electron beam based on characteristics, readout method, and uncertainty estimation [9]. In this study, the dosimetry characteristics of the Gafchromic film EBT3 on a small field electron beam have been investigated. In addition, the implementation of the dosimetry of small field on the clinical case has also been conducted in this study.

IOP Conf. Series: Journal of Physics: Conf. Series **851** (2017) 012023 doi:10.1088/1742-6596/851/1/012023

## 2. Method and Materials

### *2.1. Irradiation*
Gafchromic EBT3 film was cut into the dimension of 3.2×2.5 cm$^2$ (1.25"×1") and was irradiated using electron beam energies of 6, 8, 10, 12, and 15 MeV with dose in the range of 0 to 500 cGy. The experiment was performed using two applicators; 10×10 cm$^2$ and 6×6 cm$^2$. For irradiation, the film was inserted to the slab phantom with a dimension of (30×30×5 cm$^3$) at the $d_{max}$. Thus, irradiation EBT3 film using small field electron beams with the field size of 8×8, 5×5, 3×3, 2×2, and 1×1 cm$^2$ is illustrated in Figure 1.

### *2.2. Digitalisation*
After irradiation, Gafchromic EBT3 film was kept at the temperature between 20 to 24°C for 72 hours for developing purpose. Then, Epson V700 flatbed scanner with 72 dpi in 48 bit RGB mode was used to digitize Gafchromic EBT3 film. The digital film was saved in Tagged Image File Format (TIFF) image [7], [9]–[12].

### *2.3. Analysis*
In this study, Image-J and in-house software were used to analyze the pixel value readout. The region of interest (ROI) of EBT3 film was 0.25×0.25 cm$^2$ in each center of the image. The response of each energy became our point of interest and hence the relation between dose and net optical density (netOD). The netOD value was calculated by subtraction the pre-scan OD value from the corresponding post-scan OD value [13]. The value of netOD which was used to analyze characteristic of the EBT3 film was given according to the equation (1):

$$netOD = \log_{10}(\frac{PV_{un}}{PV_{ex}}) \quad (1)$$

where $PV_{un}$ is an unexposed pixel value, and $PV_{ex}$ is electron beam exposed pixel value [9].

Calibration curve formula in this research used the comparison of polynomial order three and trust-region described in the equation (2) [7], [10], [14]

$$D = b.netOD + c.netOD^n \quad (2)$$

where $D$ is dose, $b$ and $c$ are free parameters with $n$ is the problem-dependent parameter. Furthermore, the influenced dose and energy in each field size was analyzed. The uncertainty value was shown in equation (3) [15]

$$\sigma_{netOD} = \frac{1}{ln10}\sqrt{\frac{\sigma_{unx}}{PV_{unx}} + \frac{\sigma_{exp}}{PV_{exp}}} \quad (3)$$

where $\sigma_{unx}$ is the deviation standard of the pixel value in EBT3 film before irradiation. Moreover, the sensitivity of EBT3 film in each energy was calculated using equation (4)

$$S = \frac{\Delta netOD}{D^*} = \frac{netOD - netOD_{0cGy}}{D^*} \quad (4)$$

where $S$ is the sensitivity of EBT3 film and $D^*$ dose estimate of the EBT3 film.

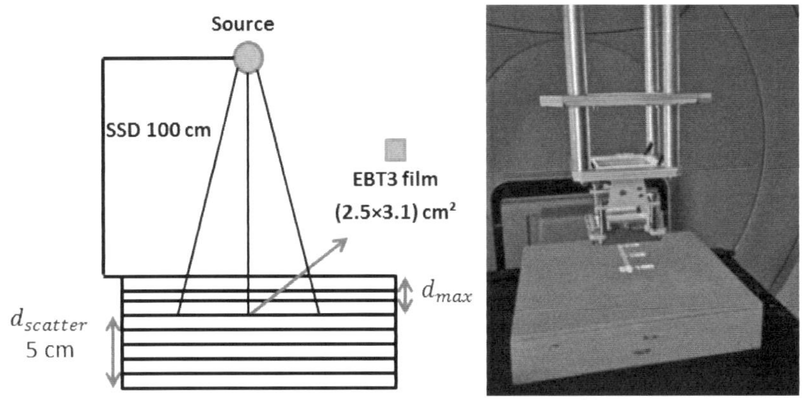


(a) (b)

**Figure 1.** (a) Illustration of EBT3 film calibration of 10×10 cm$^2$, (b) irradiation of small field electron

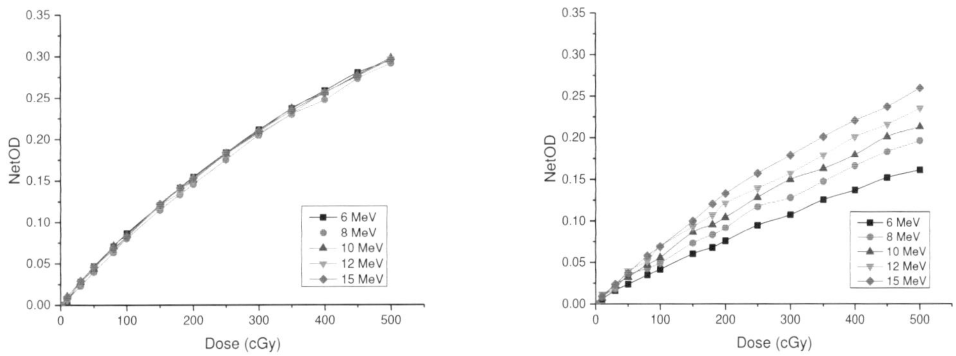


(a) (b)

**Figure 2.** Relation curve between dose and netOD of (a) 10×10 cm$^2$, (b) 1×1 cm$^2$ field size

## 3. Result and Discussion

Calibration of EBT3 film was conducted on $d_{max}$ using field size of 10×10 cm$^2$ within dose range of 0 to 500 cGy for each energy. Scanner response in each channel was studied in this study. The red channel was used to analyse since it had the highest netOD [9].

In EBT3 film calibration, the netOD value was analyzed according to Equation (1) on each calibration equation of energy as shown in Table 1. The comparison between polynomial and trust region equation for calibration indicated that the trust region equation was more accurate approximately 1%. Figure 2 (a) illustrates the graph of the relationship between the dose prescribed to the netOD in the field size of 10×10 cm$^2$. While the relationship curve between the dose and netOD for 1×1 cm$^2$ field size can be seen in Figure 2 (b). The curve tends to overlap at all energy electron beams for 10×10 cm$^2$, but it is influenced much by the energy of electron beams for the small field. For the dose and netOD relationship at various energy, the curves indicates the overlapping curve for high energy, whereas it is spread out the curve for low energy electron beams. These were caused by the scattering effect highly found in low energy than the high one.

Figure 3 (a) depicts the sensitivity of EBT3 film which has the large variation in the range of dose of 0 to 100 cGy, while the sensitivity for higher dose more than 100 cGy has low variation and tends to overlap and constant. It was in agreement with Papaconstadopoulos et al. (2014) explaining that the sensitivity of the EBT3 film was constant at the dose greater than 50 cGy [10]. Figure 3 (b) depicts the sensitivity on a 1×1 cm$^2$ field which is more influenced by the given electron beam energy. For example, the energy of 15 MeV electron beam had the greatest sensitivity in compared to other energies. The characteristic of EBT film sensitivity for various energy can be seen in Figure 3 (b).

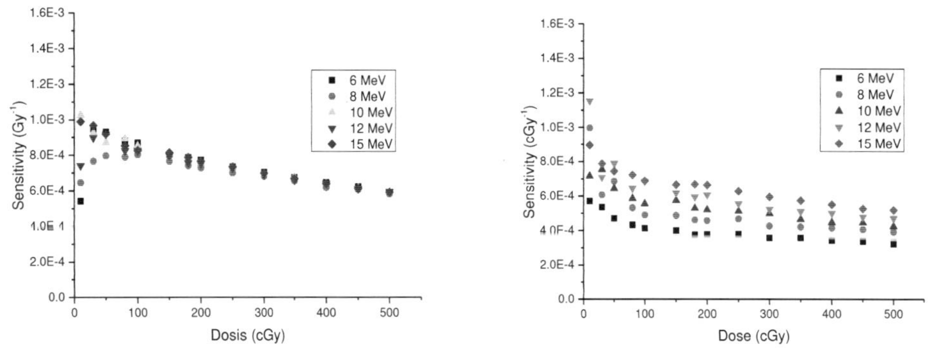


(a) (b)

**Figure 3.** Sensitivity of EBT3 film (a) 10×10 cm$^2$ (b) 1×1 cm$^2$ field size with dose variations

**Table 1.** Calibration equation in each energy electron beam of 10×10 cm$^2$ field size

| Energy | Polynomial equation | Trust-region equation | | |
|---|---|---|---|---|
| | | b | c | n |
| 6 MeV | y = 5,129.08x$^3$ + 483.14x$^2$ + 1,085.83x + 0.48 | 1120.12 | 5519.06 | 2.88 |
| 8 MeV | y = 4,349.63x$^3$ + 540.75x$^2$ + 1,182.3x + 1.45 | 1228.19 | 4704.10 | 2.84 |
| 10 MeV | y = 3,688.25x$^3$ + 893.02x$^2$ + 1,087.53x - 0.79 | 1095.44 | 3760.71 | 2.54 |
| 12 MeV | y = 2,899.38x$^3$ + 1,144.79x$^2$ + 1,088.49x + 0.35 | 1124.23 | 3421.90 | 2.49 |
| 15 MeV | y = 4,878.02x$^3$ + 531.37x$^2$ + 1,111.49x - 1.03 | 1107.93 | 4600.95 | 2.69 |

**Table 2**. Uncertainty value comparison using other research

| Parameter | *Uncertainty* (%) | | References |
|---|---|---|---|
| | *Image-J* | in-house | |
| Electron beam 10×10 cm$^2$ field size (EBT3) | 3.23 | 1.95 | |
| Electron beam 8×8 cm$^2$ field size (EBT3) | 2.26 | 1.79 | |
| Electron beam 5×5 cm$^2$ field size (EBT3) | 3.55 | 1.75 | This study |
| Electron beam 3×3 cm$^2$ field size (EBT3) | 2.93 | 1.59 | |
| Electron beam 2×2 cm$^2$ field size (EBT3) | 3.16 | 3.14 | |
| Electron beam 1×1 cm$^2$ field size (EBT3) | 4.61 | 4.36 | |
| Electron beam (EBT) 10×10 cm$^2$ field size | 4 | | Fan-Chi Su, et al. (2007) |
| Photon and electron beam 9 MeV (EBT2) 10×10 cm$^2$ field size | 2.6 | | Moylan, et al. (2013) |
| 6 MV (EBT3) 10×10 cm$^2$ field size | 4 | | Borca, et al. (2013) |
| Electron (EBT3) 10×10 cm$^2$ field size | 2 | | Sorriaux, et al. (2012) |
| Photon 6 MV and electron10×10 cm$^2$ field size | 3.7 | | Sipilä, et al. (2016) |

This study also analyzed the uncertainty output of the EBT3 film of electron beam energy according to Equation (3). The percentage uncertainty was generated by dividing the standard deviation of irradiated film to the mean of netOD and multiplied by 100% using the Image-J software. From calculation, the average of the uncertainty was circa of 3.23% of 10×10 cm$^2$ field size as indicated in Table 2. Furthermore, the table also presents that EBT3 film offers more accurate results than that previously studied by Fan-Chi Su et al.[16]. In addition, the evaluation of EBT3 film sensitivity for 8×8, 5×5, 3×3, 2×2, and 1×1 cm$^2$ field size was around of 2.26%, 3.55%, 2.93%, 3.16 % and 4.61% s, respectively. Furthermore, the EBT3 sensitivity was also evaluated using the in-house software for comparison as described in Table 2. The result showed that the EBT3 sensitivity of Image-J software is higher than the in-house software calculation. The differences of EBT3 sensitivity between Image-J and in-house software output is due to the ROI technique on the digitized film. The ROI in Image-J was generated manually while the ROI in in-house software was generated automatically.

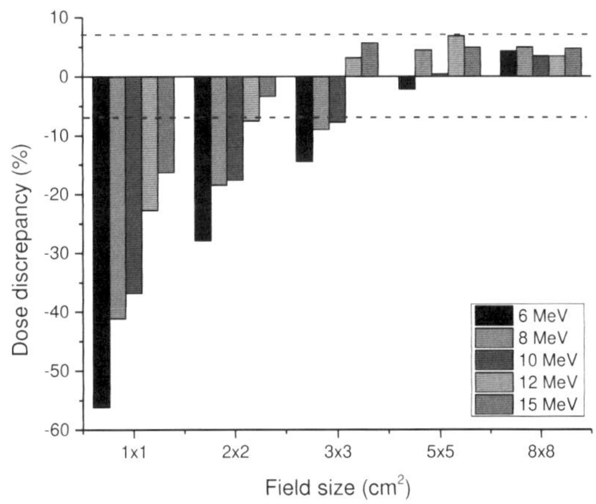


**Figure 4.** The curve of calibration range 10×10 $cm^2$ in small field electron beam (with ±7 % tolerance from the recommendation by Gibbons et al. (2014)) [17].

The implementation of small field electron beams was performed using simple planning at PrecisePLAN Release 2.16–28.76. We simulated the calculation with a prescribed dose of 200 cGy and using 100 cm SSD at gantry angle $0^o$ for the field size of 10×10 $cm^2$ to 1×1 $cm^2$. The measurements were compared to the reference field size of 10×10 $cm^2$ as illustrated in Figure 4. According to AAPM TG 25, the tolerance range of dose discrepancy of the small field electron beam is circa of 3% - 7% [18].

Figure 4 showed that the EBT3 calibration curve equation of 10×10 $cm^2$ can be implemented until 5×5 $cm^2$. This is indicated by the discrepancy of the measurements of small field size 5×5 $cm^2$ is higher more than 7%. In addition, the calibration equation can also be implemented for in the dose calculation analysis of 3×3 $cm^2$ field size for 12 MeV and 15 MeV and 2×2 $cm^2$ for 15 MeV energy.

The comparison the Trust-region algorithm and the calibration formula of polynomial order three was also conducted. This method was adapted from the work of Papaconstadopoulos et al. [10]. Other researchers such as Sorriaux et al. (2012), Devic et al. (2005), and Reinhardt et al. (2012) were also carried out this methodology for determining the calibration curve of EBT film [8], [14], [19]. The trust-region algorithm result was slightly different from polynomial order three measurements base on the discussion above.

## 4. Conclusion

Characteristics of the EBT3 film depend on dose, field size, and energy of the electron. Sensitivity value is almost similar for dose more than 100 cGy, whereas it depends on energy for radiation dose less than 100 cGy. Moreover, the highest uncertainty was found approximately of 4.6% at 1×1 $cm^2$. The uncertainty of measurements tends to increase with decreasing field size. In addition, the use of 10×10 $cm^2$ calibration curve was sufficient for calculating the dose up to 5×5 $cm^2$ field size for all energies. However, it can also be used for 3×3 $cm^2$ and 2×2 $cm^2$ for 12 MeV and 15 MeV, respectively.

## 5. References

[1] Sharma S C, Johnson M W and Gossman M S 2005 Practical considerations for electron beam small field size dosimetry *Med. Dosimetry* **30** 104
[2] Amin M N, Heaton R, Norrlinger B and Islam M K 2011 Small field electron beam dosimetry using MOSFET detector *Jounal Appl. Clin. Med. Phys* **12** 50
[3] Richley L, John A C, Coomber H and Fletcher S 2010 Evaluation and optimization of the new EBT2 radiochromic film dosimetry system for patient dose *Phys. Med. Biol* **55** 2601

[4] Butson M J, Yu P K N, Cheung T and Alnawaf H 2010 Energy response of the new EBT2 radiochromic fi lm to x-ray radiation Radiat. Measurements 45 836
[5] ISP 2010 Gafchromic ® EBT2 self-developing film radiotherapy dosimetry
[6] Devic S, Aldelaijan S, Mohammed H, Tomic N, Liang L, Deblois F and Seuntjens J 2010 Absorption spectra time evolution of EBT-2 model gafchromic™ film *Int. J. Med. Phys. Res. Pract* **37** 2207
[7] Borca V C, Pasquino M, Russo G, Grosso P, Cante D, Sciacero P, Girelli G, Porta M R L and Tofani S 2013 Dosimetric characterization and use of gafchromic EBT3 film for IMRT dose verification *Jounal Appl. Clin. Med. Phys* **14** 158
[8] Sorriaux J, Kacperek A, Rossomme S, Lee J A, Bertrand D, Vynckier S and Sterpin E, 2013 Evaluation of Gafchromic EBT3 films characteristics in therapy photon, electron and proton beams *Phys. Med* **29** 599
[9] Sipilä P, Ojala J, Kaijaluoto S and Jokelainen I 2016 Gafchromic EBT3 film dosimetry in electron beams — energy dependence and improved film read-out *J. Appl. Clin. Med. Phys* **17** 360
[10] Papaconstadopoulos P, Hegyi G, Seuntjens J and Devic S 2014 A protocol for EBT3 radiochromic film dosimetry using reflection scanning A protocol for EBT3 radiochromic film dosimetry using reflection scanning *Med. Phys* **41** 122101–1
[11] Chełminski K, Bulski W, Georg D, Bodzak D, Maniakowski Z, Oborska D, Rostkowska J and Kania M 2010 Energy dependence of radiochromic dosimetry films for use *Rep. Pract. Oncol. Radiother* **5** 40
[12] Villarreal-barajas J E and Khan R F H 2014 Energy response of EBT3 radiochromic films : implications for dosimetry in kilovoltage range *J. Appl. Clin. Med. Phys* **15** 331
[13] Ning Wen, Siming Lu, Jinkoo Kim, Yujiao Qin, Yimei Huang, Bo Zhao, Chang Liu and Indrin J. Chetty 2016 Precise film dosimetry for stereotactic radiosurgery and stereotactic body radiotherapy quality assurance using Gafchromic™ EBT3 films *Radiation Oncology* **11** 132
[14] Devic S, Seuntjens J, Sham E, Podgorsak E B and Soares C G 2005 Precise radiochromic film dosimetry using a flat-bed document scanner *Med. Phys* **32** 2245
[15] Massillon G, Chiu-Tsao S, Domingo-Munoz I and Chan M F 2012 Energy dependence of the new gafchromic EBT3 film : dose response curves for 50 kV, 6 and 15 MV X-ray beams *Int. J. Med. Physics, Clin. Eng. Radiat* **1** 60
[16] Su F, Liu Y, Stathakis S, Shi C and Esquivel C 2007 Dosimetry characteristics of gafchromic EBT film responding to therapeutic electron beams *Appl. Radiat.* **65** 1187
[17] Gerbi B J, Antolak J A, Deibel F C, Followill D S, Herman M G, Higgins P D, Huq M S, Mihailidis D N, Yorke E D, Hogstrom K R and Khan F M 2009 Recommendations for clinical electron beam dosimetry: supplement to the recommendations of Task Group 25 *Med. Phys.* **36** 3239
[18] Gibbons J P, Antolak J A, Followill D S, Huq M S, Klein E E, Lam K L, Palta J R, Roback D M, Reid M and Khan F M 2014 Monitor unit calculations for external photon and electron beams : Report of the AAPM Therapy Physics Committee Task Group No. 71 *Med. Phys.* **41** 031501-1
[19] Reinhardt S, Hillbrand M, Wilkens J J and Assmann W 2012 Comparison of Gafchromic EBT2 and EBT3 films for clinical photon *Med. Phys.* **39** 5257

**Acknowledgments**
This works was supported by National General Hospital Cipto Mangunkusumo and Hibah Publikasi Internasional Terindeks Tugas Akhir (PITTA) Universitas Indonesia 2016 with contract number: 2050/UN2.R12/HKP.05.00/2016.

# Characterisation of sensitive Ge-doped silica flat fibre-based thermoluminescence detectors for high resolution radiotherapy dosimetry

**A R Abdul Rahim[1], N A Zahaimi[1], H M Zin[2], D A Bradley[3], G A Mahdiraji[4] and A T Abdul Rahman[1, 3, 5*]**

[1]School of Physics and Material Studies, Faculty of Applied Sciences, Universiti Teknologi MARA, 40450 Shah Alam, Selangor, Malaysia.
[2]Radiotherapy & Oncology Department, Advanced Medical & Dental Institute Malaysia, Universiti Sains Malaysia, Bertam 13200 Penang, Malaysia.
[3]Centre for Nuclear and Radiation Physics, Department of Physics, University of Surrey, Guildford GU2 7XH, Surrey, United Kingdom.
[4]School of Engineering, Taylor's University, 47500 Subang Jaya, Selangor Darul Ehsan, Malaysia.
[5]Institute of Science, Universiti Teknologi MARA, 40450 Shah Alam, Selangor, Malaysia.

E-mail: ahmadtaufek@salam.uitm.edu.my

**Abstract:** Present study focuses on characterisation of $SiO_2$ optical fibers as a potential thermoluminescence (TL) system for radiation therapy dosimetry. Irradiations were made using 6 MV photon beams from a linear accelerator. Flat $SiO_2$ optical fibers of various dimensions with 8% concentration of germanium doped were used. The dimensions of the flat fibers were 270×60 μm, 360×73 μm, 100×510 μm and 160×750 μm. Flat $SiO_2$ optical fibers were characterised for TL dose response in terms of linearity, sensitivity, fading and reproducibility. The uncertainty measured was ±1 standard error of the mean and the coefficient variation was within ±4%, as required for clinical radiotherapy dosimetry. Results shown a good distribution of TL response measured by flat $SiO_2$ optical fibers with uncertainties less than 4%. Linearity of TL comes out with a coefficient of determination ($r^2$) of each fibers that is better than 99% which resulted in high percentage of confidence level. The loss of TL response due to fading, for photon irradiation at fixed energy and constant dose was found to be (20.4 ± 0.2)% over a post irradiation period of 30 days. The TL fading well, showing rapid loss in the first seven (7) days (17.8 ± 0.2)% followed by a more linear like loss subsequently the following day (3.2 ± 0.2)%. A perfect selection of fibers can enhance the accuracy of radiation dosimeter in order for better determination and measurement of radiation doses with a linear response over wide range therapeutic dose.

## 1. Introduction

Recent technological gains in radiation therapy, including intensity modulated radiation therapy (IMRT), image-guided radiation therapy (IGRT) and recent attempts to implement intensity-modulated proton therapy (IMPT) [1, 2], enable delivery of more complex and conformal treatment plans and the use of tighter margins to irradiate smaller treatment volumes with higher doses. Correctly delivered radiotherapy dose ensures the possibility to improve survival rate and reduce toxicities to normal tissues for better quality of life. Any slight miss in the dosimetric verification may defeat the whole aims for better tumour control. The challenge in advanced dosimetry is to have a system that is able to verify those complex plans. Obviously, not all dosimeters can satisfy all the characteristics, therefore, the choice of a radiation dosimeter and its reader must be made sensibly, taking into account the requirements of the measurement situation, such for beam calibrations, for the evaluation of the dose distribution or for dose verification. Therefore, study will investigate a potential TL dosimetry system for radiation therapy application. These preliminary studies will characterise the new candidate for dosimetry system of germanium doped silica flat fibers in term of its feasibility.

Characteristics of dosimetric properties of germanium doped silica glass optical fibre have been undertaken and established by a several groups for various applications in radiation therapy [3-18]. Commercially available germanium doped silica glass optical fiber has been shown to possess a number of desirable TL characteristics, in terms of dose stability, linearity of response, and sensitivity to dose. Previous studies have examined the performance of tailor-made doped silica cylindrical glass optical fibre and its dopant concentration effect on TL response due to photon and electron irradiation [4, 19]. Present study will focus on characterisation of tailor-made doped silica flat fibre at therapeutic doses.

## 2. Material and methods

### 2.1. Sample preparation

The dosimeter used is made of Ge-doped $SiO_2$ flat fibre (FF), which has been fabricated using MCVD process, with ultra-pure fused silica Suprasil F300 glass tube as the substrate (University of Malaya). The Ge-doped $SiO_2$ flat fibre has been fabricated by applying a vacuum pressure of 10 kPa from the top of the hollow preform during the drawing process [20]. The $Z_{eff}$ of the flat fiber is in the range of 11.3-11.8, based on Field Emission Scanning Microscope (FESEM) and Energy Dispersive X-ray Spectroscopy (EDXRS) analysis [21]. Five different dimensions of flat fiber have been used as shown in Table 1. Fibers were cut into approximately 5 mm length and average mas of each fiber have been measured.

**Table 1.** Physical properties of fabricated germanium doped silica flat fibers

| FLAT FIBRE | DIMENSION (um) | Ge - DOPANT (%) | MASS (gram) |
|---|---|---|---|
| FF-A | 180 X 45 um | 8 | 0.074 |
| FF-B | 270 X 60 um | 8 | 0.151 |
| FF-C | 360 X 73 um | 8 | 0.311 |
| FF-D | 100 X 510 um | 8 | 0.913 |
| FF-E | 160 X 750 um | 8 | 1.325 |

The flat fibers were annealed using a furnace (Carbolite CWF 23L Laboratory Chamber Furnaces). Fibers are placed in an alumina ceramic boat and annealed at 400°C for an hour. The fibers were then removed and allowed to cool to room temperature at the rate of 10°C/min. Thermal annealing is intended to erase accumulated radiation signal and stabilise the background of the medium such that dosimetric properties of flat fibers used are being controlled at the same state before any irradiation.

### 2.2. Irradiation Setup

The characterisation of Ge-doped $SiO_2$ flat fibres were investigated using 6 MV photon beam from an Synergy linear accelerator (Elekta, Crawley, UK) at Advances Medical and Dental Institute (IPPT), Universiti Sains Malaysia. Flat fibers were placed in a water phantom at 2.0 cm depth. A field size of 10 x 10 $cm^2$ and standard Source-Surface Distance (SSD) of 100 cm was applied.

***Uncertainty***: 100 samples of each Ge-doped $SiO_2$ flat fibres were characterised for TL sensitivity selection within a standard division of mean distribution. Provided at 3 Gy dose at constant dose rate of 400 cGy/min.

***Dose Response***: Each capsule (contained ~25 flat fibers) irradiated to 6 MV photon energy at a fixed dose rates of 400 cGy/min in the dose range of 1 to10 Gy.

***Reproducibility***: To allow a study on reproducibility and repeatability of the dosimeter, the irradiations of each group of flat fibers have been repeated for several times.

***Fading Effect***: It is significant to know how long the information of radiation deposition can be stored by the flat fibre. Fading effect of the flat fibers has been investigated up to a month beyond the time of irradiation.

### 2.3. TL Measurement

The measurement of thermoluminescence (TL) yield response of the irradiated flat fibres has been performed using a Harshaw 3500 TLD reader. Measurements being carried out under nitrogen ($N_2$) gas atmosphere. The TLD reader has been set up to a preheat temperature of 50°C for 10 s, an acquired temperature rate of 40°C/s, an acquisition time of 13.33 s and maximum anneal temperature of 400°C. This readout parameter was set up to ensure optimal data acquisition and radical sweep-out of any residual signal. The TL readings were normalised to the mass of flat fiber, to obtain results in µC/mg.

## 3. Results and discussions

Figure 1 shows the selected Ge-doped silica flat fibres are capable of producing a uniform TL response of better than 3% (1 S.D).

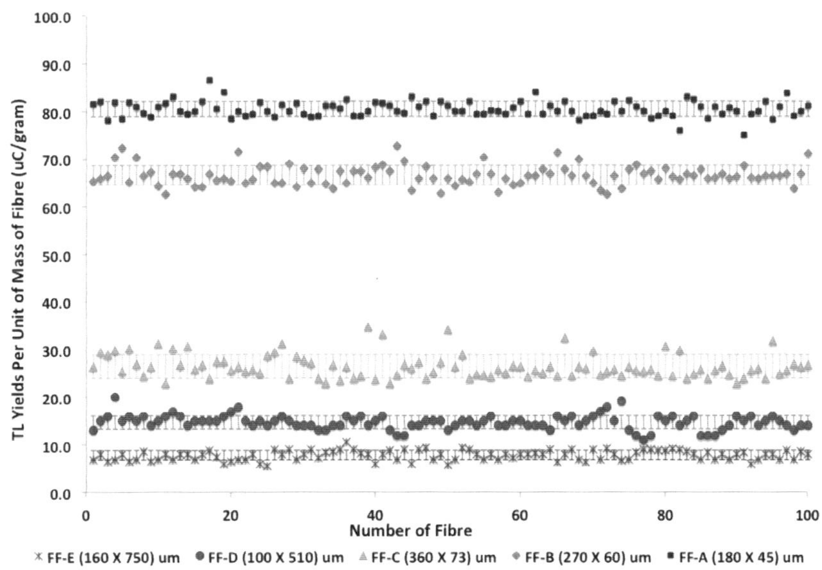


**Figure 1.** TL response of Ge-doped silica flat fibres irradiated with 3 Gy dose of 6MV photon energy at 400 cGy/min dose rate.

The above irradiation has been repeat consequently four time with the same set up of 6MV photon energy at the fixed dose delivered of 3 Gy at the dose rate of 400 cGy/min. Figure 2 shows that Ge-doped silica flat fibres are also capable of reuse, the TL response varying by about 0.3 % (1 S.D). The flat fibers dosimeter is able to shows a better uncertainty and reproducibility compared to cylindrical fibres obtained by [15, 19].

A linear response is obtained over the dose range 1 – 10 Gy, with a correlation coefficient of 0.994 or better. The least-squares straight line fits to the measured data, revealing a TL light yield (in counts per second per unit mass of fiber) corresponding to a dose-dependency of ~26 time the absorbed dose, measured in Gy, for photon irradiations. Each group of flat fibers provide the basis for sensitive dosimeter throughout this range, the TL yield increasing by on average 95 % per additional 1Gy of dose delivered for photon irradiation. The results show a small but systematic difference between the TL response at lower and higher dimension of Ge-doped silica flat fibres for photon beams, as shown by the various least-squares straight line fits in Figure 3.

IOP Conf. Series: Journal of Physics: Conf. Series **851** (2017) 012024 doi:10.1088/1742-6596/851/1/012024

The loss of TL response due to fading, for photon irradiation at fixed energy and constant dose was found to be (20.4 ± 0.2)% over a period of 30 days post irradiation. The TL fading well showing rapid loss in the first seven (7) days (17.8 ± 0.2)% followed by a more linear like loss subsequently the following day (3.2 ± 0.2) %. The fading response of Ge-doped optical fibers investigated is shown in Figure 4.

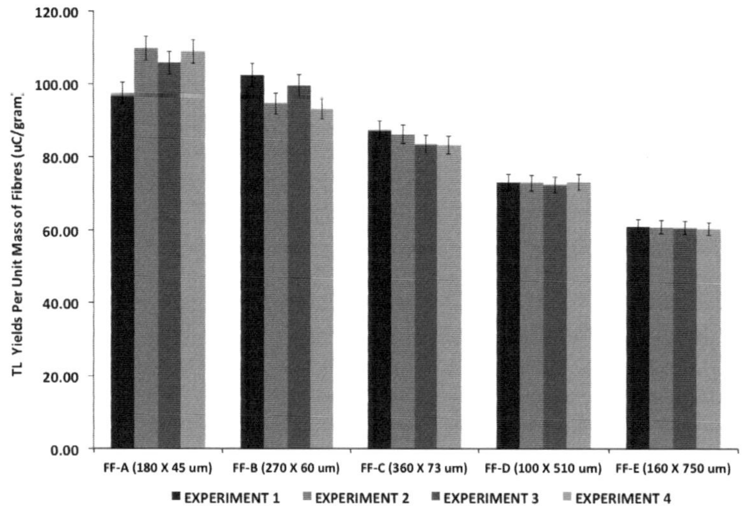


**Figure 2.** The reproducibility and repeatability of Ge-doped silica flat fibres irradiated for four times consequently at a fixed dose of 3 Gy delivered at a dose rate of 400 cGy/min of 6MV photon energy.

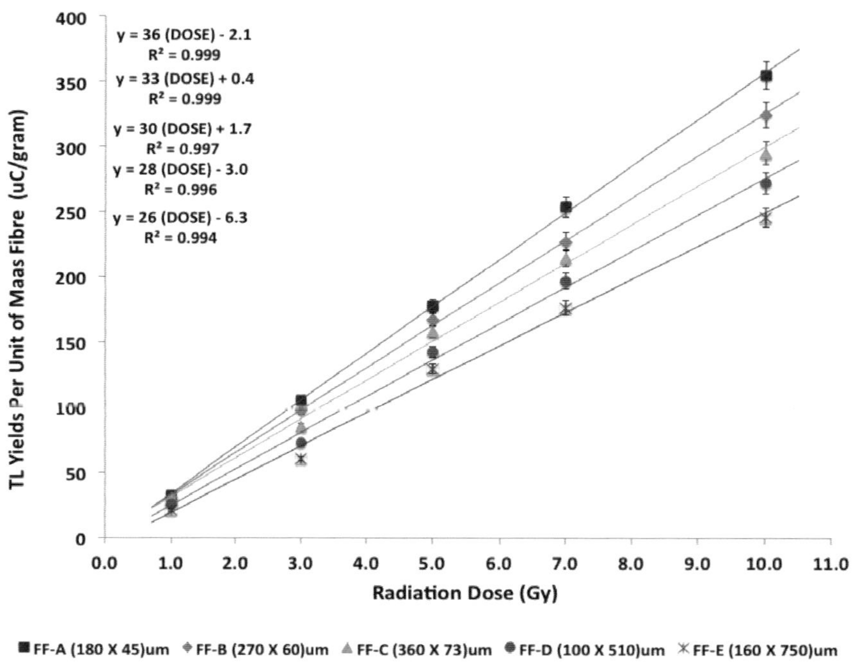


**Figure 3.** Dose response of Ge-doped optical fibers of cross-section dimensions 180 × 5 μm (FF-A), 270 × 60 μm (FF-B), 360 × 73 μm (FF-C) and 100 × 510 μm (FF-D) and 160 × 750 μm (FF-E), provided together with the standard error of the mean. The solid lines are least square fits to the data, obtaining respective correlation coefficients of better than 0.994. The irradiation has been made used of 6 MV photon energy at 400 cGy/min for a dose range 1 – 10 Gy. (Note: in some cases, the error bars are smaller than the data points).

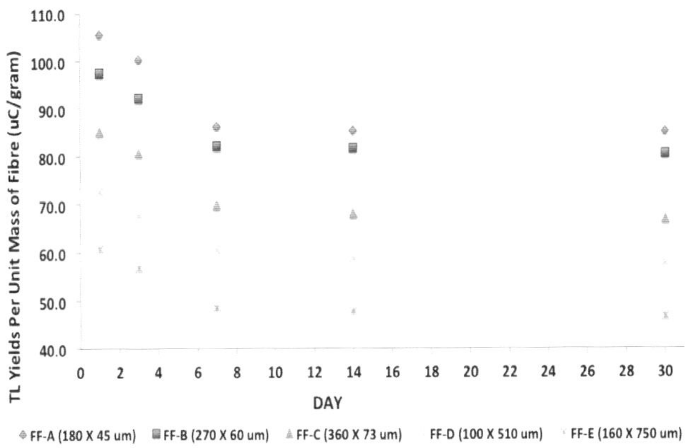


**Figure 4.** The loss of TL yield over a period of 30 days post-irradiation for the variously dimensioned Ge-doped optical fibers of cross-section dimensions 180 × 5 μm (FF-A), 270 × 60 μm (FF-B), 360 × 73 μm (FF-C) and 100 × 510 μm (FF-D) and 160 × 750 μm (FF-E). The irradiation has been made used of 6 MV photon energy at 400 cGy/min for a fixed dose 3 Gy.

## 4. Conclusions

Ge-doped $SiO_2$ flat fibers produce the greatest sensitivity of all of the fiber samples, also displaying uniform sensitivity over the full range of measured doses (1 – 10 Gy). The various features in term of high spatial resolution, impervious to water, reproducibility and low signal loss over 30 days post irradiation (0.14% per day make these tailored Ge-doped $SiO_2$ flat fibers a promising TL material for use as a dosimetric system in radiotherapy.

## 5. References

[1] Michael Baumann, Mechthild Krause, Jens Overgaard, Jürgen Debus, Søren M. Bentzen, Juliane Daartz, Christian Richter, Daniel Zips and Thomas Bortfeld, *Radiation oncology in the era of precision medicine.* Nat Rev Cancer, 2016. **16**(4): p. 234-249.

[2] Saif S Ahmad, Simon Duke, Rajesh Jena, Michael V Williams and Neil G Burnet, *Advances in radiotherapy.* BMJ : British Medical Journal, 2012. **345**.

[3] N A Zahaimi, H Zin, G A Mahdiraji, A L Abdul Rahman, D A Bradley and A T Abdul Rahman, *Characterisation of the thermoluminescence (TL) properties of tailor-made Ge-doped silica glass fibre for applications in medical radiation therapy dosimetry.* Journal of Physics: Conference Series, 2014. **546**(1): p. 012012.

[4] N A Zahaimi, M H R Ooi Abdullah, H Zin, A L Abdul Rahman, S Hashim, M I Saripan, M C Paul, D. A. Bradley and A T Abdul Rahman, *Dopant concentration and thermoluminescence (TL) properties of tailor-made Ge-doped SiO2 fibres.* Radiation Physics and Chemistry, 2014. **104**: p. 297-301.

[5] Bradley D. A. Abdul Sani Siti F, Alalawi Amani I. Jafari, S. M. Noor Noramaliza, M. Hairul Azhar, A. R. Mahdiraji, Ghafour Amouzad, Tamchek Nizam, Ghosh S, Paul M. C, Alzimami Khalid S, Nisbet A. Maah M. J, *Development of tailor-made silica fibres for TL dosimetry.* Radiation Physics and Chemistry, 2014. **104**: p. 3-9.

[6] Abdul Sani, Siti. F.; Alalawi, Amani I.; Azhar, Hairul A. R.; Amouzad Mahdiraji, Ghafour; Tamchek, Nizam; Nisbet, A.; Maah, M. J.; Bradley, D. A., *High sensitivity flat $SiO_2$ fibres for medical dosimetry.* Radiation Physics and Chemistry, 2014. **104**: p. 134-138.

[7] Hossain, I., H. Wagiran, and N.H. Yaakob, *Thermoluminescence of Ge- and Al-Doped $SiO_2$ Optical Fibers Subjected to 0.2–4.0 Gy External Photon Radiotherapeutic Dose.* Journal of Applied Spectroscopy, 2013. **80**(4): p. 620-623.

[8] Bradley DA, Hugtenburg RP, Nisbet A, Abdul Rahman AT, Issa F, Mohd Noor N, Alalawi A, *Review of doped silica glass optical fibre: Their TL properties and potential applications in radiation therapy dosimetry.* Applied Radiation and Isotopes, 2013(0).

[9] Fatma Issaa, A.T. Abdul Rahmana, Richard P. Hugtenburg, David A. Bradleya, Andrew Nisbet, *Establishment of Ge-doped optical fibres as thermoluminescence dosimeters for brachytherapy.* Applied Radiation and Isotopes, 2012. **70**(7): p. 1158-61.

[10] Abdul Rahman, A.T., Siti Fairus Abdul Sani, and D. A. Bradley, *Doped Silica telecommunication fibre as a 1-D detector for radiation therapy dosimetry.* AIP Conference Proceedings, 2012. **1423**(1): p. 347-353.

[11] Noor, Noramaliza M; Hussein, M; Bradley, D.A; Nisbet, A., *Investigation of the use of Ge-doped optical fibre for in vitro IMRT prostate dosimetry.* Nuclear Instruments and Methods in Physics Research Section A: Accelerators, Spectrometers, Detectors and Associated Equipment, 2011. **652**(1): p. 819-823.

[12] Issa, Fatma; Latip, Nur Atiqah Abd; Bradley, David A.; Nisbet, Andrew, *Ge-doped optical fibres as thermoluminescence dosimeters for kilovoltage X-ray therapy irradiations.* Nuclear Instruments and Methods in Physics Research Section A: Accelerators, Spectrometers, Detectors and Associated Equipment, 2011. **70(7), 11**

[13] Abdul Rahman, A.T., S.F.A. Sani, and D. Bradley, *Doped Silica Telecommunication Fibre as a 1-D Detector for Radiation Therapy Dosimetry*, in *IX Latin American Symposium on Nuclear Physics and Applications (IX-LASNPA 2011)*2011, AIP: Quito, Ecuador.

[14] A.T. Abdul Rahman, R.P. Hugtenburg, Siti Fairus Abdul Sani, A.I.M. Alalawi, Fatma Issa, R. Thomas, M.A. Barry, A. Nisbet, D.A. Bradley. *An Investigation of the Thermoluminescence of Ge-doped Silica Optical Fibres for Application in Interface Radiation Dosimetry.* in *8th Industrial Radiation and Radioisotope Measurement Applications (IRRMA).* 2011. Kansas City, USA.

[15] Abdul Rahman, A.T., A. Nisbet, and D.A. Bradley, *Dose-rate and the reciprocity law: TL response of Ge-doped $SiO_2$ optical fibers at therapeutic radiation doses.* Nuclear Instruments and Methods in Physics Research Section A: Accelerators, Spectrometers, Detectors and Associated Equipment, 2011. **652**(1): p. 891-895.

[16] Noor, NM, Hussein, M, Bradley, DA and Nisbet, A, *The potential of Ge-doped optical fibre TL dosimetry for 3D verification of high energy IMRT photon beams.* Nuclear Instruments and Methods in Physics Research Section A: Accelerators, Spectrometers, Detectors and Associated Equipment, 2010. **619**(1-3): p. 157-162.

[17] S. Hashim, D. A. Bradley, M. I. Saripan, A. T. Ramli, H. Wagiran, *The thermoluminescence response of doped $SiO_2$ optical fibres subjected to fast neutrons.* Applied Radiation and Isotopes, 2010. **68**(4-5): p. 700-703.

[18] D. A. Bradley, R. P. Hugtenburg, Suhairul Hashim, O. O. Okoya, A. L. Yusoff, A. Aziz Mat Hassan, A. T. Ramli, and H. Wagiran, *The Development of Doped Radiosensitive Glass.* AIP Conference Proceedings, 2007. **909**(1): p. 9-18.

[19] A.T. Abdul Rahman, R.P. Hugtenburg, Siti Fairus Abdul Sani, A.I.M. Alalawi, Fatma Issa, R. Thomas, M.A. Barry, A. Nisbet, D.A. Bradley, *An investigation of the thermoluminescence of Ge-doped $SiO_2$ optical fibres for application in interface radiation dosimetry.* Applied Radiation and Isotopes, 2012. **70**(7): p. 1436-1441.

[20] Entezam, A., Khandaker, M. U., Amin, Y. M., Ung, N. M.; Maah, J., Bradley, D. A., *Thermoluminescence response of Ge-doped $SiO_2$ fibres to electrons, X- and γ-radiation.* Radiation Physics and Chemistry, 2016. **121**: p. 115-121.

[21] S. Hashim, S. S Che Omar, S. A. Ibrahim, W. M S Wan Hassan, N. M. Ung, G. A. Mahdiraji, D. A. Bradley, K. Alzimami, *Thermoluminescence response of flat optical fiber subjected to 9 MeV electron irradiations.* Radiation Physics and Chemistry, 2015. **106**: p. 46-49.

**Acknowledgments**

The authors would like to acknowledge the Universiti Teknologi MARA (UiTM) for providing the facilities; the Advanced Medical and Dental Institute (AMDI) for their help in performing the irradiations. We also would like to thank University of Malaya - High Impact Research (UM-HIR) grant number A000007-50001 that supports this study, UM-HIR grant number UM.C/625/1/HIR/33 that supported fabrication of the Ge-doped preform and fibers, and TM-R&D and MMU MCVD group for fabricating the Ge-doped preform.

IOP Conf. Series: Journal of Physics: Conf. Series **851** (2017) 012025 doi:10.1088/1742-6596/851/1/012025

# Immobilisation precision in VMAT for oral cancer patients

**M N Norfadilah[1,2], R. Ahmad[1], S P Heng[3], K S Lam[2], A B Ahmad Radzi[2] and L S H John[2]**

[1]Diagnostic and Radiotherapy Programme, Faculty of Health Sciences, Universiti Kebangsaan Malaysia, 50300 Kuala Lumpur, Malaysia
[2]Radiotherapy Unit, Cancer Centre, Pantai Hospital Kuala Lumpur, 59100 Kuala Lumpur, Malaysia
[3]Cancer, Radiosurgery and Nuclear Medicine Centre, Sunway Medical Centre, 46150 Petaling Jaya, Selangor, Malaysia

E-mail: norfadilahmn@gmail.com

**Abstract**. A study was conducted to evaluate and quantify a precision of the interfraction setup with different immobilisation devices throughout the treatment time. Local setup accuracy was analysed for 8 oral cancer patients receiving radiotherapy; 4 with HeadFIX® mouthpiece moulded with wax (HFW) and 4 with 10 ml/cc syringe barrel (SYR). Each patients underwent Image Guided Radiotherapy (IGRT) with total of 209 cone-beam computed tomography (CBCT) data sets for position set up errors measurement. The setup variations in the mediolateral (ML), craniocaudal (CC), and anteroposterior (AP) dimensions were measured. Overall mean displacement (M), the population systematic (Σ) and random (σ) errors and the 3D vector length were calculated. Clinical target volume to planning target volume (CTV-PTV) margins were calculated according to the van Herk formula ($2.5\Sigma+0.7\sigma$). The M values for both group were < 1 mm and < 1° in all translational and rotational directions. This indicate there is no significant imprecision in the equipment (lasers) and during procedure. The interfraction translational 3 dimension vector for HFW and SYR were 1.93±0.66mm and 3.84±1.34mm, respectively. The interfraction average rotational error were 0.00°±0.65° and 0.34°±0.59°, respectively. CTV-PTV margins along the 3 translational axis (Right-Left, Superior-Inferior, Anterior-Posterior) calculated were 3.08, 2.22 and 0.81 mm for HFW and 3.76, 6.24 and 5.06 mm for SYR. The results of this study have demonstrated that HFW more precise in reproducing patient position compared to conventionally used SYR ($p<0.001$). All margin calculated did not exceed hospital protocol (5mm) except S-I and A-P axes using syringe. For this reason, a daily IGRT is highly recommended to improve the immobilisation precision.

## 1. Introduction

Novel radiotherapy technique Volumetric Modulated Arc Therapy (VMAT) allows superior conformal dose distribution to the treatment target volume within a short time interval. Thus, significant dose reduction to surrounding normal tissues can be achieved due to clinical target volume to planning target volume (CTV-PTV) margin was reduced for this advanced technique [1, 2]. However, this technique is frequently associated with a wide range of acute and chronic side effect complications causing local discomfort and pain.

The most common side effects of radiation therapy (RT) for oral cancer patients are dry mouth (xerostomia) and inflamed mucous membranes (mucositis) [3, 4]. Patients may have dental decay (caries) and death of bone tissue (osteoradionecrosis) resulting from poorly directed radiation during radiotherapy treatment due to setup uncertainties [5, 6]. Therefore, a study was conducted to identify and evaluate the ideal immobilisation device that can minimise setup uncertainties by maintaining the patient in the same position during the course of treatment. Thus, we comparing two immobilisation devices; syringe 10 ml (conventional approach) and customized HeadFIX® mouthpiece moulded with wax (a sub-device initially invented for stereotactic radiotherapy procedure).

## 2. Materials and methods

### *2.1 Patient selection*

Ethical approval from research committee members of Universiti Kebangsaan Malaysia (JEPUKM) with ethic number NN-166-2014 and permission from management of Pantai Hospital Kuala Lumpur were granted for this study. Permission from patients and oncologists involved was consented after researcher explained risks and benefits of the research. Local setup accuracy was analysed for 8 oral cancer patients receiving radiotherapy. The patient that included in this study were consecutively assigned to one of the two immobilisation devices which were HeadFIX® mouthpiece moulded with wax (HFW) and 10 ml/cc syringe barrel (SYR). Inclusion criteria include patients with one type of oral cancer. Stages and types of the oral cancers are verified from histopathology and radiology reports. The patient was planned to receive radical radiotherapy treatment using VMAT technique and received a total dose of 60-70 Gy for 6-7 weeks. Patients who failed to finish the treatment as prescribed by the oncologist are excluded. Furthermore, patients who have excessive salivary secretion are also excluded as subjects because it will influence the immobilisation device setup. The characteristics of the patient are shown in Table 1.

**Table 1**. Patient characteristics.

| Characteristics | No. of patients (%) |
|---|---|
| **Patients (n)** | 8 |
| Men | 2 (25.0) |
| Women | 6 (75.0) |
| **Age (y)** | |
| Mean | 57 |
| Range | 23-83 |
| **Tumour location (n)** | |
| Hard Palate | 1 (12.5) |
| Buccal Mucosa | 1 (12.5) |
| Sublingual gland | 1 (12.5) |
| Submandibular gland | 1 (12.5) |
| Tongue | 4 (50.0) |
| **Medical History** | |
| Post-surgery | 2 (25.0) |
| Concurrent chemotherapy | 2 (25.0) |
| Post-surgery & concurrent chemotherapy | 3 (37.5) |
| None | 1 (12.5) |

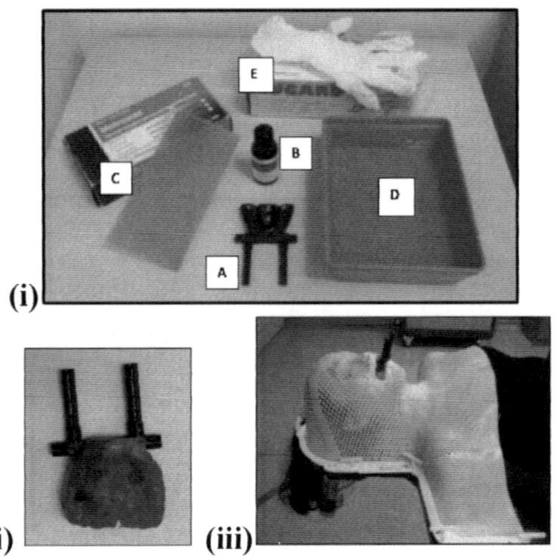


(i) (ii) (iii)

**Figure 1.** Preparation for fabrication of HeadFIX® mouthpiece moulded by wax (i) Apparatus; A: HeadFIX® mouthpiece, B: adhesive material, C: a blend of paraffin & micro-crystalline waxes, D: hot water and E: glove. (ii) HFW after fabrication process. (iii) Patient in supine position while HFW threaded into a small hole of thermoplastic mask.

### *2.2 Radiotherapy simulation*

HFW was prepared and fabricated prior to simulation as shown in Figure 1. No special fabrication needed for SYR. Type-S thermoplastic mask of head-neck-shoulder was used to maintain head, neck and shoulder's patient during treatment. Then, the immobilisation device HFW or SYR was threaded through a small hole as the mask was placed over the head as Figure 1 (iii). CT scans with contrast media were obtained in 2 mm slice thickness using 16 slice Brilliance™ CT Big Bore (Philips, Netherlands). Coverage of the scan was from frontal sinus to carina. The reconstructed CT images were imported to Monaco treatment planning system version 3.20.02 for planning purposes.

### *2.3 Radiotherapy treatment planning*

Clinical oncologist contoured the tumour volume and medical physicist contoured organ-at-risk (OAR) e.g. spinal cord, parotid glands, and brain stem. Generally, CTV encompasses the entire gross tumour

IOP Conf. Series: Journal of Physics: Conf. Series **851** (2017) 012025 doi:10.1088/1742-6596/851/1/012025

volume (GTV) with an additional margin of 5 mm. PTV was created with addition of 5 mm margin on CTV. The secondary target was the metastatic cancer at high-risk lymph nodes group around the neck area. PTV to this lymph nodes group was the CTV plus 3 mm margin. A lower dose was given to low-risk lymph nodes. Next, the optimisation and dose calculation of CT images was performed with Monte Carlo algorithm. Treatments were planned by medical physicists with 6 megavolt (MV) photon with two arch VMAT technique. This followed by quality assurance routine to ensure that the dosage was as planned. The treatment planning was considered successful when the target volume dose of treatment exceeds 95% of the prescription dose.

### *2.4 Cone-Beam Computed Tomography (CBCT) volume acquisition*
After radiotherapy planning has been approved, patients will be scheduled for radiotherapy treatment. On each treatment day, patients were positioned on the radiotherapy couch in supine position and instructed to lie motionless in their respective fixation. Prior to radiotherapy treatment, image guided radiotherapy (IGRT) was performed using CBCT scanner integrated with Elekta Synergy linear accelerator (Elekta, Crawley, UK). The parameters used were 100 kVp, 10 mA and 10 ms for nominal frame, collimator and filter F0 was S20. One full rotation of the head and neck cases was about 200 degrees starting from an angle of 100° to 250°. Approximately 361 projections been done. These projections were undergone reconstruction process to form a three-dimensional (3D) image in medium resolution. CBCT scanning was done by using a volumetric imaging system kV X-rays (XVI, Elekta Ltd., Crawley, UK). This process takes about two minutes. CBCT images were processed and displayed in coronal, sagittal and axial dimension.

### *2.5 Images registration*
Registration of the images from treatment planning CT and acquired CBCT were processed by an automatic bone value algorithm. This was followed by manual adjustment based on target volume and bone near the PTV especially maxilla, mandible, and first until third cervical vertebrae [7]. Once the registration satisfied, the position of the treatment table was corrected based on the resulting position error. There were two categories of reading error; translational and rotational errors. Translational errors represent the position of the subject in three degrees of freedom (3 DoF); lateral (left-right), longitudinal (superior-inferior) and vertical (anterior-posterior). Whereas, rotational errors represent roll, yaw and pitch position. The ability of adjustment in translational and rotational errors in simultaneously means the system capable to correct errors in 6 DoF. The image registration process must be done carefully so that the radiation dose could be delivered as per prescribed.

### *2.6 Statistical analysis*
Each position setup error measurement represents the total variation in patient positioning for the treatment session. The mean value (M), standard deviation, systematic ($\Sigma$) and random ($\sigma$) errors of all translational and rotational errors were calculated and compared for these 2 different immobilisation devices. CTV-PTV margin was calculated according to the formula $PTV = 2.5\Sigma + 0.7\sigma$ [8]. A three-dimensional (3D) vector was calculated using the following formula: $v = \sqrt{(x^2 + y^2 + z^2)}$. The hypothesis of a normal distribution was rejected for the majority of variables. Hence, the Mann-Whitney U test was used to determine if there was a significant difference between HFW and SYR results. All p-values were calculated with two-tailed tests, and statistical significance was defined as $p < 0.001$ with IBM SPSS statistic version 23.

## 3. Results
A total of 209 CBCT images were acquired and analysed from the 8 oral cancer patients; 120 for HFW and 89 for SYR.

*3.1 Setup accuracy and reproducibility between immobilisation devices*

Table 2 summarizes the isocentric setup variations for oral cancer patients HFM and SYR. The M values for both group were <1 mm and <1° in all translational and rotational directions. This indicate there is no significant imprecision in the equipment (lasers) and during procedure [9]. Figure 2 shows percentage distribution of setup errors in all 6DoF for HFW (represented in areas) and SYR (represented in line). All error toward zero value, referred as minimum error. If the number away from zero neither positive nor negative values, it means the error become bigger. Setup error in HFM was statistically significant lower than SYR for A-P, pitch and yaw directions with $p<0.001$. Meanwhile, SYR significantly ($p<0.001$) shows lower setup errors in R-L direction.

From Table 2, HFW has lower values (<1 mm) of systematic errors in all translational and in yaw directions (< 1°) compared to SYR. In contrast, SYR only shows lower values (< 1°) of systematic errors in all rotational directions. All random errors values in translational and rotational directions were lower in HFM compared to SYR.

In addition, comparative analysis for interfraction 3D vector and average rotational error reveal that there was a significant difference ($p<0.001$) in setup accuracy between the two immobilisation devices studied. By referring Table 3, the interfraction translational 3D vector for HFW and SYR were 1.93±0.66mm and 3.84±1.34mm, respectively. The interfraction average rotational errors were 0.00°±0.65° and 0.34°±0.59°, respectively.

**Table 2.** Summary of the isocentric setup variations for oral cancer patients HFM and SYR.

| **Immobilisation devices** | **HFW** | | | **SYR** | | |
|---|---|---|---|---|---|---|
| Translational errors (mm) | **R-L** | **S-I** | **A-P** | **R-L** | **S-I** | **A-P** |
| Min | -2.10 | -2.30 | -3.00 | -4.50 | -5.20 | -6.20 |
| Max | 3.20 | 3.40 | 2.00 | 2.70 | 8.90 | 4.50 |
| M | 0.53 | -0.12 | -0.76 | -0.48 | 0.31 | 0.18 |
| Σ | 0.99 | 0.64 | 0.09 | 1.09 | 1.70 | 1.52 |
| σ | 0.87 | 0.87 | 0.85 | 1.47 | 2.86 | 1.80 |
| CTV-PTV margin (mm) | 3.08 | 2.22 | 0.81 | 3.76 | 6.24 | 5.06 |
| Rotational errors (°) | **Pitch** | **Roll** | **Yaw** | **Pitch** | **Roll** | **Yaw** |
| Min | -2.50 | -2.20 | -1.50 | -1.40 | -3.00 | -1.90 |
| Max | 2.90 | 2.80 | 1.40 | 3.30 | 2.70 | 3.00 |
| M | -0.13 | 0.12 | 0.01 | 0.32 | 0.10 | 0.41 |
| Σ | 1.12 | 1.09 | 0.39 | 0.60 | 0.51 | 0.43 |
| σ | 0.51 | 0.64 | 0.49 | 0.62 | 0.89 | 0.75 |

*Abbreviations*: R-L=right-left; A-P=anterior-posterior; S-I=superior-inferior; min=minimum, max=maximum, M=population mean setup errors, Σ=population systematic error; σ=population random errors; CTV-PTV= clinical target volume-planning target volume.

**Table 3.** 3D vector, and average rotational error for 2 different immobilisation device.

| Immobilisation device | 3-dimensional vector (mm) | | | | Average rotational error (°) | | | |
|---|---|---|---|---|---|---|---|---|
| | Min | Max | Mean | SD | Min | Max | Mean | SD |
| HFM | 0.51 | 3.86 | 1.93 | 0.66 | -1.17 | 2.17 | 0.00 | 0.65 |
| SYR | 1.14 | 9.08 | 3.84 | 1.34 | -1.10 | 2.60 | 0.34 | 0.59 |

*3.2 CTV-PTV margins*

The margins calculated for each axis using the two immobilisation device were also shown in Table 2. For the HFW, the margin calculated in R-L axis was the largest (3.08 mm), while the A-P axis was the smallest (0.81 mm). However, for SYR, the margin calculated in R-L axis was the smallest (3.76 mm) and the S-I axis had the largest margin (6.24 mm).

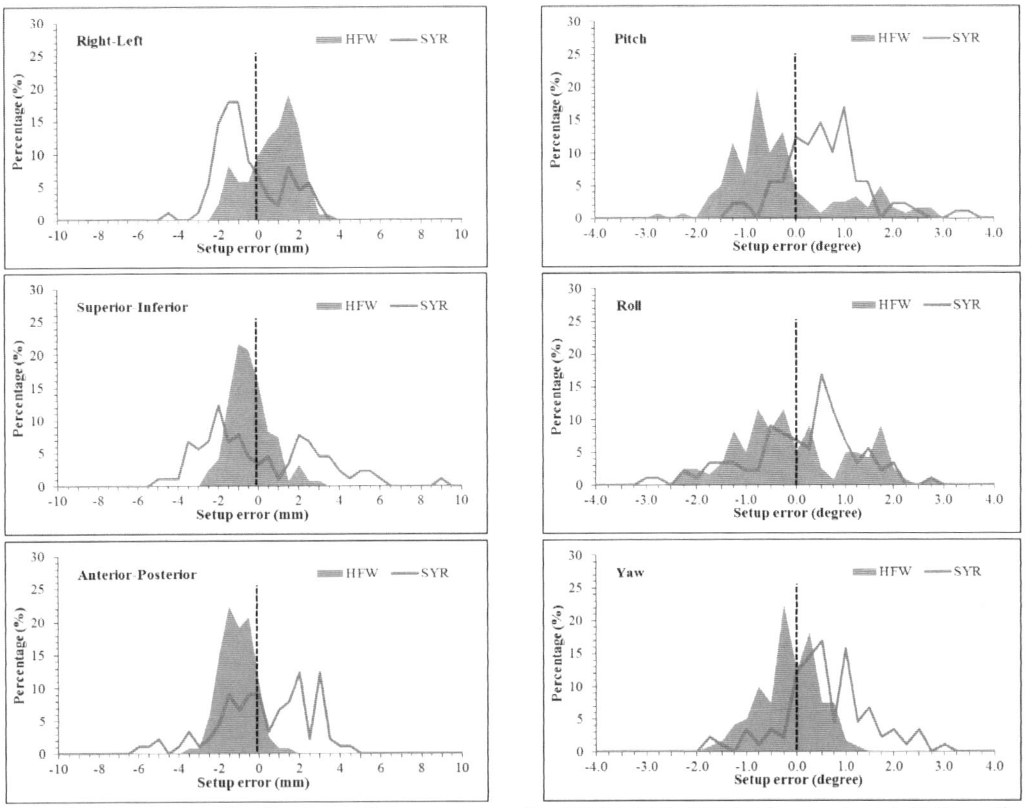


**Figure 2.** Percentage distribution of setup errors for HFW (120 CBCT images) and SYR patients (89 CBCT images) in right-left, superior-inferior, anterior-posterior, pitch, roll and yaw directions.

## 4. Discussion

It has been widely observed that immobilisation devices are important in positioning reproducibility. In this study, the relative impact of HFW and SYR on the reproducibility during oral cancer radiotherapy was evaluated by using CBCT images. CBCT images provide an on-line 3D patient positioning assessments precision in sub-millimetre [9]. In order to evaluate the precision of the interfraction setup for immobilisation devices throughout the treatment time, population random and systematic errors must be determined. Systematic errors that occur in the same magnitude and direction each fraction throughout the treatment course was identified and minimised using image guided correction for every patient. During this study, systematic errors may be introduced during a patient's treatment at the localisation, planning or treatment delivery phases. Meanwhile, random errors were influenced by patient compliance, department protocols and the immobilisation system itself [10].

HFW shows small values in overall of systematic and random errors described higher reproducibility of the treatment preparation performed compared to SYR. Thus, it improves setup accuracy and reduces PTV movement. With higher precision of customized immobilisation device, it could minimise the adverse effects of radiation especially oral mucositis [11,12,13]. It also increased the distance between the maxilla and mandible, focusing the radiation dose more precisely on the target volume (dosimetry) and immobilising the mandible [13,14,15].

Besides that, all margin calculated did not exceed hospital protocol (5mm) except S-I and A-P axes using SYR. However, since the VMAT treatment plans generated by steep dose gradients between PTV and OAR, the precise targeting and delivering of radiation therapy on a daily basis is imperative. Therefore, in some special situations, such as re-irradiation or the close proximity of OAR with high-dose regions and lower (i.e., 3 mm) margins could benefit from daily image guidance correction.

## 5. Conclusion
HFW shows smaller position setup errors in overall isocentric shift indicates higher reproducibility and thus more precise than the SYR. Even though the value were smaller, but this errors can be furthered minimised by careful attention to immobilisation and patient preparation techniques.

## 6. References
[1] Lee T F, Ting H M, Chao P J and Fang F M 2012 Dual arc volumetric-modulated arc radiotherapy (VMAT) of nasopharyngeal carcinomas: A simultaneous integrated boost treatment plan comparison with intensity-modulated radiotherapies and single arc VMAT *Clin. Oncol.* **24**(3), 196-207.

[2] Teoh M, Clark C H, Wood K, Whitaker S and Nisbet A 2011 Volumetric modulated arc therapy: A review of current literature and clinical use in practice *Br. J. Radiol.* **84** (1007), 967-996.

[3] Nagler R M and Baum B J 2003 Prophylactic treatment reduces the severity of xerostomia following radiation therapy for oral cavity cancer *Arch Otolaryngol. Head Neck Surg.* **129**(2), 247-250.

[4] Ch'ng S, Oates J, Gao K, Foo K, Davies S, Brunner M and Clark J R 2014 Prospective quality of life assessment between treatment groups for oral cavity squamous cell carcinoma *HeadNeck* **36**, e834-e840.

[5] Niewald M, Mang K, Barbie O, Fleckensteuin J, Holtmann H, Spitzer W J and Rube C 2014 Dental status, dental treatment procedures and radiotherapy as risk factors for infected osteoradionecrosis (IORN) in patients with oral cancer – a comparison of two 10 years' observation periods *Springerplus* **3**(263), 1-12

[6] Ben-David M A, Diamante M, Radawski JD, Vineberg K A, Stroup C, Murdoch-Kinch C, Zwetchkenbaum S R and Eisbruch A 2007 Lack of osteoradionecrosis of the mandible after IMRT for head and neck cancer: likely contributions of both dental care and improved dose distributions *Int. J. Radiat. Oncol. Biol. Phys.* **68**(2), 396-402.

[7] Simon V K, Suzanne V B, Coen R, Marcel V K and Jan J S 2009 Setup uncertainties of anatomical sub-regions in head-and-neck cancer patients after offline CBCT guidance *Int. J. Radiat. Oncol. Biol. Phys.* **73**(5), 1566–73.

[8] Herk V M, Remeijer P, Rasch C and Lebesque J V 2000 The probability of correct target dosage: Dose-population histograms for deriving treatment margins in radiotherapy *Int. J. Radiat. Oncol. Biol. Phys.* **47**(4), 1121-35

[9] Herk V H 2004 Errors and margins in radiotherapy *Semin. Radiat. Oncol.* **14(**1), 52–64

[10] Hong T S, Tomé W A, Chappell R J, Chinnaiyan P, Mehta M P, and Harari P M 2005 The impact of daily setup variations on head-and-neck intensity-modulated radiation therapy *Int. J. Radiat. Oncol. Biol. Phys*. **61**(3), 779–788

[11] Kaanders J H A M, Fleming T J, Ang K K, Maor M H and Peters L J 1992 Devices valuable in head and neck radiotherapy *Int. J. Radiat. Oncol. Biol. Phys.***23**(3), 639-645.

[12] Verrone J R, Alves F D A, Prado J D, Boccaletti K W, Sereno M P, Silve M L G and Jaguar G C 2013 Impact of intraoral stent on the side effects of radiotherapy for oral cancer *HeadNeck* **35**, e213-217.

[13] Johnson B, Sales L, Winston A, Liao J, Laramore G and Parvathaneni U 2013 Fabrication of customized tongue-displacing stents *Journal of American Dental Association* **144**(6), 594-600.

[14] Bodard A G, Racadot S, Salino S, Pommier P, Zrounba P and Montbarbon X 2009 A new, simple maxillary-sparing tongue depressor for external mandibular radiotherapy: A case report *Wiley InterScience: Head & Neck* **31**, 1528-30

[15] Sales L R, Liao J, Johnson B, Winston A, Laramore G and Parvathaneni U 2011 Customized tongue-displacing dental stents for oral mucosal sparing and immobilisation in head and neck radiotherapy *Proceedings of the American Society for Radiation Oncology 53rd Annual Meeting* S493.

IOP Conf. Series: Journal of Physics: Conf. Series **851** (2017) 012026 doi:10.1088/1742-6596/851/1/012026

# Bladder filling variation during conformal radiotherapy for rectal cancer

**S Sithamparam[1], R Ahmad[1], A Sabarudin[1], Z Othman[2], M Ismail[2]**

[1] Diagnostic Imaging and Radiotherapy Programme, Faculty of Health Sciences, 50300 Kuala Lumpur, Malaysia
[2] Radiotherapy and Oncology Department, Hospital Kuala Lumpur, Malaysia

E-mail: santhekumari@gmail.com

**Abstract** Conformal radiotherapy for rectal cancer is associated with small bowel toxicity mainly diarrhea. Treating patients with a full bladder is one of the practical solutions to reduce small bowel toxicity. Previous studies on prostate and cervix cancer patients revealed that maintaining consistent bladder volume throughout radiotherapy treatment is challenging. The aim of this study was to measure bladder volume variation throughout radiotherapy treatment. This study also measured the association between bladder volume changes and diarrhea. Twenty two rectal cancer patients were recruited prospectively. Patients were planned for treatment with full bladder following departmental bladder filling protocol and the planning bladder volume was measured during CT-simulation. During radiotherapy, the bladder volume was measured weekly using cone-beam computed tomography (CBCT) and compared to planning bladder volume. Incidence and severity of diarrhea were recorded during the weekly patient review. There was a negative time trend for bladder volume throughout five weeks treatment. The mean bladder volume decreased 18 % from 123 mL (SD 54 mL) during CT-simulation to 101 mL (SD 71 mL) on the 5th week of radiotherapy, but the decrease is not statistically significant. However, there was a large variation of bladder volume within each patient during treatment. This study showed an association between changes of bladder volume and diarrhea (P = 0.045). In conclusion bladder volume reduced throughout radiotherapy treatment for conformal radiotherapy for rectal cancer and there was a large variation of bladder volume within patients.

## 1. Introduction

Radiotherapy combined with chemotherapy or chemo-radiotherapy (CRT) is established as a standard treatment for locally advanced rectal cancer and improves local control [1]. A significant number of patients treated with radiotherapy for rectal cancer experience gastrointestinal acute and late toxicity. Acute and late gastrointestinal toxicity such as pain, bloating, nausea, fecal urgency, diarrhea, and rectal bleeding have a significant impact on patient's quality of life. In a study by Gérard et al. [2], 8-40% of patients were reported to have acute toxicity (grade $\geq$ 3) and 24 % late toxicity. The volume of irradiated small bowel and the radiation dose were reported as major factors affecting small bowel toxicity [3].

Therefore, non-invasive, effective and clinically practical approaches were needed to reduce the amount of small bowel in an irradiated field. Kavanagh et al. [4] and Fiorino et al. [5] described several methods to minimise the amount of small bowel in the treatment volume using surgical and non-surgical methods. Surgical techniques are procedures such as clip placement in high-risk areas, pelvic reconstruction, re-peritonealisation of the pelvic floor, and placement of an omental sling, retroversion of the uterus and by placing a synthetic prosthesis under the small bowel (a removable pelvic spacer). The irradiated small bowel volume can also be minimalized by radiotherapy techniques such as 3D conformal radiotherapy; intensity modulated radiotherapy (IMRT), adaptive radiotherapy, customised shielding, a shrinking field technique, bladder distension and optimal irradiation position such as supine, prone or by using a belly board.

Many institutions combined bladder distension with radiotherapy technique such as using belly board as a method to reduce the amount of small bowel in the irradiated field because it is more practical. Kim et al. [6] reported that bladder distension is more effective in reducing the irradiated small bowel compared to using belly board. A distended bladder pushes the small bowel superiorly and away from radiotherapy treatment field [7, 8, and 9]. However, maintaining consistent bladder volume throughout radiotherapy treatment is challenging.

Previous studies have shown that consistent bladder volume throughout radiotherapy treatment is difficult to reproduce. Most studies showed a reduction in the bladder volume towards the end of treatment particularly in prostate and cervical cancer [10], [11], [12] and [13]. Reduction or variation in bladder volume leads to the inconsistent amount of small bowel being irradiated. There are limited studies on efforts to maintain bladder volume for rectal cancer treatment. Therefore, the implementation of an effective bladder filling protocol is required for rectal cancer patients. Effective bladder filling protocol will also result in greater treatment accuracy, particularly with the advancement in IMRT in future and reduced incidence of diarrhea, which will eventually improve patient's quality of life during and after radiotherapy.

This is a preliminary study to assess the effectiveness of existing bladder filling protocol in maintaining bladder volume consistency in our radiotherapy department. This study assesses the bladder volume variation, trend of bladder volume changes throughout radiotherapy treatment and the association of bladder volume changes and incidence of diarrhea in rectal patients treated with conformal radiotherapy in Radiotherapy and Oncology Department, Hospital Kuala Lumpur.

## 2. Materials and method

### *2.1. Patient population*

This study enrolled 22 rectal cancer patients who were scheduled for radical pre- and post-operative pelvic radiotherapy based on stratified random sampling. Informed consent was obtained from all the patients. Patients receiving palliative radiotherapy, radical short course pre-operative radiotherapy, had a functional disorder in their bladder, had cystectomy were excluded from this study. Those patients, who showed a strong aversion to retaining urine, had previous lower gastrointestinal radiotherapy and patients with the Eastern Cooperative Oncology Group (ECOG) performance status of 2 and lower were also excluded. This study was conducted with Malaysian Medical Research and Ethics Committee approval NMRR-15-807-24333(IIR) from October 2015 until May 2016. The clinical profiles of patients are shown in Table 1.

**Table 1.** Clinical profiles of patients.

| Criteria | No (%) |
|---|---|
| Age , average (range) | 55 (27 – 71) |
| Male | 16 (72) |
| Female | 6 (27) |
| Pre-operative | 11 (50) |
| Post-operative | 11 (50) |
| Concurrent chemo radiotherapy | 15 (68) |
| Radiotherapy dose (Gy) , average | 45 |
| Radiotherapy fractions | 25 |

*2.2. CT-simulation*
According to departmental bladder filling protocol, patients were required to empty the bladder, drink 200ml of water and wait for 30 minutes before CT-simulation. CT-simulation was carried out using Brilliant Big Bore Philips CT-simulator. A radio-opaque marker was placed on the anal verge. Patients were positioned supine with both hands resting on chest. CT scan was performed from the level of third lumbar vertebrae until the middle of the femur with 5 mm slice thickness.

*2.3. Radiotherapy treatment*
Patients are treated using 10-MV Varian Medical System- Silhouette Edition linear accelerator and Varian Clinac RapidArc linear accelerator. Dosage and fractionation for rectal cancer radiotherapy was 45 Gy in 25 daily fractions over five weeks. The superior border: 1.5 cm superior to sacral promontory, which correlates to L5-SI inner space. The lateral borders: 2 cm lateral to pelvic brim and inlet to include iliac lymph nodes. The inferior border: includes the entire obturator foramina, although this may vary depending on the location of the tumor. The recommended inferior margin is 3 to 5 cm from the gross tumor pre-operatively or below the most distal extent of dissection post-operatively. The posterior field border: 1.5 to 2 cm behind the anterior bony sacral margin. Anterior field border: the anterior edge of the femoral heads to ensure coverage of internal iliac nodes. Organs at risk are small bowel, femoral head and neck, bladder, and rectum.

*2.4. Bladder volume measurement*
During planning, bladder volume was measured from CT planning images. Planning bladder volume was used as the reference to measure the variation of bladder volume during treatment. Treatment bladder volume was derived every week from offline treatment Cone Beam CT (CBCT) images. In this study the both the CT-simulation and treatment bladder volume were measured using calculation based on ellipsoid formula. This is because bladder volume could not be measured directly from CBCT images. The accuracy of ellipsoid formula were validated by comparing with bladder volume derived from bladder contour by physician on CT-simulation images. Bladder width, height, and depth were used to calculate bladder volume using ellipsoid formula. Two senior radiation therapists measure the bladder dimension independently, and the average measurement was taken. Each patient has 5 CBCT images.

*2.5. Assessment on diarrhea*
The incidence and severity of diarrhea were assessed weekly during patient's weekly review by attending medical officer or oncologist. A standardised form derived from National Cancer Institute's Common Terminology Criteria for Adverse Events (CTCAE version 3.0) Events: Diarrhea was used to scoring diarrhea.

*2.6. Statistical test*
The accuracy of bladder volume measurement using ellipsoid formula and bladder contour in CT-simulation images was determined using Pearson correlation coefficient. Variation of mean bladder volume during CT-simulation and fifth week of radiotherapy was compared using paired t-test. Spearman's rank test was used to determine the association between changes in bladder volume with incidence and severity of diarrhea. All the statistical test was carried out using SPSS version 23.0 (IBM, Armouk, NY, USA).

## 3. Results and discussion

*3.1. Validation of ellipsoid formula*
The mean difference between the bladder volume measurement using ellipsoid formula and bladder volume measurement from bladder contour on CT-simulation images is 1.45 mL ± 10.74 mL. There is a strong correlation between bladder volume measurement using ellipsoid formula and bladder volume

measurement from bladder contour on CT-simulation images, $r = 0.98$, $P < 0.01$ (Figure 1). The accuracy of ellipsoid formula for measuring bladder volume was earlier validated by Miralbell et al. [13] ($r^2 = 0.9859$) and Dicuio et al. [14].

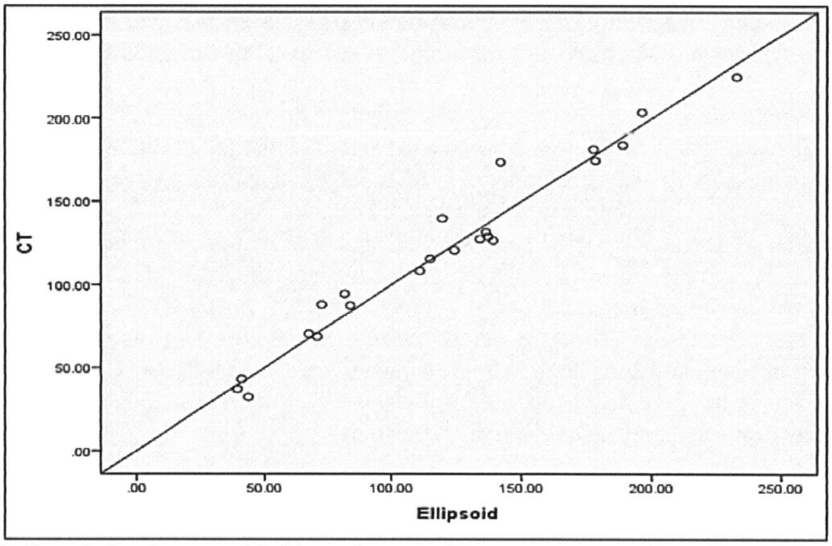


**Figure 1.** Correlation between bladder volume (mL) measurement using ellipsoid formula and bladder volume (mL) measurement from bladder contour on CT-simulation images.

*3.2. Bladder volume variation*

The population mean bladder volume variation for 22 patients throughout the five weeks radiotherapy treatment was compared with the bladder volume during CT-simulation as shown in figure 2. A total of 110 images were used in this study. Mean bladder volume during CT-simulation was 123 ± 54 mL During week four and week five, treatment bladder volume reduced 10% to 111 ± 85 mL and 18% to 101 ± 71 mL respectively from CT-simulation bladder volume. However, the population mean bladder volume reduction is not statistically significant ($P = 0.104$). The reduction of bladder volume during the last two weeks of radiotherapy treatment could be due to patients becoming less attentive in the following bladder filling instructions towards the end of radiotherapy treatment or due to increased frequency of diarrhea which makes it difficult for patients to control micturition.

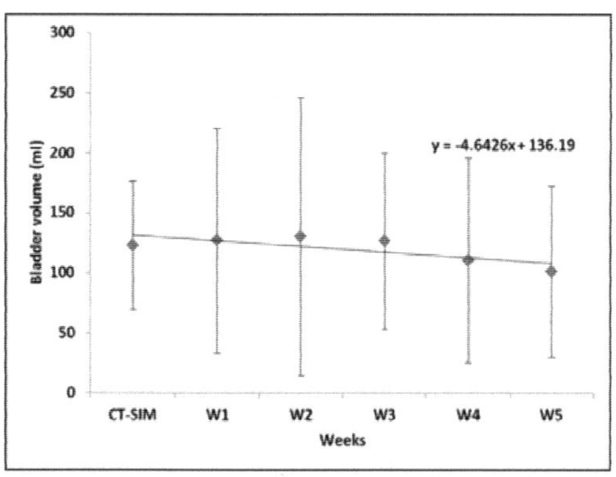


**Figure 2.** The change and variation in bladder volume during treatment compared to bladder volume during planning (CT-sim).

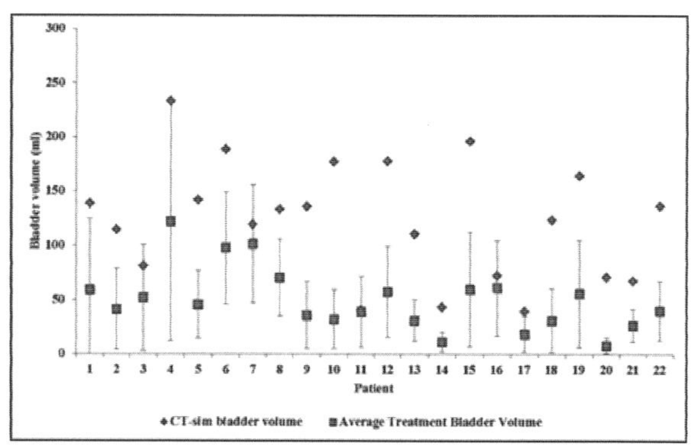


**Figure 3.** Variation of bladder volume within patient. Diamond: Bladder volume during CT simulation. Square: Mean Bladder volume for each patient during treatment.

Similar studies on bladder volume variation for prostate and cervix cancer patients reported a significant reduction in bladder volume [10], [11], [12] and [13]. Patients enrolled in all the above studies had larger mean bladder volume at CT-simulation compared to this study, ranging from 219 mL – 378 mL It has been reported that maintaining consistency of large bladder volume throughout radiotherapy treatment is more difficult than smaller bladder volume.

However, in this study bladder volume variation within each patient was large (Figure 3). During the first week of radiotherapy treatment, 50% of patients were having more than 50% change in bladder volume and on the fifth week of treatment 64 % of patients are having more than 50% change in bladder volume compared to planning bladder volume (Table 2). Such large variation of bladder volume within patients during radiotherapy treatment could lead to significant change in the volume of small bowel in the irradiated field. Joost J Nuyttens et al. [15] studied the implication of variation of small bowel in rectal cancer radiotherapy and had found that variation in bladder volume cause shift the clinical target volume (CTV) by 7 mm cranially and 2.5 cm caudally. Therefore, large variation in bladder volume could affect the accuracy of treatment and dose volume histogram of organ at risk particularly of the small bowel. Small bowel toxicity was shown to be primarily dependent on the volume of small bowel receiving over 15 Gy, therefore reducing the volume of the small bowel may have an impact on the tolerance to therapy.

**Table 2.** Number and percentage (%) of patients having bladder volume changes of more than 50% and 100% throughout five weeks radiotherapy treatment.

| Percentage of bladder volume change | Week 1 | Week 2 | Week 3 | Week 4 | Week 5 |
|---|---|---|---|---|---|
| >50% | 11/22 (50.0%) | 10/22 (45.6) | 4/22 (25.0%) | 9/22 (41.0%) | 14/22 (64.0%) |
| >100% | 2/22 (9.1%) | 4/22 (18.2%) | 2/22 (9.1%) | 2/22 (9.1%) | 1/22 (4.5%) |

There was a correlation between bladder volume changes and incidence and severity of diarrhea (Spearman's rank correlation $r_s$ = -0.432, P = 0.045). The highest score for diarrhea is 2 which indicates an increase of 4–6 stools/day over baseline; moderate increase in ostomy output compared with baseline. During the third week of radiotherapy 10/22 (45%) of patients were reported having score 2 for diarrhea and was later controlled with medication and diet modification. Future research is required to reduce the variation of bladder volume within patients throughout radiotherapy treatment to improve treatment accuracy and minimise acute and chronic small bowel toxicity.

The standardized protocol in this study was not successful in ensuring a consistent bladder volume within patient. This could be because patients were given instruction on the importance for having full bladder only on the first day of treatment and there were no written information given to patients. There were no tools used to measure patient's bladder volume daily prior to treatment. Written bladder filling instructions and measurement of bladder volume using ultrasound was reported to reduce bladder volume variation within patients [11]. Therefore, future study need to include written bladder filling instruction and measurement of bladder volume using ultrasound prior to treatment in bladder filling protocol in order to reduce bladder volume variation within patients.

## 4. Conclusion

There was 18% reduction of population mean bladder volume from CT-simulation to the fifth week of radiotherapy (123 ± 54 mL to 101 mL ± 71 mL) (p = 0.104), however within patient bladder volume variation was large. Further study using a more effective protocol is required to reduce variation of bladder volume within patient. The reduction in bladder volume was correlated with incidence and severity of acute diarrhea.

## 5. References

[1] Roh, M. S., Colangelo, L. H., O'Connell, M. J., Yothers, G., Deutsch, M., Allegra, C. J., Kahlenberg, M. S. et al. 2009. Preoperative multimodality therapy improves disease-free survival in patients with carcinoma of the rectum: NSABP R-03. *Journal of clinical oncology*, 27(31), 5124–30. doi:10.1200/JCO.2009.22.0467

[2] Gérard, J. P., Conroy, T., Bonnetain, F., Bouché, O., Chapet, O., Closon-Dejardin, M. T., Untereiner, M. et al. 2006. Preoperative radiotherapy with or without concurrent fluorouracil and leucovorin in T3-4 rectal cancers: Results of FFCD 9203. *Journal of Clinical Oncology*, 24(28), 4620–4625.

[3] Stacey, R. & Green, J. T. 2014. Radiation-induced small bowel disease: latest developments and clinical guidance. Therapeutic advances in chronic disease, 5(1), 15–29. doi:10.1177/2040622313510730

[4] Kavanagh, B. D., Pan, C. C., Dawson, L. A., Das, S. K., Li, X. A., Ten Haken, R. K. & Miften, M. 2010. Radiation dose-volume effects in the stomach and small bowel. *International journal of radiation oncology, biology, physics, 76(3 Suppl)*, S101–7. doi:10.1016/j.ijrobp.2009.05.071

[5] Fiorino, C., Valdagni, R., Rancati, T. & Sanguineti, G. 2009. Dose-volume effects for normal tissues in external radiotherapy: Pelvis. *Radiotherapy and Oncology*.

[6] Kim, T. H., Chie, E. K., Kim, D. Y., Park, S. Y., Cho, K. H., Jung, K. H., Kim, Y. H. et al. 2005. Comparison of the belly board device method and the distended bladder method for reducing irradiated small bowel volumes in preoperative radiotherapy of rectal cancer patients. *International journal of radiation oncology, biology, physics*, 62(3), 769–75. doi:10.1016/j.ijrobp.2004.11.015

[7] Nuyttens, J. J., Robertson, J. M., Yan, D. & Martinez, a. 2001. The position and volume of the small bowel during adjuvant radiation therapy for rectal cancer. *International journal of radiation oncology, biology, physics*, 51(5), 1271–80.

[8] Ahmad, R., Hoogeman, M. S., Quint, S., Mens, J. W., de Pree, I. & Heijmen, B. J. M. 2008. Inter-fraction bladder filling variations and time trends for cervical cancer patients assessed with a portable 3-dimensional ultrasound bladder scanner. *Radiotherapy and oncology : journal of the European Society for Therapeutic Radiology and Oncology*, 89(2), 172–9. doi:10.1016/j.radonc.2008.07.00

[9] Tournel, K., De Ridder, M., Engels, B., Bijdekerke, P., Fierens, Y., Duchateau, M., Linthout, N. et al. 2008. Assessment of intrafractional movement and internal motion in radiotherapy of rectal cancer using megavoltage computed tomography. *International journal of radiation oncology, biology, physics*, 71(3), 934–9. doi:10.1016/j.ijrobp.2008.02.032

[10] Nakamura, N., Shikama, N., Takahashi, O., Ito, M., Hashimoto, M., Uematsu, M., Hama, Y. et al. 2010. Variability in bladder volumes of full bladders in definitive radiotherapy for cases of localized prostate cancer. *Strahlentherapie und Onkologie : Organ der Deutschen Röntgengesellschaft* ... [et al], 186(11), 637– 42. doi:10.1007/s00066-010-2105-6

[11] Stam, M. R., van Lin, E. N. J. T., van der Vight, L. P., Kaanders, J. H. a M. & Visser, A. G. 2006. Bladder filling variation during radiation treatment of prostate cancer: can the use of a bladder ultrasound scanner and biofeedback optimize bladder filling? *International journal of radiation oncology, biology, physics*, 65(2), 371–7. doi:10.1016/j.ijrobp.2005.12.039)

[12] Chang, J. S., Yoon, H. I., Cha, H. J., Chung, Y. & Cho, Y. 2013. Bladder filling variations during concurrent chemotherapy and pelvic radiotherapy in rectal cancer patients : early experience of bladder volume assessment using ultrasound scanner. *Radiation Oncology Journal*, 31(1), 41–47.

[13] Miralbell, R., Nouet, P., Rouzaud, M., Bardina, A., Hejira, N. & Schneider, D. 1998. Radiotherapy of Bladder Cancer: Relevance of Bladder Volume Changes in Planning Boost Treatment. International Journal of Radiation Oncology*Biology*Physics 41(4): 741-746.

[14] Dicuio, M., Pomara, G., Fabris, F. M., Aleis, V., Dahlstrand, C. & Morelli, G. 2005.

Measurements of Urinary Bladder Volume: Comparison of Five Ultrasound Calculation Methods in Volunteers. Arch Ital Urol Androl 77(1): 60-62.
[15] Nuyttens, J. J., Robertson, J. M., Yan, D. & Martinez, A. 2002. The variability of the clinical target volume for rectal cancer due to internal organ motion during adjuvant treatment. *International journal of radiation oncology, biology, physics*, 53(2), 497–503.

**Acknowledgement**
The authors would like to thank Radiotherapy and Oncology staff of Hospital Kuala Lumpur for supporting this study.

# Motion effects in stereotactic body radiotherapy (SBRT) dosimetry of bone metastases: case study using homogeneous and inhomogeneous CIRS phantoms

**Y.S. Asril[1], W.E. Wibowo[2], and S.A. Pawiro[1]**

[1]Department of Physics, Faculty of Mathematics and Natural Sciences, Universitas Indonesia, Depok, West Java, 16424, Indonesia;
[2]Department of Radiotherapy, Cipto Mangunkusumo General Hospital, Jakarta, 10430, Indonesia;

E-mail: supriyanto.p@sci.ui.ac.id

**Abstract**. The Stereotactic Body Radiotherapy (SBRT) for bone metastases has been implemented in radiotherapy center in Indonesia. In this study, we simulated and explored the effect of target motion in SBRT on bone metastases using the homogeneous (002 H9K) and inhomogeneous (002 LFC) CIRS phantom to simulate the presence of inhomogeneous medium near the target, with the holder for chamber. Both phantoms have the interchangeable rod for ionisation chamber, while for TLD and gafchromic EBT3, a holder was devised using Teflon material. In order to evaluate the impact of target motion, the measurements were performed in static and superior-inferior motion with the amplitude of 5, 10, and 20 mm. The measurement in the static condition has the standard deviation of <1.5 for gafchromic film EBT3 and <0.2 for PTW N30013. For the measurement of superior-inferior dynamic motion, we obtained a decrease in the dose of the target volume with increasing amplitudes of the movements. In addition, the standard deviation of dynamic measurement results was in the range of 1.13 to 11.7, 9.5 to 28.6, and 0.05 to 7.21 for gafchromic film EBT3, exradin A16 and PTW N30013, respectively.

## 1. Introduction

Greco et al stated that 13% of deaths were caused by cancer, and bone metastases is a common complication that occurs in more than 40% of oncology patients [1] and approximately 70% will develop spinal metastases [2]. Bone metastases and primary cancer may cause pain and functional disorder of the patient itself [3]. Stereotactic body radiotherapy (SBRT) is a technique deemed appropriate for bone metastases for its ability to deliver a high radiation dose to a small volume with a very tight margin [4]. The advantages of radiotherapy using small field is to improve dose accuracy in target volume and to decrease dose impact in organ at risk (OAR) around the tumor [5].

In terms of Quality Assurance (QA), target motion present specific challenges due to a regular displacement of the target during dose delivery, causing undesired dose differences in the target volume [6]. The motion itself is affected by pressure in the lungs, whose volume can change together with the movement of the diaphragm and ribs [7]. Since the main respiratory system is supported by the diaphragm, the movement was largely on superior-inferior (SI) direction with averaged displacement of the spine in superior direction is around 4.7 to 19.1 mm in normal patients [8]. The commercially available CIRS Dynamic Platform Model PL 008 is able to simulate the specific patient movement in SI direction. The movement of the spine and dose differences owing to it were not well described in existing literature. In terms of the several types of motion, it is better to know and understand how much the differences of dose between static and moving (dynamic) position. Two CIRS phantom (002 H9K and 002 LFC) were employed in this study to observe dose differences in static and dynamic target under homogeneous and inhomogeneous situations.

## 2. Methodology

### *2.1. Target Contouring*
Contouring process on both CIRS phantoms was performed to determine the dimension and volume on the target. In treatment planning, the target was simulated by a cylindrical object at the spine point area with volume and dimensions of the target is around 7 cc in both of CIRS phantom. The volume and dimension being determined and adjusted with respect to the clinical references of SBRT techniques.

### *2.2. Treatment planning and delivery*
Treatment plans were generated in Pinnacle$^3$® (Philips Radiation Oncology Systems, Fitchburg, WI) TPS, by using adaptive cone-convolution algorithm. Treatment was performed using 6 MV photon beam produced by Synergy-S (Elekta AB, Stockholm, Sweden). Prescribed dose and number of fields was determined based on the clinical requirement and thirteen irradiation fields were applied in both of homogeneous and inhomogeneous phantom. The results of optimization from the overall plan was evaluated by observing the Dose Volume Histogram (DVH) to check if the percentage of dose received by the PTV in the SBRT success criteria.

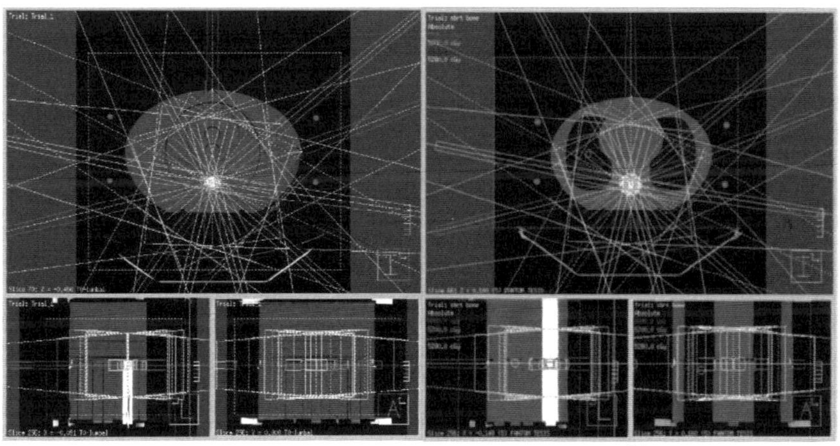

(a) (b)

**Figure 1.** Treatment Plans on (a) Homogeneous (002 H9K) and (b) Inhomogeneous (002 LFC) CIRS Phantoms

### *2.3. Dosimeter preparation*
Calibration was performed on TLD and gafchromic EBT3 film dosimeters. Gafchromic EBT3 film was cut into a small square with a size of 2 x 2 cm$^2$ and 6 x 1 x 1 mm$^3$ sized of TLD were also prepared accordingly. Both dosimeters were calibrated using high range dose (500 to 1800 cGy), with 10 x 10 cm$^2$ field size at 5 cm depth in solid water phantom.

### *2.4. Measurements*
In order to evaluate the impact of target motion, the measurements were carried out in static and SI movement with the amplitude of 5, 10, and 20 mm. In target dose measurement, there were four dosimeters employed: PTW N30013, Exradin A16, TLD, and gafchromic EBT3 film. Dose measurements using TLD and film were performed using holders which have made from teflon. Holders were designed resembling the original interchangeable rod from the CIRS phantom.

### 2.5. *Result Analysis*
The evaluation was performed by calculating the magnitude of dose differences (discrepancy) between the value of measured dose ($D_{\text{measured}}$) with a value of TPS calculated dose ($D_{\text{planned}}$) using the following equation:

$$\text{Discrepancy } (\Delta\%) = \left(\frac{D_{measured} - D_{planned}}{D_{planned}}\right) x\ 100$$

## 3. Result and Discussion

### 3.1. *Static and Dynamic Measurements*
In static measurement, four dosimeters were employed: PTW N30013, Exradin A16, TLD, and gafchromic EBT3 film. Figure 2 shown the results of static measurements where all measured dose were lower than planned dose (underestimate) except for gafchromic EBT3 film. The gafchromic EBT3 film was characterised by high spatial resolution, linear response to a given dose and not being affected by the effects of volume averaging. PTW N30013 and TLD showed the lowest accuracy. PTW ionisation chamber with 0.6 $cm^3$ active volume is not suitable to be used in small field measurements because it indicates volume averaging effect occured, and TLD is a detector with uncertainty measurement up to 15% [9]. The results obtained by Exradin A16 is also sufficient to prove the ability of Exradin A16 in high-dose measurement on small field, being a type of microchamber with an active volume of 0.007 $cm^3$. Moreover, the standard deviations on both phantoms were less than 1.5 for gafchromic EBT3 film and less than 0.2 for PTW N30013.

In dynamic measurement, we used three dosimeters employed: PTW N30013, Exradin A16, and gafchromic EBT3 film. Measurement on both phantoms was performed with the phantom moving 5, 10, and 20 mm in SI direction. The dynamic situation generated significant difference and being unsynchronized between the 5, 10 and 20 mm motion. An increase in amplitude of the phantom movement generated a drastic dose reduction from the average doses received by target bone.

### 3.2. *Dosimetric Comparison*
Figure 3 shown the results of all measurements in static and dynamic conditions. Measured dose decreased with the increase of movement amplitude. This dose reduction is due to the fact that the objects are regularly moving away from the field during the irradiation. PTW N30013 and TLD is the most perturbative dosimeter in 5, 10, and 20 mm motion. This is due to the volume averaging effect, in which the area of the target radiation has a nearly equal size to the size of the cavity of a PTW N30013.

On measuring targets in motion, Exradin A16 has a slightly different pattern with other dosimeters, with the measured dose being greater than the dose of static target. This is because of the increase in the attenuation of the primary beam radiation affecting the distribution of the scattering. Moreover, relatively rapid movement hinders the electrometer from recombining of ion since the measurement points are away from the radiation field, resulting in high fluctuation with the value being not stable at electrometer readings.

In addition, the standard deviation of dynamic measurement results was in the range of 1.13 to11.7, 9.5 to 28.6, and 0.05 to 7.21 for gafchromic EBT3 film, Exradin A16 and PTW N30013, respectively. Dose target evaluation of homogeneous and inhomogeneous phantom between static and dynamic, yielded differences in doses with a dose range of 0.62 to 347.44 cGy. The dose profile measurements result obtained proved that an increased amplitude of phantom movement from 5 mm, 10 mm and 20 mm resulted in significant decrease in the dose on target volume. This finding is in line with the finding of Seco *et al.* [10] and Siebenthal [7] that motion effect correlate with decreasing dose on target volume.

### 3.3. *Dose Profil of Measurements*
Dose profile of each motion in a homogeneous and inhomogeneous phantom was analyzed to observe reduced dose on movements. Figure 4 shown the normalized graph of homogeneous and inhomogeneous phantom, showing that the position of 90% to 100% dose being away from the point of measurement.

Whereas the dose was decreased at the point of measurement, the highest dose point is obtained during movement in superior direction (i.e. cranially).

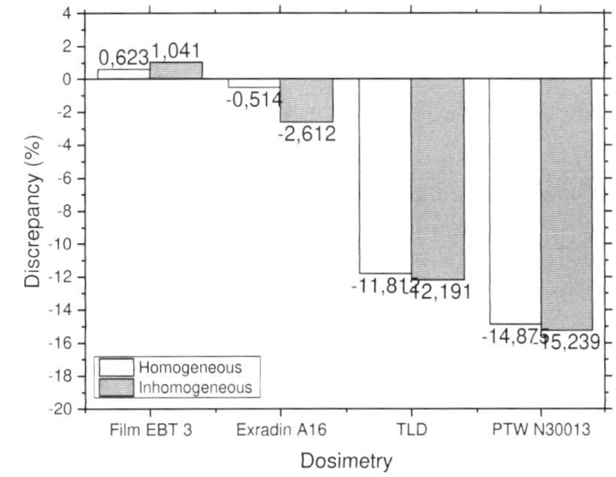


**Figure 2.** Illustration of Static Measurements

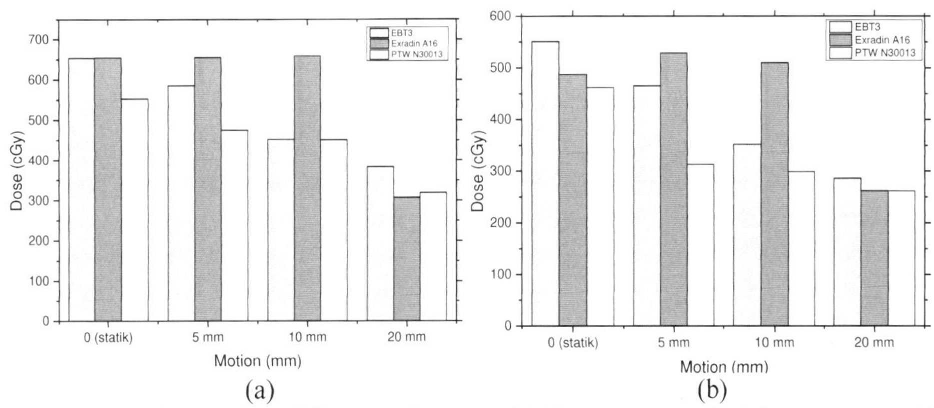


**Figure 3.** Illustration of dose differences between (a) Homogeneous (b) Inhomogeneous CIRS Phantom

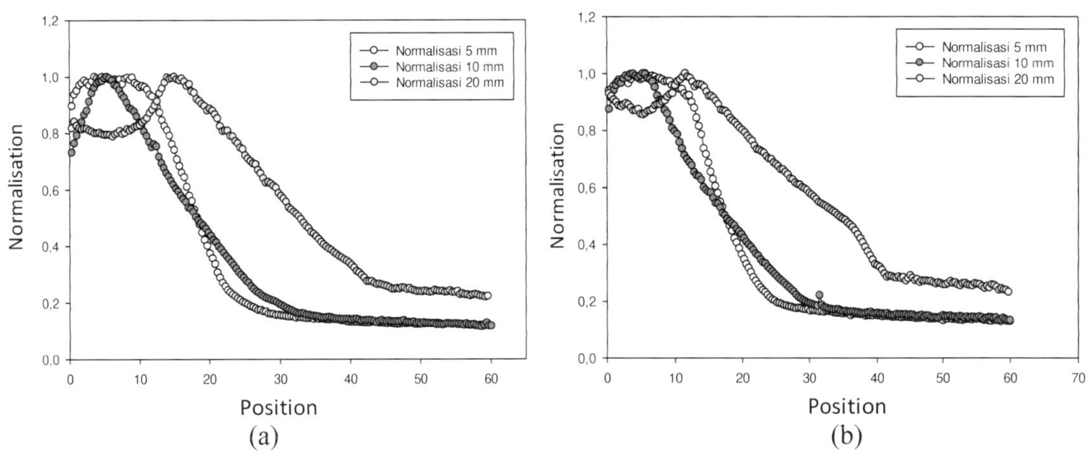


**Figure 4.** Dose Normalization in (a) Homogeneous (b) Inhomogeneous CIRS Phantom

## 4. Conclusion

In general, gafchromic EBT3 film shown the best performance for measurement of static target. However, the dose profile measurements result obtained proved that an increase of motion amplitude from 5 mm, 10 mm and 20 mm resulted drastic decrease in the dose on the target volume in phantom motion.

## 5. References

[1] C. Greco, O. Pares, N. Pimentel, E. Moser, V. Louro, X. Morales, B. Salas, and Z. Fuks, Spinal metastases: From conventional fractionated radiotherapy to single dose SBRT, Reports Pract. Oncol. Radiotherapy. 20 (2015), p. 1–10.
[2] G. M. Berardino De Bari, Filippo Alongi, Gianluca Mortellaro, Rosario Mazzola, Schiappacasse Luis, Spinal Metastases: Is Stereotactic Body Radiation Therapy Supported by evidence, Crit. Rev. Oncology/Hematology (2015), p. 1–12.
[3] G. Bedard and E. Chow, The failures and challenges of bone metastases research in radiation oncology, J. Bone Oncol. 2 (2013), p. 84–88.
[4] Bhattacharya I.S. et al., Stereotactic Body Radiotherapy for Spinal and Bone Metastases, Journal Clinical Oncology. 27 (2015), p. 298-306.
[5] J. U. Wuerfel, Dose measurements in small fields, Medical Physics Int. 1 (2013), p. 81–90.
[6] Keall P.J. et al., The Management of Respiratory Motion in Radiation Oncology Report of AAPM Task Group 76, Medical Physics 33 (2006), p. 3874-3900.
[7] Siebenthal Von M., Analysis and Modelling of Respiratory Liver Motion using 4DMRI, Dissertation (2008), ETH Zurich.
[8] Leong J.C.Y. et al., Kinematics of the Chest Cage and Spine During Breathing in Healthy Individuals and in Patients with Adolescent Idiopathic Scoliosis, Journal of Spine, 24 (199), p. 1310-1315.
[9] Fitriandini A. et al., Comparison of dosimeter response: ionisation chamber, TLD, and Gafchromic EBT2 film in 3D-CRT, IMRT, and SBRT technique for lung cancer, IOP Journal of Physics: Conference Series 694 (2016), p. 1-5.
[10] Joao Seco, Greg C.S., Ziji W. et al., Dosimetric Impact of Motion in Free-Breathing and Gated Lung Radiotherapy: A 4D Monte Carlo Study of Intrafraction and Interfraction Effects, Medical Physics Journal, 35 (2008), p. 356-366.

## Acknowledgements

This study was funded by Hibah PITTA Universitas Indonesia 2016 (contract number 2051/UN2.R12/HKP.05.00/2016). The authors would also like to acknowledge the help and assistance during the preparation and measurement process from medical physicists in the Department of Radiotherapy, Cipto Mangunkusumo General Hospital.

# The characteristics of dose at mass interface on lung cancer Stereotactic Body Radiotherapy (SBRT) simulation

**I H Wulansari[1], W E Wibowo[2] and S A Pawiro[1]**

[1]Departement of Physics, Faculty of Mathematics and Natural Sciences, Universitas Indonesia, Depok, West Java, 16424, Indonesia
[2]Radiotherapy Departement, National General Hospital Cipto Mangunkusumo, Jakarta, 10430, Indonesia.

E-mail: supriyanto.p@sci.ui.ac.id

**Abstract**. In lung cancer cases, there exists a difficulty for the Treatment Planning System (TPS) to predict the dose at or near the mass interface. This error prediction might influence the minimum or maximum dose received by lung cancer. In addition to target motion, the target dose prediction error also contributes in the combined error during the course of treatment. The objective of this work was to verify dose plan calculated by adaptive convolution algorithm in Pinnacle[3] at the mass interface against a set of measurement. The measurement was performed using Gafchromic EBT 3 film in static and dynamic CIRS phantom with amplitudes of 5 mm, 10 mm, and 20 mm in superior-inferior motion direction. Static and dynamic phantom were scanned with fast CT and slow CT before planned. The results showed that adaptive convolution algorithm mostly predicted mass interface dose lower than the measured dose in a range of -0,63% to 8,37% for static phantom in fast CT scanning and -0,27% to 15,9% for static phantom in slow CT scanning. In dynamic phantom, this algorithm was predicted mass interface dose higher than measured dose up to -89% for fast CT and varied from -17% until 37% for slow CT. This interface of dose differences caused the dose mass decreased in fast CT, except for 10 mm motion amplitude, and increased in slow CT for the greater amplitude of motion.

## 1. Introduction

In radiotherapy planning, it is difficult for the Treatment Planning System (TPS) to predict the dose at or near mass interface. These are caused by the loss of lateral or longitudinal electronic equilibrium [1]. The TPS commonly uses dose correction methods to compensate the dose change in photon fluence, but the changes in charged particle transport to calculate the inhomogeneity is not taken into account [2]. TPS inaccuracies in predicting the interface dose causes a discrepancy between TPS dose calculations and the experimental results; there was an underdose at mass lung interface [3].

Measurements of lung SBRT doses during the procedure is deemed complicated to be performed owing to the inclusion of some factors such as tissue density differences, charged particle disequilibrium in the small field, and motion of the lungs. Moreover, dosimetry complexity is also caused by the beam passing through the difference of material densities [4]. In addition, the effect of the materials density difference was exacerbated by the presence of charged particle disequilibrium (CPD) on a small field in SBRT [5]. However, Treatment planning system (TPS) with a simple calculation algorithm generally did not calculate this effect [3]. This can lead to overdose or underdose in contrast to the actual dose received by patients. In addition, the motion of the lungs might also influence the accuracy of the radiation beam delivery [6], [7].

In this work, dose measurement in mass lung interface with a static and dynamic condition is presented. The lung mass was irradiated with 6 MV photon beam and compared to TPS Pinnacle[3] with adaptive convolution algorithm. The aim of this work was to identify the TPS tendency of predicting dose of mass interface and to identify underdose degree in the interface region.

## 2. Materials and Methods

Simulation of lung cancer has been created using CIRS phantom 002FLC with cylindrical acrylic of 2.9 cc volume to simulate the cancer target. The simulated target cancer was introduced to the left lung as seen in Figure 1. The phantom, with target, was scanned using helical CT with fast and slow CT techniques. The scanning time were 1 second per slice and 2 seconds per slice for fast and slow CT techniques, respectively. The phantom was scanned in a static and dynamic setup in the superior-inferior direction with the amplitudes of the phantom motion were of 5 mm, 10 mm and 20 mm to represent the lung motion. The dynamic setup was managed by CIRS phantom 002FLC using phantom motion control 008PL. The acquired images were sent to the TPS to simulate SBRT forward planning. Contouring of simulated cancer mass as clinical traget volume (CTV) was performed on both images from fast and slow CT scanning in a static and dynamic setup. The gross target volume (GTV) has a margin of 2 mm for CTV, whereas the planing target volume (PTV) was arranged with an error margin of 2 mm from GTV. 15 beams were arranged around the cancer mass with 80% prescription dose of 6000 cGy in 12 fractions on GTV. Afterwards, the planning was exported to Synergy S linear accelerator for 6 MV photon beam delivery. To ensure the positioning accuracy of the measurements, isocenter position was verified using CBCT and room laser.

For measurements, the dose in the interface cancer was measured using Gafchromic EBT3. It was calibrated with source axis distance (SAD) 100 cm, filed size of 10 x10 cm$^2$ in solid water phantom with dose in the range of 0 to 1800 cGy. The irradiated films were scanned using EPSON V700 of 72 dpi resolution and saved in .TIF format. In order to determine the dose distribution and point dose of interface cancer, the scanned film was read using in-house dose verification (IDV) software. From Equation 1, the tendency of TPS Pinnacle[3] in predicting the interface doses were found, in addition to the obtained dose reduction degree on interface mass of static and dynamic phantom. For comparison, Equations 2 and 3 expresses the degree of underdose on mass interface as adopted below from Taylor et al [3];

$$\Delta D = \left(\frac{D_{measured} - D_{TPS}}{D_{TPS}}\right) x\ 100\ \% \quad (1)$$

$$R = \frac{D_{boundary}}{D_{ref}} \quad (2)$$

$$DRF = \frac{1}{4D_{ref}}(D_F + D_D + D_B + D_H) \quad (3)$$

where $\Delta D$ is dose discrepancy of interface point measurement ($D_{measured}$) to the TPS calculation ($D_{TPS}$). $R$ is a ratio of the dose in the mass interface ($D_{boundary}$) to the dose at center mass ($D_{ref}$). R value less than 1 indicates that the interface areas is underdose to the center mass, while the R value of greater than 1 indicates that interface area is overdose to the center mass. Dose reduction factor ($DRF$) is the average of dose reduction in the mass interface. $D_F$, $D_D$, $D_B$, and $D_H$ are interface dose point, shown in Figure 2.

## 3. Results and Discussion

### 3.1. Dose Distribution

The EBT3 film outputs using IDV software were shown in Figure 3 and Figure 4 for fast and slow CT technique, respectively. The white line described the GTV area of the simulated lung cancer which is also the interface area between cancer and lung tissue (Figure 2). The differences in the size of white lines in Figure 3 and 4 was caused by the difference in film size. Red color indicated the area of high doses. *Figure 3a, Figure 4a, Figure 4b, Figure 4c* and *Figure 4d* show that all of GTV area received a high dose, while not all areas of the GTV receiving high doses in dynamic phantom scanned using the fast CT technique (Figure 3b, 3c and 3d)). The range of doses for each distribution is displayed in the histogram in each image.

Figure 3 shown that the GTV area of simulated cancer in a static phantom scanned using a fast CT technique received a dose in the range of 500 to 700 cGy, while the GTV area of cancer in a dynamic phantom received doses within range of 60 to 450 cGy. Such dose distribution caused the overall interface cancer of static phantom to receive high dose (in red area), while only upper cancer received high doses for the dynamic phantom. These dose distributions caused by the moving phantom effect that gave consequence to the accuracy of radiation beams to hit the target. The radiation beam can not be in the GTV region continuously. Figure 3*b*, *c*, and *d* shown that in the dynamic setup the radiation beam was more often exposed the upper part than the lower part of the mass.

For slow CT scanning technique, Figure 4 displayed the dose distribution in GTV area for static and dynamic run. In these Figures, it can be observed that the whole GTV area of the target is in the red region. The GTV area from slow CT technique received doses in the range of 500 to 750 cGy and 480 to 750 for static and dynamic setup, respectively. On slow CT technique, the overall interface of GTV has covered the high doses both for static and dynamic phantom condition. The situation is influenced by the cancer size differences of each scanning result. The cancer size of slow CT technique is larger compared to fast CT technique, especially for dynamic phantom. The result agrees with previous works; that images resulting from slow CT technique has larger target size than the target image from fast CT techniques [8], [9],[10], [11]. Larger target size kept the phantom inside the radiation beam along the dynamic movement. This, in turn, caused the cancer from slow CT technique to receive relatively high dose in all GTV area.

### *3.2. Interface Point Dose*

Doses at the mass interface point were verified with TPS Pinnacle[3] to identify the tendency of estimating interface dose. Table 1 and Table 2 shown the discrepancy of interface point dose for fast and slow CT techniques, respectively.

For fast CT scanning mode, Table 1 shown that TPS Pinnacle[3] tended to estimate dose lower than the measured values for the static phantom condition, estimating higher interface doses than the measurement for all of phantom motion amplitudes for the dynamic phantom. On the other hand, slow CT scanning technique presented Table 2 which described that TPS tend to estimate dose lower than the measurements for static phantom condition, while the TPS tend presented the variation dose for each interface region at dynamic phantom.

For the phantom motion with amplitudes of 5 mm and 10 mm, the TPS estimates lower interface dose against measured values at the upper target; and higher estimation at the lower target area. In addition, the TPS estimated the point dose at interface area to be lower than measurement results for 20 mm amplitude of motion. From these result in *Tables 1* and *2*, it is seen that interface effect on TPS calculation was only identified in static phantom but it could not be identified in dynamic phantom. In dynamic phantom, motion effect was found to be more dominant than the interface effect.

In this work, an adaptive convolution algorithm estimated the interface dose to be lower than the measurement for both fast and slow CT for static phantom (0 mm motion amplitude). Previous studies by Shahine (1998) and Rosa (2010) reported that a simple dose calculation algorithm, like BATHO and ETAR algorithm, estimate the dose at interface area lower than the measurement up to 55% [1], [12]. However, Monte Carlo and collapsed cone convolution (CCC) showed a better result for estimating the dose at interface region [13]. Jones reported that CCC algorithm tend to estimate interface dose higher than Monte Carlo simulation results [14].

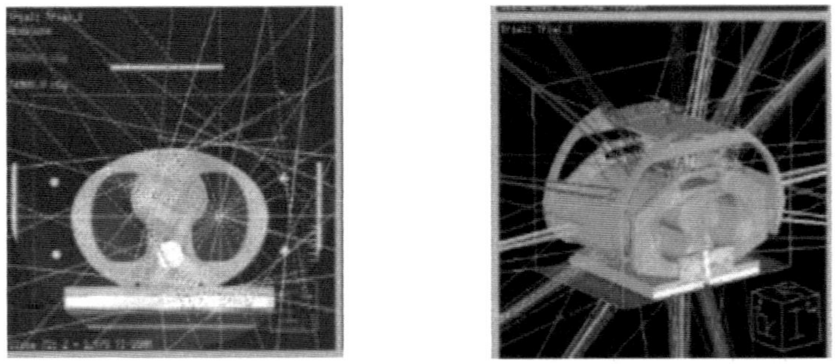

**Figure 1.** The position of lung cancer and SBRT planning

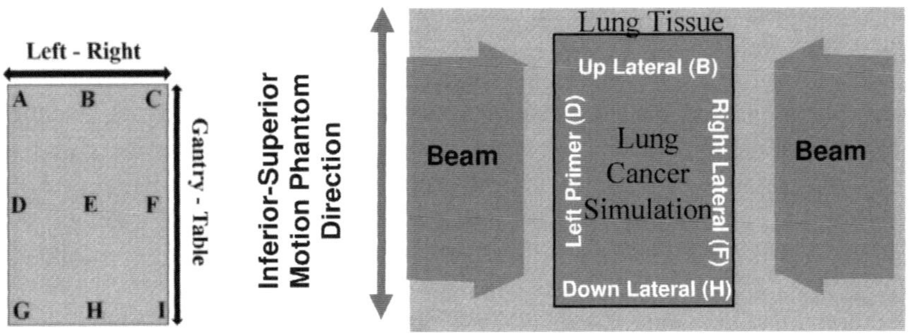


**Figure 2.** Illustration lung mass interface point measurement, beam direction and phantom motion direction

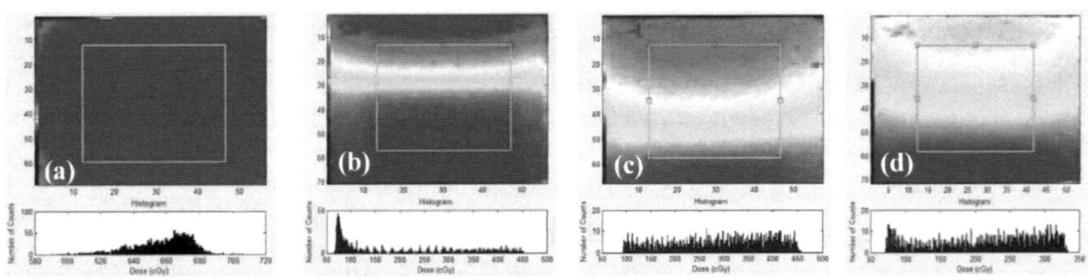


**Figure 3.** Dose distribution of lung mass (a) static and dynamic (b) 5 mm (c) 10 mm (d) 20 m phantom amplitude fast CT technique

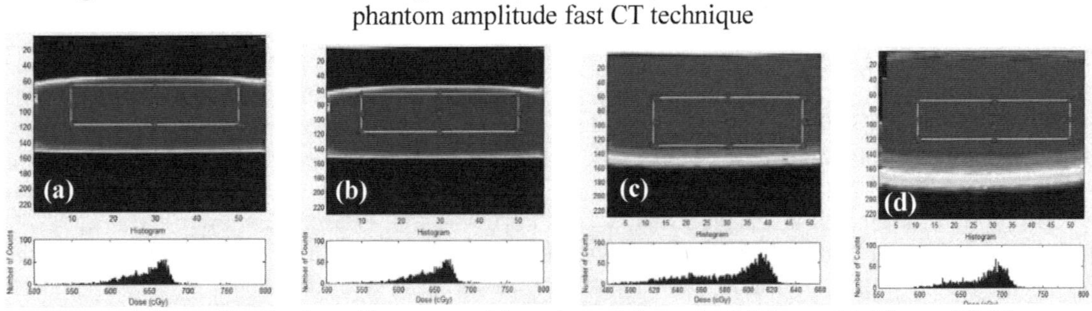


**Figure 4.** Dose distribution of lung mass (a) static and dynamic (b) 5 mm (c) 10 mm (d) 20 mm phantom amplitude slow CT technique

### *3.3.* Dose *Reduction Factor (DRF)*

The dose of lung cancer interface decreased significantly, reaching up to 16% as seen in Table 3. Reduction of interface dose mostly occurred in the upper and lower lateral region of the simulated lesion.

IOP Conf. Series: Journal of Physics: Conf. Series **851** (2017) 012028 doi:10.1088/1742-6596/851/1/012028

In the fast CT scanning, the *R* value of dynamic phantom was more than unity (especially on the upper side of the lateral position), indicating that the area of lung mass interface received higher dose than the center of the mass area. The average dose reduction in the interface lung cancer was shown in Table 4. The relation between the amplitude of motion phantom and the average dose reduction was not found in this work.

**Table 1.** Discrepancy of Interface Point Dose for Fast CT Scanning

| **Interface Point** | **Discrepancy (ΔD) (%)** | | | |
|---|---|---|---|---|
| | 0 mm | 5 mm | 10 mm | 20 mm |
| **A** | 2.86 | -12.47 | -25.82 | -43.39 |
| **B** | 1.40 | -17.35 | -23.01 | -45.19 |
| **C** | -1.67 | -7.29 | -13.25 | -38.73 |
| **D** | 9.14 | -78.10 | -49.91 | -62.63 |
| **E** | 1.27 | -83.17 | -49.86 | -62.79 |
| **F** | 5.60 | -78.44 | -45.54 | -60.86 |
| **G** | 3.14 | -85.32 | -81.11 | -86.60 |
| **H** | 0.69 | -86.52 | -83.63 | -88.42 |
| **I** | 5.04 | -84.33 | -79.59 | -86.87 |

**Table 2.** Discrepancy of Interface Point Dose for Slow CT Scanning

| **Interface Point** | **Discrepancy (ΔD) (%)** | | | |
|---|---|---|---|---|
| | 0 mm | 5 mm | 10 mm | 20 mm |
| **A** | 3.55 | 7.18 | 6.22 | 35.59 |
| **B** | 13.85 | 5.76 | 7.22 | 37.48 |
| **C** | 4.05 | 8.59 | 19.41 | 36.14 |
| **D** | 0.45 | 5.56 | 0.43 | 9.81 |
| **E** | 0.16 | -2.53 | -1.68 | 13.52 |
| **F** | 1.22 | -6.94 | -1.99 | 7.57 |
| **G** | 7.80 | -16.51 | -9.58 | 15.21 |
| **H** | 11.85 | -11.48 | -5.47 | 15.56 |
| **I** | 12.42 | -5.18 | -1.52 | 15.20 |

**Table 3.** The ratio of interface dose to the centre mass dose (R)

| **Interface Point** | **Scanning Technique** | | **Interface Point** | **Scanning Technique** | |
|---|---|---|---|---|---|
| **Right Primer Dose (F)** | Fast CT | Slow CT | Up Lateral Dose (B) | Fast CT | Slow CT |
| **0 mm** | 0.95 | 0.92 | 0 mm | 0.94 | 0.86 |
| **5 mm** | 1.12 | 0.87 | 5 mm | 4.45 | 0.92 |
| **10 mm** | 0.97 | 0.89 | 10 mm | 1.46 | 0.99 |
| **20 mm** | 0.95 | 0.89 | 20 mm | 1.44 | 0.98 |
| **Left Primer Dose (D)** | Fast CT | Slow CT | Down Lateral Dose (H) | Fast CT | Slow CT |
| **0 mm** | 0.97 | 0.90 | 0 mm | 0.87 | 0.92 |
| **5 mm** | 1.19 | 0.96 | 5 mm | 0.66 | 0.74 |
| **10 mm** | 0.91 | 0.91 | 10 mm | 0.28 | 0.86 |
| **20 mm** | 0.94 | 0.89 | 20 mm | 0.30 | 0.88 |

**Table 4.** Dose Reduction Factor (DRF)

| Phantom Motion Amplitude (mm) | Fast CT | Slow CT |
|---|---|---|
| 0 | 0.93 | 0.90 |
| 5 | 1.85 | 0.87 |
| 10 | 0.90 | 0.91 |
| 20 | 0.91 | 0.91 |

## 4. Conclusion

This work shown that the interface dose characteristics are influenced by scanning technique and phantom motion. For static phantom, fast CT scanning technique has smaller interface dose discrepancy than slow CT technique, while the slow CT scanning mode has smaller interface dose discrepancy than fast CT technique for dynamic phantom. On the other hand, calculation results from Pinnacle[3] TPS implies that the dose of fast CT scanning was higher than measurement, with dose variation for slow CT scanning in dynamic phantom. Dose reduction of the interface cancer were up to 12% and 16% for static and dynamic phantom, respectively.

## 5. References

[1] B. H. Shahine, M. S. . Al-Ghazi, and E. El-Khatib, "Exerimental evaluation of interface doses in the presence of air cavities compared with treatment planning algorithms," *Med. Phys.*, vol. 26, no. 3, pp. 350–355, 1998.

[2] T. R. Mackie, E. El Khatib, J. Battista, J. Scrimger, J. Van Dyk, and J. R. Cunningham, "Lung dose corrections for 6 - and 15 - MV x rays," *Med. Phys.*, vol. 12, no. 327, 1985.

[3] M. Taylor, L. Dunn, T. Kron, F. Height, and R. Franich, "Determination of peripheral underdosage at the lung-tumor interface using Monte Carlo radiation transport calculations," *Med. Dosim.*, vol. 37, no. 1, pp. 61–66, 2012.

[4] A. I. Saito, J. G. Li, C. Liu, K. R. Olivier, and J. F. Dempsey, "Accurate heterogeneous dose calculation for lung cancer patients without high-resolution CT densities.," *J. Appl. Clin. Med. Phys.*, vol. 10, no. 2, p. 2847, 2009.

[5] M. Miften, S. K. Das, I. J. Chetty, and D. Westerly, "Treatment Planning for Stereotactic Body Radiation Therapy," in *Stereotactic Body Radiation Therapy*, vol. 39, no. 1, 2012, pp. 91–113.

[6] W. Van Elmpt, M. Öllers, M. Velders, K. Poels, B. Mijnheer, D. De Ruysscher, A. Dekker, P. Lambin, and L. Boersma, "Transition from a simple to a more advanced dose calculation algorithm for radiotherapy of non-small cell lung cancer (NSCLC): Implications for clinical implementation in an individualized dose-escalation protocol," *Radiother. Oncol.*, vol. 88, no. 3, pp. 326–334, 2008.

[7] R. Boopathy, S. Padmanaban, V. Nagarajan, P. Sukumaran, P. Jeevanandam, S. Kumar, D. Rajasekaran, and L. A. Venkataraman, "Effects of Lung Tumor Motion on Delivered Dose Distribution during RapidArc Treatment Technique," *J. Med. Biol. Eng.*, vol. 30, no. 3, pp. 189–192, 2010.

[8] D. Koste, F. J. Lagerwaard, R. H. Schuchhard-schipper, J. R. Van So, M. R. J. Nijssen-visser, P. W. J. Voet, S. S. Oei, and S. Senan, "Dosimetric consequences of tumor mobility in radiotherapy of stage I non-small cell lung cancer – an analysis of data generated using ' slow ' CT scans," *Radiother. Oncol.*, vol. 61, pp. 93–99, 2001.

[9] J. R. V. S. De Koste, F. J. Lagerwaard, H. C. J. De Boer, M. R. J. Nijssen-Visser, and S. Senan, "Are multiple CT scans required for planning curative radiotherapy in lung tumors of the lower lobe?," *Int. J. Radiat. Oncol. Biol. Phys.*, vol. 55, no. 5, pp. 1394–1399, 2003.

[10] S. Seki, E. Kunieda, A. Takeda, T. Nagaoka, H. M. Deloar, T. Kawase, J. Fukada, O. Kawaguchi, M. Uematsu, and A. Kubo, "Differences in the definition of internal target volumes using slow CT alone or in combination with thin-slice CT under breath-holding conditions during

IOP Conf. Series: Journal of Physics: Conf. Series **851** (2017) 012028 doi:10.1088/1742-6596/851/1/012028

the planning of stereotactic radiotherapy for lung cancer," *Radiother. Oncol.*, vol. 85, no. 3, pp. 443–449, 2007.

[11] K. Wurstbauer, H. Deutschmann, P. Kopp, and F. Sedlmayer, "Radiotherapy planning for lung cancer : Slow CTs allow the drawing of tighter margins," *Radiother. Oncol.*, vol. 75, pp. 165–170, 2005.

[12] L. A. R. Rosa, S. C. Cardoso, L. T. Campos, and V. G. L. Alves, "Percentage depth dose evaluation in heterogeneous media using thermoluminescent dosimetry," *J. Appl. Clin. Med. Phys.*, vol. 11, no. 1, pp. 1–11, 2010.

[13] M. R. Arnfield, C. H. Siantar, J. Siebers, P. Garmon, L. Cox, R. Mohan, C. H. Siantar, and L. Cox, "The impact of electron transport on the accuracy of computed dose," vol. 1266, no. 2000, 2006.

[14] A. O. Jones and I. J. Das, "Comparison of inhomogeneity correction algorithms in small photon fields," *Med. Phys.*, vol. 32, no. 3, pp. 766–776, 2005.

**Acknowledgements**

This study was funded by Hibah PITTA Universitas Indonesia 2016 (contract number 2051/UN2.R12/HKP.05.00/2016). The authors would also like to acknowledge the help and assistance during the preparation and measurement process from medical physicists in the Department of Radiotherapy, Cipto Mangunkusumo General Hospital.

IOP Conf. Series: Journal of Physics: Conf. Series **851** (2017) 012029 doi:10.1088/1742-6596/851/1/012029

# Quantitative image quality evaluation for kV cone-beam CT-based IGRT

**S Y Lim and Hafiz M Zin**

Advanced Medical and Dental Institute, Universiti Sains Malaysia, Bertam 13200 Kepala Batas, Pulau Pinang, Malaysia

E-mail: sylim617@gmail.com

**Abstract.** The objective of this study is to quantitatively evaluate the image quality of a kV cone-beam CT-based IGRT system (Elekta, XVI) using two commercial CT image quality phantoms, Catphan-600 and CIRS-062QA. Both phantoms consist of similar image quality test modules (uniformity, CT linearity and spatial resolution) but each phantom has different diameter and test pattern design. Each test module was imaged separately using an optimised cone-beam CT imaging parameter. The quality metrics of the reconstructed images were analysed using algorithms developed with MatLab. The image uniformity and the spatial resolution measured with Catphan were of 4% and 40% greater respectively, compared to those measured with CIRS phantom. The differences were due to the beam scattering and hardening originated from the CIRS phantom holder. The contrast-to-noise ratio (CNR) values measured with CIRS phantom were at least 2% higher than that of Catphan. The diameter of CIRS phantom is smaller and resulted in lower beam attenuation. The quantitative image quality assessment algorithms developed for both phantoms provided a phantom-specific set of reference values for a cone-beam CT imaging system as recommended by AAPM TG-142. Further investigation will be performed to resolve beam hardening issue arising from the CIRS phantom holder.

## 1. Introduction

Kilovoltage (kV) cone-beam computed tomography (CT) based image guided radiotherapy (IGRT) system is a useful tool for target localisation and patient setup verification during advanced radiotherapy treatment [1]. The kV cone-beam CT is able to distinguish soft tissues and provides volumetric images of the patient using low radiation doses [2–4]. The imaging system provides a tool to monitor patient positioning error and internal organ motion [2, 3] for immediate corrections of patient setup, in order to reduce the uncertainties of radiotherapy treatment. The performance of kV cone-beam CT system is dependent on the quality of the images obtained by the imaging system [5, 6]. Poor image quality, such as poor soft tissue contrast, will affect the accuracy in image registration between the CT reference image and the cone-beam CT image. This will result in patient setup error that lead to inaccurate treatment delivery.

The American Association of Physicist in Medicine (AAPM) task group (TG) 142 and 179 have recommended a set of quality assurance (QA) programmes to assure the kV cone-beam CT system has a consistent performance [7, 8]. TG-142 report provides a brief guideline of the QA programme and the frequency of performing the QA test. The details of the QA programme and the calculation of the image quality parameters were discussed in TG-179. The reports proposed that the value of each image quality parameter measured during commissioning of the system is served as the baseline value for the periodical QA tests. Both reports did not provide any specific tolerance limits for the baseline values measured. This was due to the wide variety of commercially available kV cone-beam CT systems for IGRT. Each system is of different design and provides different imaging performance. Hence it is impossible to provide a universal quantitative tolerance limits for every image quality parameter. Besides, the quantitative metrics are also affected by the image quality phantom used to evaluate the quality metrics. To address the issues, a system-specific and phantom-specific tolerance limits are required for each IGRT system in the clinic.

In the year 2008, Bissonnette *et al.* established a set of system-specific tolerance limits for two kV cone-beam CT system according to TG-142 [3]. However, to the best of author knowledge, no research has been done in developing a phantom-specific tolerance limits for kV cone-beam CT system. To resolve the issue, this study was performed to evaluate the image quality of a kV cone-beam CT-based IGRT system quantitatively using two commercial CT image quality phantoms. The phantoms were scanned with an optimised cone-beam CT imaging parameters. A set of algorithms was developed to analyse the image quality metrics of the reconstructed images.

## 2. Materials and Methods

### *2.1. kV cone-beam CT system*

The study was performed on an on-board kV cone-beam CT system, X-ray volumetric imaging (XVI) system (Elekta Oncology System, Crawley, UK), mounted on an Elekta Synergy linear accelerator platform. The XVI system consists of a kV x-ray tube and an amorphous silicon flat-panel imager that are attached on a retractable arm perpendicular to the megavoltage beam treatment head. The flat-panel imager has an active area of 41 x 41 $cm^2$. It comprises of 1024 x 1024 detector elements of 16-bit pixel depth and a primary filtration of 6.8 mm Al equivalence. In this study, the phantoms were scanned with an optimised imaging protocol, which captured 660 projection images in a single rotation at a frame rate of 5.5 frames per second. The field-of-view (FOV) collimator selected was S10 (27.68 cm x 13.53 cm) and the beam energy was set at 120 kVp and 0.4 mAs per projection. The scanned images were then reconstructed using Feldkamp 3D filtered back-projection reconstruction algorithm.

### *2.2. Phantoms*

The imaging performance of the kV cone-beam CT system was evaluated using Catphan-600 (The Phantom Laboratory, Salem, NY) and CIRS-062QA (CIRS, Norfolk, VA) image quality phantom. Catphan-600 is a cylindrical phantom with a diameter of 20.0 cm and a length of 19.5 cm (Figure 1a). It is constructed of Acrylic and consists of five test modules, including CT linearity test modules, beads geometry test module, high-contrast spatial resolution test module, low contrast spatial resolution test module and uniformity test module. Whereas, CIRS-062QA phantom is smaller than Catphan-600 with a diameter of 18.0 cm and a length of 10.0 cm (Figure 2a). It also consists of all the test modules in Catphan-600 except the beads geometry test modules. Figure 1(b-f) and 2(b-e) show the reconstructed CT slice of each test module for both phantoms. Each CT slice was used to assess a specific image quality parameter. For instances, uniformity metric was evaluated with the uniformity test module, contrast-to-noise ratio (CNR) and low contrast visibility were assessed with the CT linearity test module, and the high-contrast spatial resolution was assessed with the high-contrast spatial resolution test module.

In this study, each test module was scanned separately. The phantom was positioned on the treatment couch at the gantry end. The alignment marker of the specific test module was aligned with the room laser to position the test module at the centre of the imaging system. The scanned images were reconstructed using an optimised reconstruction algorithm and exported as DICOM files. A set of algorithms was developed using MatLab (The MathWorks, Natick, MA) to evaluate the image quality metrics of the reconstructed images.

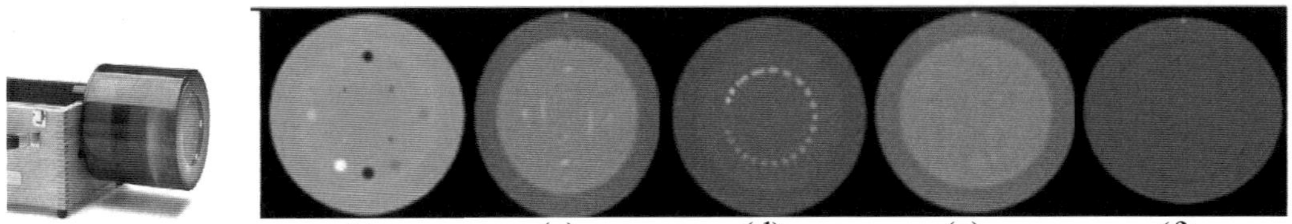

(a) (b) (c) (d) (e) (f)

**Figure 1.** (a) Catphan-600 phantom and the reconstructed image of each module: (b) CT number linearity test module (c) bead geometry test module, (d) high-contrast spatial resolution test module, (e) low-contrast spatial resolution test module and (f) uniformity test module.

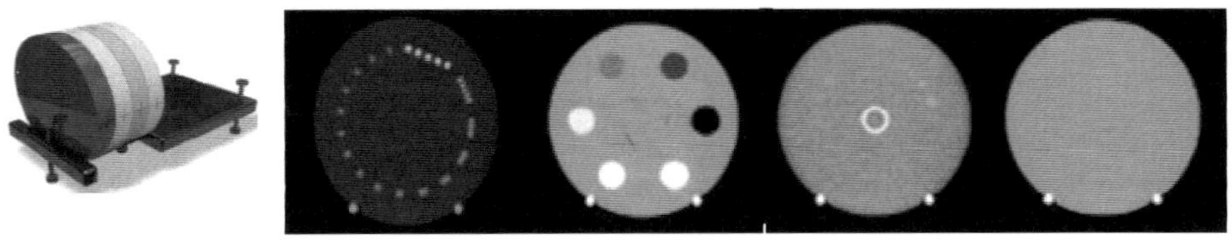

(a) (b) (c) (d) (e)

**Figure 2.** (a) CIRS-062QA image quality phantom and the reconstructed image of each test modules: (b) high-contrast spatial resolution test modules, (c) CT number linearity test modules, (d) low-contrast spatial resolution test modules, and (e) uniformity test modules.

### *2.3. Image evaluation*

Image uniformity was determined based on pixel values across the centre of the uniformity CT slice in the vertical and the horizontal directions. A region of interest (ROI) of a height of 10.0 mm and a width that covered 80% of the phantom from the centre point was selected for both vertical and horizontal directions. The mean pixel values, $\bar{x}$, across the columns of the ROIs were calculated and a uniformity profile was plotted for both vertical and horizontal axes. The mean pixel values were also used to determine the integral non-uniformity values using equation (1).

$$\text{Integral non-uniformity} = \left(\frac{\max(\bar{x})-\min(\bar{x})}{\max(\bar{x})+\min(\bar{x})}\right) \quad (1)$$

The contrast-to-noise ratio (CNR) and low contrast visibility were evaluated using the CT number linearity module. In Catphan-600, the CT linearity test module consists of several cylindrical inserts made of different materials, such as Polystyrene, low-density polyethylene (LDPE), Acrylic, Delrin, Teflon, air and polymethylpentene (PMP), of known electron densities and CT numbers. Whereas, for CIR-062QA phantom, the test module contains all the inserts in Catphan-600, except PMP, in a greater diameter of 25.4 mm (Figure 2c). A ROI of 3.5 x 3.5 mm$^2$ was selected within each insert. The mean pixel values, $\overline{x}_i$ , and the standard deviation, $\overline{\sigma_i}$, of each ROIs was calculated. Another ROI of the same size as the insert ROI was selected for each insert to calculate the mean pixel value of background, $\overline{x}_{bg}$, and the corresponding standard deviation, $\overline{\sigma}_{bg}$. The CNR value was calculated using equation (2). The CNR is defined as the difference of mean pixel values between the insert and the corresponding background over the averaged noise [11].

$$\text{CNR} = 2 \times \left[\frac{|\overline{x}_i-\overline{x}_{bg}|}{\overline{\sigma}_i+\overline{\sigma}_{bg}}\right] \quad (2)$$

Low contrast visibility was determined using the Polystyrene insert and the LDPE insert. The low contrast visibility value was computed based on the corresponding mean and pixel value of the inserts using equation (3) [5, 6],

$$\text{Low contrast visibility} = \frac{(CT_{poly}-CT_{LDPE})/10}{(\overline{x}_{poly}-\overline{x}_{LDPE})/(\frac{1}{2})(\sigma_{poly}-\sigma_{LDPE})} = \frac{2.75(\sigma_{poly}-\sigma_{LDPE})}{\overline{x}_{poly}-\overline{x}_{LDPE}} \quad (3)$$

IOP Conf. Series: Journal of Physics: Conf. Series **851** (2017) 012029 doi:10.1088/1742-6596/851/1/012029

where $CT_{poly}$ and $CT_{LDPE}$ represents the CT number of Polystyrene and LDPE inserts, $\overline{x}_{poly}$ and $\overline{x}_{LDPE}$ represent the corresponding mean pixel values, while $\sigma_{poly}$ and $\sigma_{LDPE}$ represent the corresponding standard deviation values of each insert. The factor of 2.75 is the calculated difference between Polystyrene and LDPE in CT numbers.

The spatial resolution of the kV cone-beam CT system was determined using the high-contrast spatial resolution test module. In Catphan-600, the spatial resolution test module consists of aluminium bar patterns of spatial frequencies ranging from 1-21 lp/cm, whereas, in CIRS-062 QA phantom, the range of spatial frequencies are from 1-16 lp/cm. Besides, the thickness of the aluminium bar in CIRS-062QA phantom is 10.0 mm greater than that in Catphan-600 of only 2 mm. The spatial resolution was evaluated in terms of modulation transfer function (MTF) values. The MTF value of each bar pattern was calculated by using the Droege and Morin method [10]. With the obtained results, an MTF curve was plotted. Among the MTF values, *f*50 and *f*10 are the important values because *f*50 indicates the point where human perceive an image to be sharp; while *f*10 is the highest resolution limit which the human eye can possibly differentiate the bar patterns. The spatial frequencies corresponding to *f*50 and *f*10 were determined from the MTF curve.

## 3. Results and discussion

Figure 3 shows the uniformity profiles in both horizontal and vertical directions for Catphan-600 and CIRS-062QA phantom. The results obtained with Catphans-600 show a greater image uniformity compared to CIRS-062QA. This is reflected in the integral non-uniformity obtained, where the integral non-uniformity of Catphan-600 was 0.0163 and the corresponding results for CIRS-062QA was 0.0504. The lower the integral non-uniformity values, the better the image uniformity. Besides, there is a presence of cupping artefact in the image acquired with CIRS-062QA phantom, where the mean pixel value is the lowest in the central region of the phantom (Figure 3).

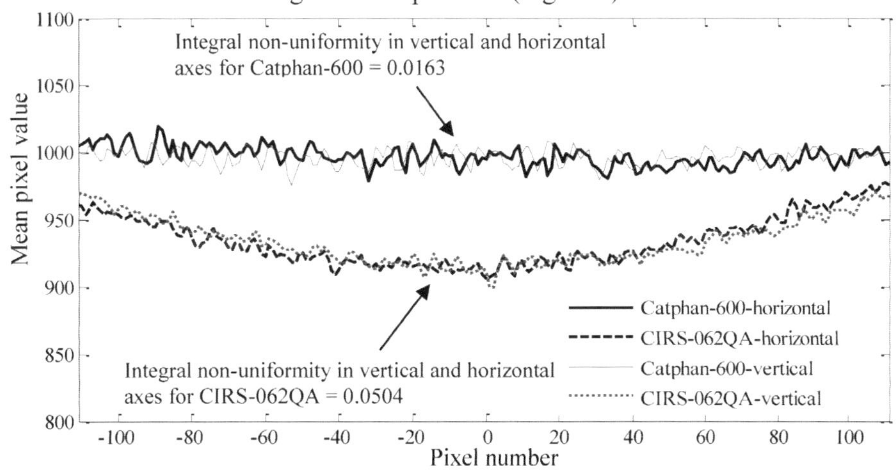


**Figure 3.** The uniformity profiles across the images in the horizontal and the vertical axis for both phantoms.

The cupping artefact is due to the beam scattering and hardening yielded from the phantom holder. According to the manufacturer, the phantom holder is made of a non-water equivalent material, which has a CT number higher than water. Moreover, due to the limitation of the collimator setting and the extension of the phantom holder, the captured projection images of the uniformity test module had included some parts of the phantom holder, such as the screw (Figure 4a), and hence resulted in greater scattering and beam hardening that in turn caused the presence of the cupping artefact, as shown in figure 4b.

In the evaluation of the contrast metrics, CNR values of each insert in CIRS-062QA are of 2% - 184% higher compared to those of Catphan-600 (Figure 4a). The mean pixel values obtained within each insert in CIRS-062QA phantom are of 8% - 59% greater than those that measured with Catphan-

600. While, the standard deviation values from CIRS-062QA phantom are 1%-40% lower than Catphan-600. These have resulted in lower CNR values in CIRS that contributed from the size of CIRS-062QA phantom that is 1.4 time smaller than Catphan-600. The smaller the phantom size, the lower the beam attenuation and the image noise and in turn increased the CNR value.

CIRS-062QA phantom yielded a lower low contrast visibility value of 1.83 compared to that of Catphan-600 of 2.42. CIRS-062QA phantom is smaller than Catphan-600, and resulted less attenuation of beam and reduction of image noise. Low contrast visibility is directly proportional to the image noise. The lower value indicates the lesser presence of the image noise and better visualisation of soft tissue for CIRS-062QA phantom compared to Catphan-600.

The MTF results obtained with both phantoms are plotted in Figure 5b. The spatial frequencies obtained with Catphan-600 at *f*50 and *f*10 were 5.26 lp/cm and 9.24 lp/cm, respectively. The corresponding results for CIRS-062QA phantom were 4.08 lp/cm for *f*50 and 6.63 lp/cm for *f*10. The spatial resolution of a CT imaging system is dependent on the focal spot size, the voxel size, the type of x-ray scintillator and the number of the projection images [11]. In this work, both phantoms were scanned using the same imaging parameters to eliminate the dependency to the aforementioned factors. The discrepancy in spatial resolution might be because of the spatial resolution test module design and the material used in constructing the test module. The difference in aluminium bar thickness and background material formulation used for both Catphan-600 and CIRS-062QA phantom generate different level of beam attenuation and image visibility and hence affect the spatial resolution of the reconstructed CT image.

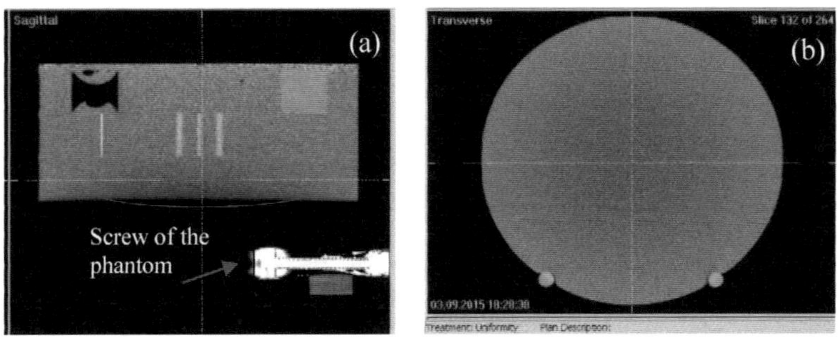


**Figure 4.** Photo of the reconstructed image (a) in the sagittal view and (b) in the transverse view. The circled regions show a higher grayscale value due to the beam hardening originated from the screw and the phantom holder.

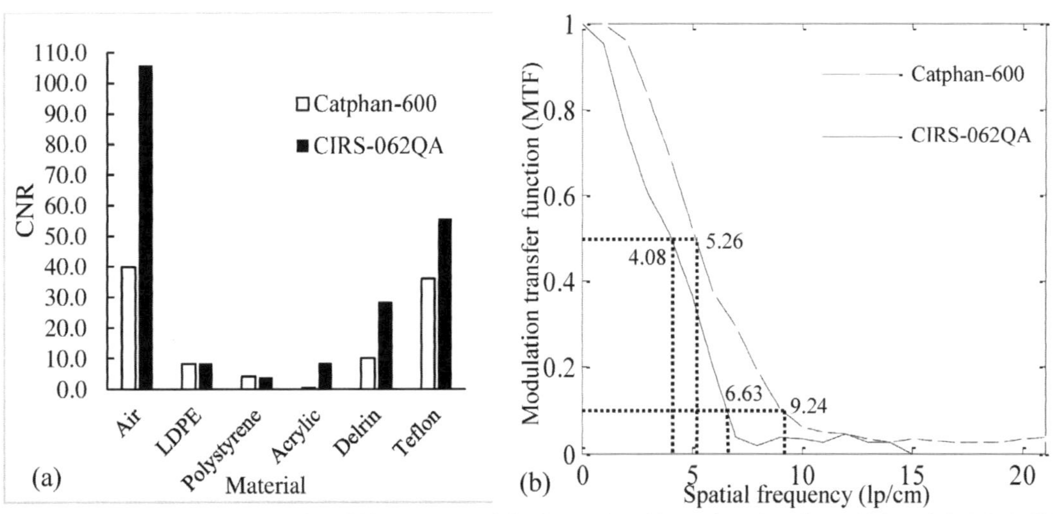


**Figure 5.** a) The CNR values for different materials determined by using Catphan-600 and CIRS-062QA phantom. (b)The modulation transfer function (MTF) obtained with both Catphan-600 and CIRS-062QA phantom.

Overall, the image quality results obtained with Catphan-600 are within the tolerance of the imaging system specification and consistent with that from Bissonnette *et al.* (2008) [3]. According to Bissonette et al. (2008) study, the reported integral non-uniformity value for Catphan-600 is ranging from 0.009-0.039; while the tolerance limit of spatial frequency at *f*10 is between 4.6 lp/cm and 9.9 lp/cm.

In summary, this research work has provided a phantom-specific baseline values for each image quality metric of the kV cone-beam CT system. Apart from image uniformity, the difference of other image quality parameters between both phantoms are due to the size and the design of the phantoms. The image uniformity measured with the CIRS-062QA phantom can be improve by eliminating the beam scattering and hardening originated from the screws and the phantom holders. Further investigation on resolving this issue will be performed.

## 4. Conclusion

Image quality metrics of a kV cone-beam CT system were assessed using two commercial phantoms. The algorithms developed for the quantitative image quality evaluation provides a phantom-specific set of reference value for a kV cone-beam CT system. The variations of image quality metrics between the phantoms were due to the difference in the size and the geometric design of the phantoms. Future work will focus on improving the setup of the CIRS-062QA phantom, in order to exclude the screws and the phantom holder from the field-of-view of the collimator during scanning.

## 5. References

[1] Stützel J, Oelfke U and Nill S 2008 A quantitative image quality comparison of four different image guided radiotherapy devices *Radiother. Oncol.* **86** 20–4.

[2] Bissonnette J, Moseley DJ, Jaffray D A 2008 A quality assurance program for image quality of cone-beam CT guidance in radiation therapy *Med. Phys.* **35** 1807–15.

[3] Garayoa J, Castro P 2013 A study on image quality provided by a kilovoltage cone-beam Computed tomography *J. Appl. Clin. Med. Phys.* **14** 239–57.

[4] Alaei P, Spezi E 2015 Imaging dose from cone beam computed tomography in radiation therapy *Phys. Med.* **31** 647-58

[5] Nakahara S, Tachibana M, Watanabe Y 2016 One-year analysis of Elekta CBCT image quality using NPS and MTF *J. Appl. Clin. Med. Phys.* **17** 6047.

[6] Stock M, Pasler M, Birkfellner W, Homolka P, Poetter R and Georg D 2009 Image quality and stability of image-guided radiotherapy (IGRT) devices: A comparative study *Radiother. Oncol.* 2009 **93** 1–7.

[7] Klein E, Hanley J, Bayouth J, Yin F, Simon W, Dresser S, Serago C, Aguirre F, Ma L, Arjomandy B, Liu C, Sandin C and Holmes T 2009 Task Group 142 report: quality assurance of medical accelerators *Med. Phys.* **36** 4197–4212.

[8] Bissonnette J, Balter P, Dong L, Langen K, Lovelock D, Miften M, Moseley D, Pouliot J, Sonke J and Yoo S 2012 Quality assurance for image-guided radiation therapy utilizing CT-based technologies: A report of the AAPM TG-179 *Med. Phys.* **39** 1946-63.

[9] Droege RT and Morin RL 1982 A practical method to measure the MTF of CT scanners *Med. Phys.* **9** 758–760.

[10] Shaw CC 2014. *Cone Beam Computed Tomography* (New York: CRC Press)

[11] U. Elstrøm, L. Muren, J. Petersen and C. Grau 2011 "Evaluation of image quality for different kV cone-beam CT acquisition and reconstruction methods in the head and neck region", *Acta Oncologica* **50** 908-917.

## Acknowledgments

Lim Siew Yong studentship is funded by MyBrain15scholarship. This research is funded by Fundamental Research Grant Scheme, Ministry of Education Malaysia, 203/CIPPT/6771383. The authors would like to thank the radiotherapists in the Radiotherapy & Oncology Unit, AMDI, USM for the help given in this research.

IOP Conf. Series: Journal of Physics: Conf. Series **851** (2017) 012030 doi:10.1088/1742-6596/851/1/012030

# Assessing the 3D accuracy of consumer grade distance camera measurement of rcspiratory motion

**M Samir**[1], **E Golkar**[1] **and A A Abd Rahni**[1]

[1]Department of Electrical, Electronic & Systems Engineering, Faculty of Engineering and Built Environment, Universiti Kebangsaan Malaysia, 43600, Selangor, Malaysia

E-mail: ashrani@ukm.edu.my

**Abstract**. Recently range imagers or distance camera systems have garnered interest for measuring respiratory motion without using markers, which can then be used as a surrogate in diagnosis and treatment for example in diagnostic imaging or radiotherapy. However, their use may have limitations, especially among lower cost systems, whereby their accuracy decrease greatly with the distance of the patient from the camera. This is considering the fact that the motion amplitude of the anterior surface of the body in normal breathing is typically around 1 cm or less, which is at the limit of accuracy of these systems. This accuracy limitation is even more pertinent when the fact that the 1 cm accuracy is desired over the whole anterior surface that is image and not just an average measurement of distance. We study this limitation in a low cost system i.e. the Microsoft Kinect™, using both version 1 and version 2 of the sensor. The 3D accuracy of both versions is compared with an alternative method of respiratory motion measurement i.e. a respiratory belt, at a distance of around 1.35 m. This study can be a guide for the design and application of range imaging systems in the clinical setting.

## 1. Introduction

Recently range imagers or distance camera systems have garnered interest for measuring respiratory motion without using markers, which can then be used as a surrogate in diagnosis and treatment for example in diagnostic imaging or radiotherapy. A variety of range imaging systems have been proposed for this use, from medically approved systems such as that from VisionRT to industrial systems such as the MESA SwissRanger, to consumer level cameras such as the Microsoft Kinect. The Microsoft Kinect™ has been proposed for respiratory motion tracking due to it being commercially available and is one of the low cost depth cameras with acceptable accuracy [1-3]. However, their use may have limitations, whereby their accuracy decrease greatly with the distance of the patient from the camera. This is considering the fact that the motion amplitude of the anterior surface of the body in normal breathing is typically around 1 cm or less, which is at the limit of accuracy of these systems. This limits the maximum range of these cameras to around 1 m from the patient which may not be entirely possible in a clinical setting. This accuracy limitation is even more pertinent when the fact that the 1 cm accuracy is desired over the whole anterior surface that is imaged and not just an average measurement of distance.

In this paper, we extend our previous comparison of the averaged (1D) depth measurement using both versions of the Kinect against a respiratory belt measurement of the anterior respiratory motion of the torso [4]. We assess how well the 3D motion detected across the torso correlates to the respiratory belt measurement.

IOP Conf. Series: Journal of Physics: Conf. Series **851** (2017) 012030 doi:10.1088/1742-6596/851/1/012030

This is in contrast to previous uses of the Kinect™ (version 1) for this purpose [1-3], which mainly look at the comparison with other respiratory motion surrogates with average depth measurements from the Kinect.

## 2. Materials and Evaluation Methodology

### *2.1. Kinect™ Cameras and Respiratory Belt Hardware*

*2.1.1. Kinect™ v1 and v2 Cameras.* For version 1 (v1) of the Kinect™ camera we use the one released for the XBox 360, whereas for version 2 (v2) we used the one released for Windows. The Kinect™ v1 camera utilises a structured light approach to computed depth by utilising two classic computer vision techniques: depth from focus and depth from stereo. The depth map is constructed by analysing a speckle pattern generated by infrared laser light. On the other hand, the Kinect™ v2 camera uses a Time-of-Flight approach to calculate the depth map [5].

The depth sensor of Kinect™ v1 has a field of view of 57° horizontally and 43° vertically [6, 7]. The depth image size is 640 × 480 pixels and has a rate of 30 frames per second. The depth sensor is able to function most properly for a distances roughly ranging from 1 m to 3 m with different degrees of accuracy. [6].

The Kinect™ v2 camera on the other hand has a wider field of view of 70° horizontally and 60° vertically [8]. The depth image size is 640 × 480 pixels and has a rate of 30 frames per second [9]. In comparison to Kinect™ v1, the v2 can measure depth to a greater range of 8 m [6], although again the accuracy is dependent on the depth to the sensor. Table 1 shows the overall comparison of the major specifications between Kinect™ v1 and v2.

**Table 1.** General comparison of the major specifications between Kinect™ v1 and v2.

| Specification | Kinect™ v1 | Kinect™ v2 |
|---|---|---|
| Colour Resolution | 640 × 480 | 1920 × 1080 |
| Depth Resolution | 640 × 480 | 512 × 424 |
| Max Depth Distance | 4 m | 8 m |
| Min Depth Distance | 0.6 m | 0.5 m |
| Horizontal field of view | 57 degrees | 70 degrees |
| Vertical field of view | 43 degrees | 60 degrees |
| USB Standard | 2.0 | 3.0 |

*2.1.2. Respiratory Belt.* In order to validate the accuracy of the respiratory measurement from both versions of the Kinect™, an alternative device is used for simultaneously measuring respiratory motion. We choose a respiratory belt, the Sleep Sense® from Sleep Lab Products. The respiratory belt is worn by the volunteer as shown in figure 1. The respiratory belt is interfaced using a g.USBamp amplifier (g.tec medical engineering GmbH). Together, the belt and the amplifier allows for accurate measurement of respiration and acts as the ground truth.

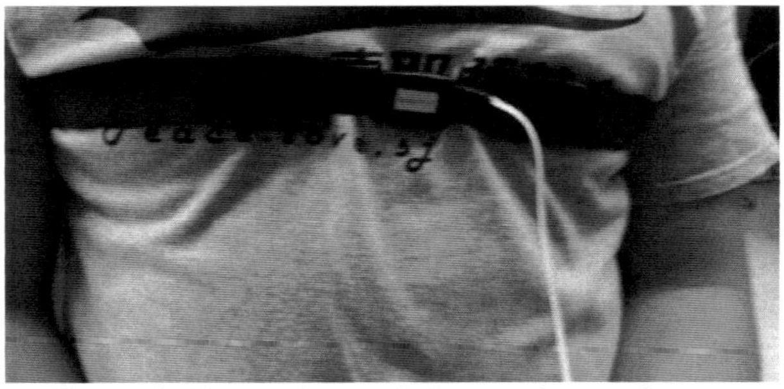

**Figure 1.** The respiratory belt worn by a volunteer.

### *2.2. Software Interface and Data Acquisition*

For interfacing the Kinect cameras, we used the Kinect™ for Windows SDK 1.8 for Kinect™ v1 and the Kinect™ for Windows SDK 2.0 for Kinect™ v2. The data acquisition application for both Kinect™ v1 and v2 was run in Windows 8 (a requirement for the SDK 2.0). The data acquisition application [10] was developed in C# and its GUI is as shown in Figure 2.

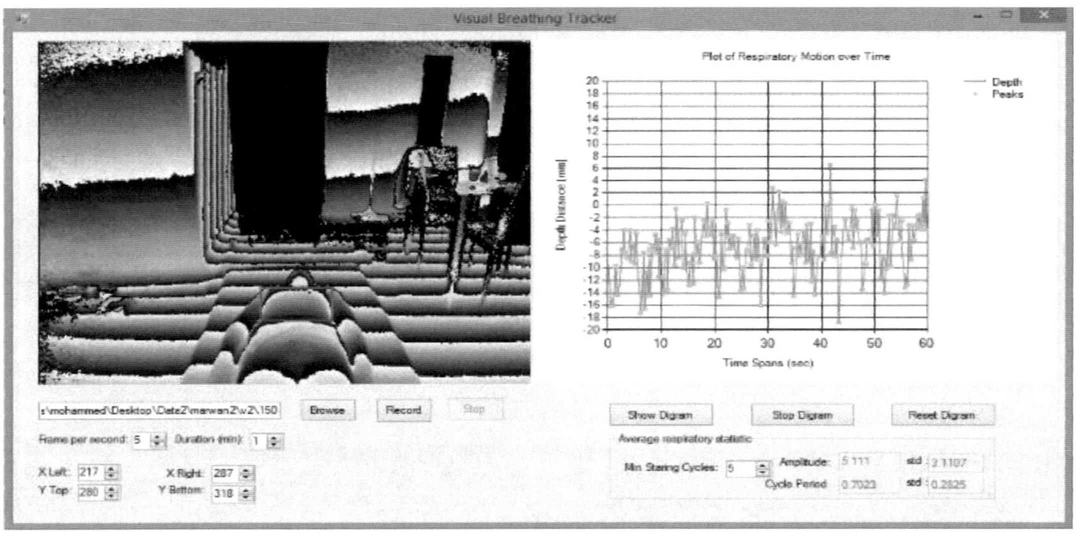


**Figure 2.** Previously developed data acquisition application for acquiring data from both versions of the Kinect™.

The belt is sampled at 32 Hz while both versions of the Kinect™ were run at around 5 Hz. We chose a low sampling rate as the belt data acquisition is performed in Windows 7 run on a virtual machine. These were deemed sufficient as respiration is a low frequency cyclic process (averaging around 1 breath every 3 seconds or having a rate of 0.33 Hz). Data from both the belt and Kinect™ were recorded simultaneously for around 60 s for further processing. The measurements were performed while the volunteer was lying down at around a distance of 1.35 m from the camera to simulate the situation of the intended application of respiratory motion tracking during radiotherapy or imaging. Further analysis of the acquired data was performed in MATLAB.

### *2.3. Methodology*

*2.3.1. 3D Calibration* A 10×10×10 cm$^3$ wooden cube is used as the calibration object. For calibration, the cube is placed around 1 m from the Kinect™ camera on a table. From the full depth image, a region of interest (ROI) containing the cube and the table is first selected. Background pixels which have depth values of more than 1 m or less than 0.8 m are removed. From the remaining foreground, the outline (six sides) of the cube is selected. A visualisation of the calibration process for both versions of the Kinect$^{TM}$ is shown in figure 3.

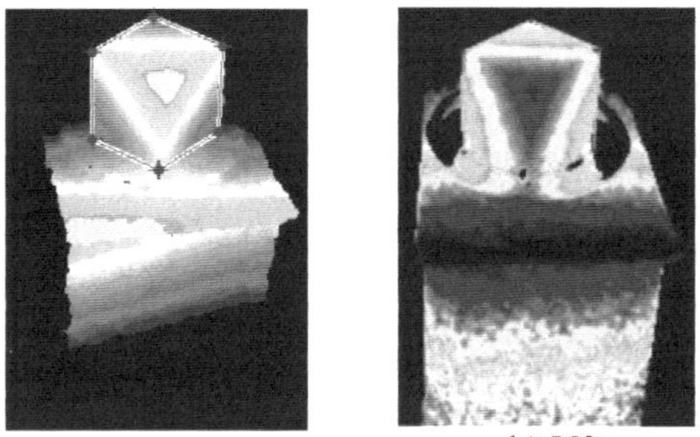

a) V1 b) V2

**Figure 3.** Visualisation of the calibration process for both versions of the Kinect$^{TM}$ cameras. (a) V1, (b) V2

As each side is known to be 10 cm in length, the focal length $f$ can be found from each side, between vertices $\mathbf{p}_i$ and $\mathbf{p}_{i+1}$, using the perspective projection formula (equation (1)) and the Euclidean distance between vertices (equation (2)):

$$\frac{c_x}{f} = \frac{p_x}{p_z}, \frac{c_y}{f} = \frac{p_y}{p_z} \tag{1}$$

$$d = ||\mathbf{p}_i - \mathbf{p}_{i+1}|| \tag{2}$$

where **c** is the projected coordinates and $d$ is the distance between adjacent vertices. For simplification we assume the same focal length along the horizontal and vertical axes of the camera. The average value of the focal length $f$ is found from all six pairs of adjacent vertices. From the average focal length, the point cloud can be found from the depth image using inverse perspective mapping (i.e. inverting equation (1)). The overall 3D calibration process is shown in figure 4 below.

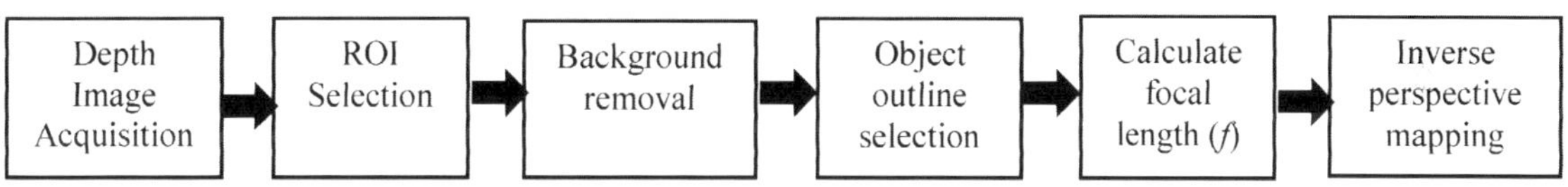


**Figure 4.** Flowchart of the 3D calibration process.

*2.3.2. 3D Motion Assessment.* For assessing the motion in 3D, similar to previous work [4] [10], we select a rectangular ROI in the depth image for further processing. However after inverse perspective mapping (using the focal length $f$ from the calibration process above) and 3D rotation so that the floor is horizontal, this initially selected ROI is not rectangular in 3D, hence a subsequent ROI, corresponding to the anterior surface of the torso, is selected based on 3D coordinates in the horizontal plane (x-, y-coordinates). Lastly, as the generated 3D point clouds for each frame do not have correspondence, we fit a mesh defined on a grid with a regular spacing of 10 mm in the x- and y-axes. Only motion in the z-axis direction is thus detected. The z-axis motion at each interpolated grid point is then compared with the 1D signal from the respiratory belt (Figure 5).

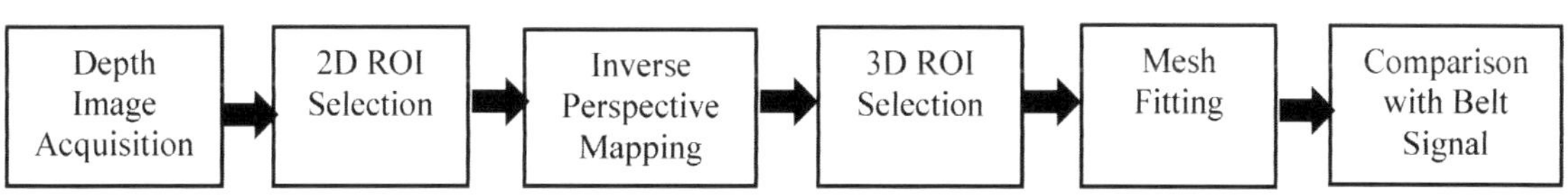


**Figure 5.** Flowchart of the 3D motion assessment process.

IOP Conf. Series: Journal of Physics: Conf. Series **851** (2017) 012030 doi:10.1088/1742-6596/851/1/012030

## 3. Results and Discussion

### *3.1. 3D Calibration Results*
The calculated mean focal length with the standard deviation found from the calibration process is shown in Table 2 below. The values found are not far from that found in other works, for e.g. [5], considering the simplified camera model adopted here.

**Table 2.** Mean focal length found for Kinect™ v1 and v2.

| | Kinect™ v1 | Kinect™ v2 |
|---|---|---|
| Mean focal length (unit pixels) | 562±24 | 369±63 |

### *3.2. 3D Motion Assessment Results*
The 3D motion assessment is performed for a single volunteer. Following the flowchart in shown in figure 4, the interpolated grid points of the fitted mesh can be primarily analysed in two different ways. Firstly we look at the maximum amplitude in the generally anterior-posterior direction (z-axis). Secondly, we look at the correlation of the grid points with the respiratory belt. This is both represented graphically in figure 6. The colour bar of the amplitude plot is in mm while the correlation plot shows the Pearson correlation coefficient at each grid point.

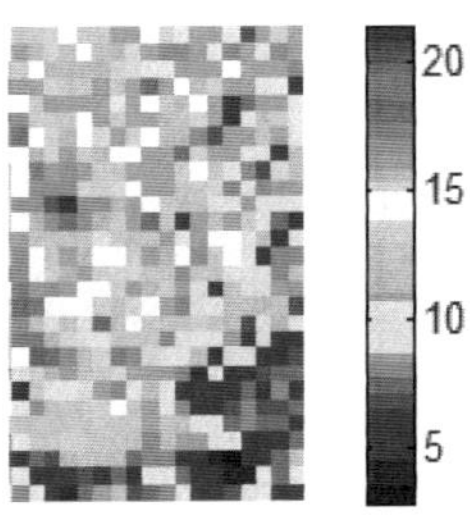

a) Maximum amplitude for Kinect™ v1

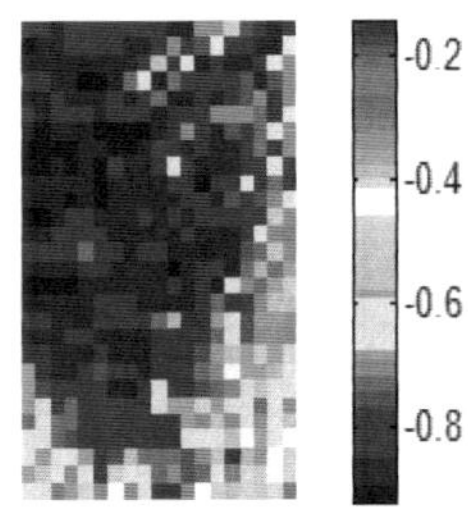

b) Correlation for Kinect™ v1

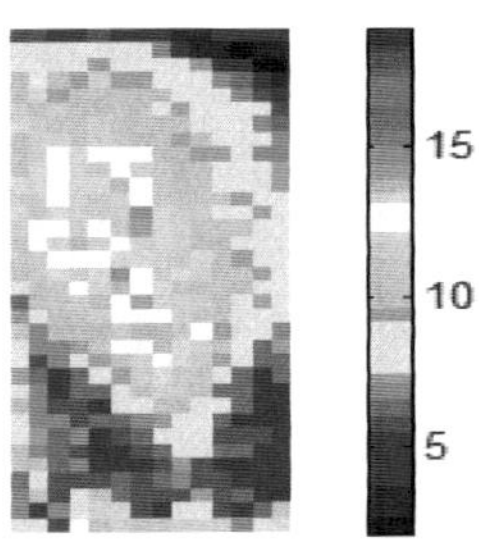

c) Maximum amplitude for Kinect™ v2

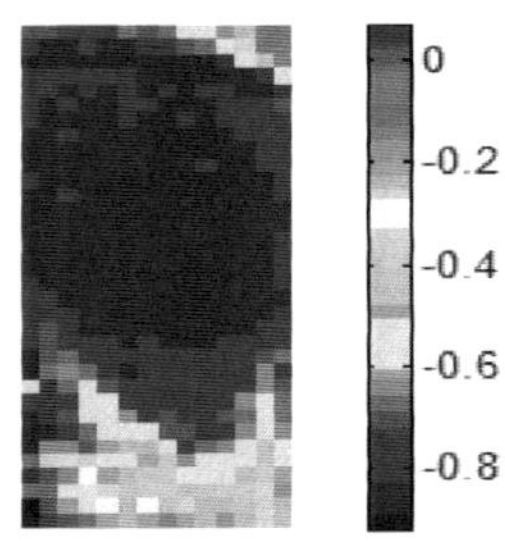

d) Correlation for Kinect™ v2

**Figure 6.** Visualisation of the maximum amplitude and correlation with the respiratory belt for both versions of the Kinect™ camera.

From figure 6, it can be seen in general, over the chosen ROI, that the Kinect™ v2 is more consistent in its measurement of the motion amplitude and likewise its correlation with the respiratory belt. The average maximum amplitude and absolute correlation over the chosen ROI is shown in Table 3.

Based on the results shown in Table 3, we note that the average amplitude with the standard deviation measured by both versions of the Kinect™ are similar, although the acquisitions are done separately. Additionally, the average correlation found supports our previous findings on 1D accuracy [4], that the Kinect™ v2 is slightly more accurate than the Kinect™ v1. We also note that the average absolute correlation over the whole ROI is lower than that of the corresponding 1D signal from the average distance to the camera over the ROI (0.94 and 0.92 for Kinect™ v1 and v2 respectively). This can be attributed to the differing correlations at different locations in the ROI.

**Table 3.** Average amplitude and correlation found for Kinect™ v1 and v2.

| | Kinect™ v1 | Kinect™ v2 |
|---|---|---|
| Average maximum amplitude (mm) | 7.91±2.75 | 8.87±2.73 |
| Average absolute correlation | 0.68 | 0.73 |

## 4. Conclusion

Based on the results, we show that the 3D respiratory motion measurement of the volunteer using both versions of the Kinect™ are consistent with each other. Accuracy over the ROI varies although it is perhaps acceptably accurate for the intended application when it is possible to locate at around 1m to 2m distance from the patient. On average the accuracy of the Kinect™ v2 is better than v1. More measurements at different distances are needed for a more conclusive assessment.

## 5. References

[1] Noonan P J, Howard J, Tout D, Armstrong I, Williams H A, Cootes T F, Hallett W A and Hinz R 2012 *Accurate Markerless Respiratory Tracking for Gated Whole Body PET using The Microsoft Kinect IEEE NSS/MIC* 3973-74

[2] Xia J and Siochi R A 2012 *A real-time Respiratory Motion Monitoring System using Kinect: Proof of concept Med Phys* **39** 2682-85

[3] Tahavori F, Alnowami M and Wells K 2015 *Marker-less Respiratory Motion Modeling using The Microsoft Kinect for Windows SPIE Medical Imaging: Image Processing* **9036**

[4] Samir M, Golkar E and Abd. Rahni A A 2015 *Comparison Between The Kinect™ V1 and Kinect™ v2 for Respiratory Motion Tracking ICSIPA* 150-5.

[5] Sarbolandi H, Lefloch D and Kolb A2015 *Kinect Range Sensing: Structured-Light Versus Time-of-Flight Kinect Comput. Vis. Image Underst* **139**

[6] Microsoft developer network kinect for windows programming guide.

[7] Dutta T 2012 *Evaluation of The Kinect™ Sensor for 3-D Kinematic Measurement in The Workplace Applied Ergonomics* **43** 645-9

[8] Yang L, Zhang L, Dong H, Alelaiwi A and El Saddik A 2015 *Evaluating and Improving The Depth Accuracy of Kinect for Windows v2 IEEE Sensors Journal* **15** 4275-85

[9] Wasenmuller O and Stricker D 2016 *Comparison of kinect v1 and v2 depth images in terms of accuracy and precision ACCV Computer Vision Workshop*

[10] Lim S H, Golkar E and Abd. Rahni A A 2014 *Respiratory Motion Tracking using The Kinect Camera IECBES* 797-800

## Acknowledgement

This work was funded by the Ministry of Higher Education Malaysia under grant no. ERGS/1/2013/TK02/UKM/03/2.

# Optimisation of 12 MeV electron beam simulation using variance reduction technique

**J Jayamani[1], N A S Mohd Termizi[2], F N Mohd Kamarulzaman[2], M Z Abdul Aziz[1]**

[1] Oncological and Radiological Science Cluster, Advance Medical and Dental Institute, Universiti Sains Malaysia, 13200 Kepala Batas, Penang, Malaysia
[2] School of Physics, Universiti Sains Malaysia, 11800 USM, Penang, Malaysia

E-mail: mohdzahri@usm.my, mr.zahri@gmail.com

**Abstract**. Monte Carlo (MC) simulation for electron beam radiotherapy consumes a long computation time. An algorithm called variance reduction technique (VRT) in MC was implemented to speed up this duration. This work focused on optimisation of VRT parameter which refers to electron range rejection and particle history. EGSnrc MC source code was used to simulate (BEAMnrc code) and validate (DOSXYZnrc code) the Siemens Primus linear accelerator model with the non-VRT parameter. The validated MC model simulation was repeated by applying VRT parameter (electron range rejection) that controlled by global electron cut-off energy 1,2 and 5 MeV using 20 x $10^7$ particle history. 5 MeV range rejection generated the fastest MC simulation with 50% reduction in computation time compared to non-VRT simulation. Thus, 5 MeV electron range rejection utilized in particle history analysis ranged from 7.5 x $10^7$ to 20 x $10^7$. In this study, 5 MeV electron cut-off with 10 x $10^7$ particle history, the simulation was four times faster than non-VRT calculation with 1% deviation. Proper understanding and use of VRT can significantly reduce MC electron beam calculation duration at the same time preserving its accuracy.

## 1. Introduction

Monte Carlo (MC) simulation is an algorithmic calculation uses a random sampling method to model a system with its specific geometry [1]. In radiotherapy application, MC is a fundamental tool to simulate a virtual model of radiation source and radiation transport in the system [2]. Advantage of MC is able to carry out dosimetry study in a challenging experiment setup which is not easily accessible in actual situation during treatment delivery [3, 4]. There are plenty of MC codes that have been used for radiotherapy applications, including Monte Carlo N-Particle (MCNP) [5], Geometry and Tracking (GEANT)[6], Electron Gamma Shower (EGS)[7] and PENELOPE[8]. However, the common MC code used for linear accelerator simulation and patient dose calculation are BEAMnrc and DOSXYZnrc where both codes were based on EGSnrc which is widely used for radiotherapy applications[9, 10]. The advantage of using EGSnrc code includes its ability for simulation three to five times faster than PENELOPE and GEANT because of its simple geometry[11].

In radiotherapy, the dose distribution in electron beams is more complex than that in photon beam[12]. Furthermore, the MC simulation of electron beam requires a long simulation time as compared to photon beam. Due to its prolonged computation time, an algorithm called VRT in BEAMnrc code was studied in this work to speed up calculation duration to run the MC simulation. The VRT parameters in EGSnrc code are range rejection, bremsstrahlung cross section involvement, Russian roulette, Bremsstrahlung splitting, interaction forcing [1, 9, 13]. VRT parameters suggested in this study are electron range rejection and particle histories. This study uses VRT parameters on evaluating simulation time on Siemens Primus linac treatment head.

## 2. Materials and Methods

### 2.1. Monte Carlo Modelling and Simulation

The treatment head of Siemens Primus linear accelerator was generated by using precise geometry provided by Siemens Medical System, USA. The accelerator head was comprised of eight component modules, which are primary scattering foil, primary collimator, secondary flattening foil, ionisation chamber, mirror, upper and lower jaws and applicator, as shown in Figure 1(a). A 12 MeV electron beam model was developed from BEAMnrc with 10 x 10 $cm^2$ field size and source-to-surface distance (SSD) of 100 cm by using 20 x $10^7$ particle history.

Simulation runs with default EGSnrc transport parameter (non-VRT method) and used AE = ECUT = 0.7 MeV for electron and AP = PCUT = 0.01 MeV for photon energy by using 700icru.pegs4 data file. AP and AE represent the low-energy threshold for secondary Bremsstrahlung photons generation. PCUT and ECUT are defined as the global cut-off energy for electron and photon transport.

A phase space file generated from BEAMnrc simulation comprised of all the data of particles, which exited from the treatment head. This file was used as an input radiation source file to DOSXYZnrc simulation for dose calculation in homogeneous water phantom. The DOSXYZnrc was used to simulate the deposited dose (central axis depth dose) in voxelised 3-D water phantom with a volume of 14 x 14 x 12 $cm^3$ and the voxel size is 0.5 x 0.5 x 0.2 $cm^3$.

Figure 1 (b) shows the model of 3-D water phantom with voxel X, Y, Z axis. The volume of 3-D voxel phantom was proportional to the reference phantom. In DOSXYZnrc, 5.0 x $10^7$ particles history was used to simulate the electron transport in 3-D water phantom. The output file from DOSXYZnrc used STATDOSE interface to extract the dose information. This data was analysed and normalized based on PDD calculation as indicated in the following equation:

$$PDD = \frac{D_d}{D_{d_0}} \text{ x } 100\,\% \qquad (1)$$

where, $D_d$ is the absorbed dose at any depth and $D_{d_0}$ dose at reference depth [14]. Ion chamber measurements were collected and compared with calculated PDD value from the MC model.

### 2.2. Variance Reduction Technique

MC simulation required the calculation of efficiency in the VRT application. The efficiency, $\eta$ is defined as:

$$\eta = 1/(\sigma^2 T) \qquad (2)$$

where, $\sigma^2$ is variance and $T$ is the total CPU time used for processing $N$ history number [13]. The variance is proportional to the inverse of the $N$, thus, efficiency of VRT is not reliable to the number of histories. However, it increases CPU computation time per history.

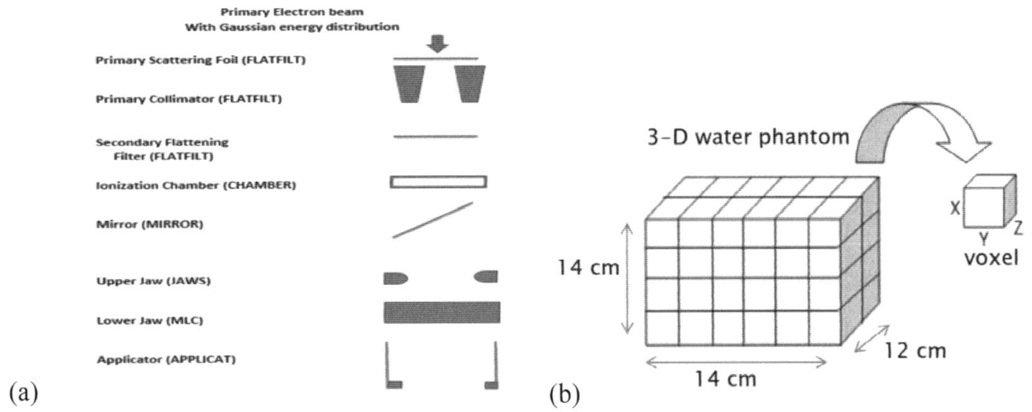


**Figure 1.** (a) Schematic diagram of Siemens Primus linear accelerator model. (b) Model of 3-D water phantom

*2.3 Range Rejection of VRT*

The VRT technique applied in this study was the range rejection for 12 MeV electron beam. The fundamental of range rejection is to eliminate electron history below a selected energy level to speed up MC simulation. Furthermore, VRT can be applied in conjunction with other techniques to improve the efficiency of MC calculation. Thus, the range rejection in this study was initialised in conjunction with number of particle history to obtain the best simulation time for EGSnrc without influencing the outcome.

The electron range rejection was turned on in BEAMnrc code by applying ESAVE global electron cut off as 1, 2 and 5 MeV. This parameter was governed in BEAMnrc to determine the energy of the electron. CPU computation time for BEAMnrc was then recorded. The phase space file generated from this simulation was used in DOSXYZnrc code for dosimetry analysis. DOSXYZnrc code runs simulation with 5.0 x $10^7$ particle history to calculate dose in 3-D voxel phantom as shown in Figure 1(b). Simulation time was recorded and 3ddose output file was analysed using STATDOSE interface. PDD evaluation was performed to ensure that range rejection has no influence on dosimetry output. This result was compared to ionisation chamber measurement in 3-D water phantom. The range rejection that provided shorter simulation time with unbiased result was chosen to be in conjunction with particle history. The particle history values was reduced from 20.0 x $10^7$ to 7.5 x $10^7$ and the analysis was done on computation time and PDD.

## 3. Results and Discussion

*3.1. Phase space file validation*

The phase space generated from BEAMnrc was evaluated in beamdp.gui by using spectral distribution and particle scatters plot. The spectral distribution from the treatment head simulation at 11.5 MeV in Figure 2(a) was found to be the closest to represent 12 MeV electron beams. Also, Figure 2(b) shows maximum intensity of scatter particle at range between –5 cm to +5 cm in x and y-axis agrees well with treatment head simulation defined at 10 x 10 $cm^2$ field size. Therefore, these results had acknowledged the validation of phase space file for 12 MeV electron beam simulation and further analysed in DOSXYZnrc code.

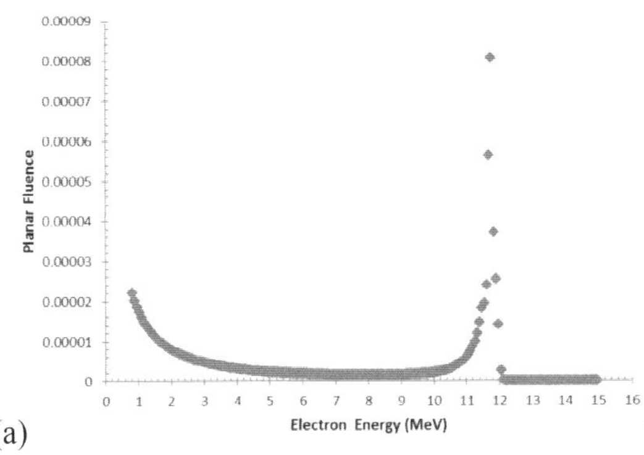

(a)

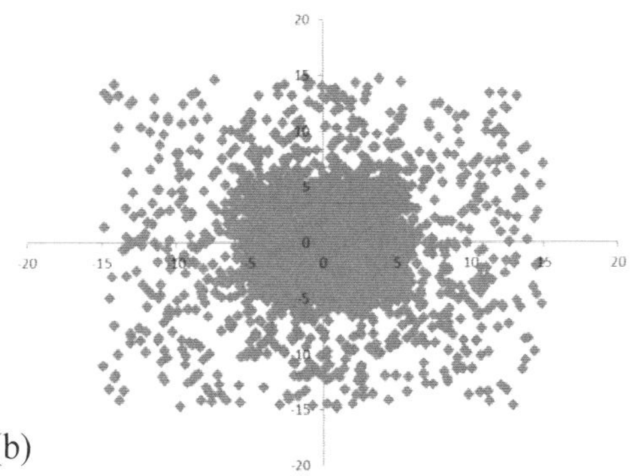
(b)

**Figure 2.** (a) Spectral Distribution of 12 MeV Electron Beam. (b) Scatter distribution of 12 MeV electron beam for 10 x 10 $cm^2$ field size

*3.2. Dose Calculation Validation*

The phase space file generated from BEAMnrc code was used further in DOSXYZnrc simulation and the results were compared with experimental measurement using ionisation chamber. The STATDOSE interface was utilised to derive PDD value for voxel phantom created in DOSXYZnrc. The MC calculation result matched well with ion chamber measurement with less than 2% error in buildup region as shown in Figure 3. This signifies that linac model for 12 MeV beam was simulated accurately using EGSnrc code compared with non-VRT method.

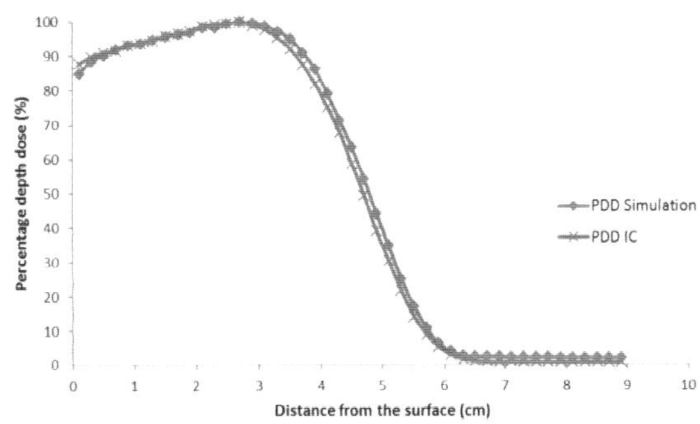


**Figure 3.** 12 MeV electron beam validation for MC calculation using PDD measurement

*3.3 Electron Range Rejection Measurement*

The non-VRT simulation time observed for BEAMnrc was 26.8 hours and DOSXYZnrc was 0.8 hours by using 20 x $10^7$ particle history and result a total computation time of 27.6 hours., The results showed that the 5 MeV electrons cut-off energy able to reduce the total simulation time of BEAMnrc and DOSXYZnrc from 27.6 hours to 13.8 hours when the electron range rejection was applied. Figure 4(a) showed that total simulation time decreased as the electron cut-off energy value increased. The range rejection showed no substantial difference on DOSXYZnrc simulation time. The result showed that the 5 MeV electrons cut-off energy was the best range rejection to reduce the computation time. The PDD values show good agreement and explain well that the output did not influence the electron range rejection as shown in Figure 4(b). The interactions occurred (as in figure 2 (a)) below the electron cut-off energy (5 MeV in this case) were not simulated and hence reduced calculation time expressively by using VRT method. The uncertainties between both simulations were not affected as cut-off energy value 5 MeV is lesser to simulate electron energy of 12 MeV due to monoenergetic peak electron beam profile generated by linac as shown in Figure 2.

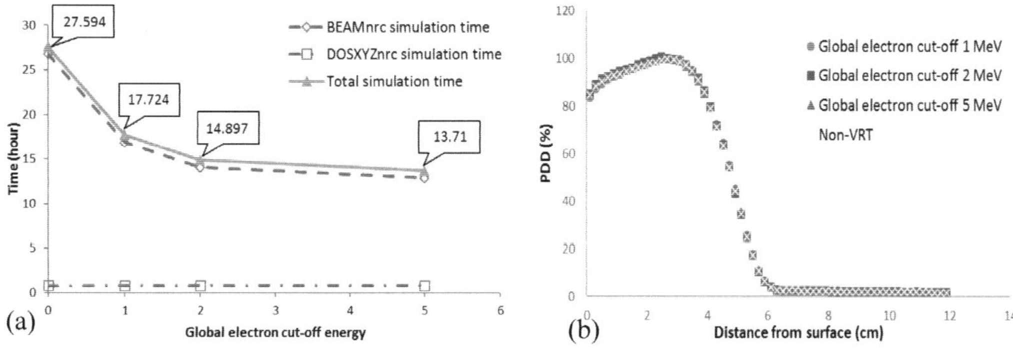


**Figure 4.** (a) Simulation time using different electron cut-off energy shows 5 MeV reduce simulation time to 13.7 hours. (b) PDD of simulation using different electron cut-off energy with less than 1% errors was observed.

### *3.4. Number of Particle History Analysis*

The electron- range rejection at 5 MeV from the previous study was referred to investigate the effect of particle history on total computation time. The results showed that the simulation time decreases by lowering the number of particle history in Figure 5 (a). Particle history of 7.5 x $10^7$ yields the lowest simulation time of 5.8 hours. However, PDD analysis in Figure 5 (b) showed that 7.5 x $10^7$ particle history deviated by 7.4% from the measured PDD result and this outcome is unacceptable as the error is more than 2%. The number of histories used in BEAMnrc must be more than 5.0 x $10^7$ to achieve 1 % statistical uncertainty in dose calculation for 6 MeV electron beam[15]. On the other hand, this result showed that minimum number of particle history that to be chosen in this study is 10 x $10^7$ with a total simulation time of 7.3 hours to produce a PDD output with a 1% dose error for 12 MeV electron beam. The cut-off range must be corresponded to required spatial resolution to conserve the final outcome. This indicated that the simulation time depended on both electron range rejection and particle history.

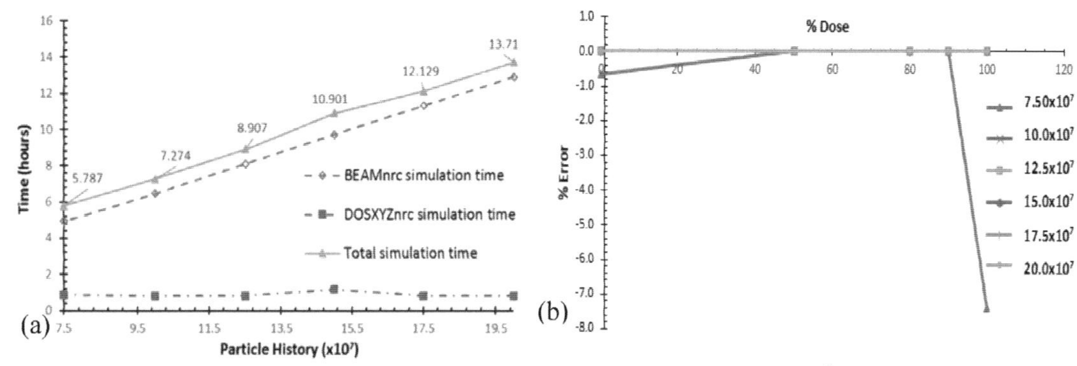


**Figure 5.** (a) Simulation time using different particle history, where 7.5 x $10^7$ gives shorter simulation time. (b) Percentage error of different particle history with non-VRT calculation, shows 7.5 x $10^7$ particle history having unacceptable dose discrepancy of 7.4%

## 4. Conclusion

The electron range rejection for Siemens Primus linear accelerator 12 MeV electron beam energy was optimised. The electron range rejection together with number of particle history used in BEAMnrc code able to give accurate results and good computation time. 5 MeV global electron cut-off in range rejection for 10 x$10^7$ particle history yields the fastest simulation time of 7.3 hours with unbiased outcome in radiation interaction. Limitation in particle history with lower than 10 x $10^7$ influences the output from treatment head and deviates PDD result. In addition, this study also proved that the use of electron range rejection did not influence time required to run DOSXYZnrc. Future work to investigate ESAVE more than 5 MeV for different electron energies is suggested.

## 5. References

[1] Seco J and Verhaegen F 2013 *Monte Carlo Techniques in Radiation Therapy* CRC Press Taylor and Francis Group Boca Raton London New York DOI:10.1017/CBO9781107415324.004.

[2] Jabbari K 2011 Review of Fast Monte Carlo Codes for Dose Calculation in Radiation Therapy *J Med Signals Sens* **1** 73–86

[3] Jabbari K, Anvar HS, Tavakoli MB and Amouheidari A 2013 Monte Carlo Simulation of Siemens ONCOR Linear Accelerator with BEAMnrc and DOSXYZnrc Code. *J Med Signals Sens* **3** 172–9

[4] Kleijnen JPC, Ridder AAN and Rubinstein RY 2010 Variance Reduction Techniques in Monte Carlo Methods CenteR Discussion Paper Series No 2010-177 Available at SSRN: http://dx.doi.org/10.2139/ssrn.1715474)

[5] Briesmeister JF, Carter L L, Cox L J, Favorite J A, Forster R A and McKinney G W 2000 Present and future capabilities of MCNP. *Appl Radiat Isot* **53** 857–861.

[6] Agostinelli S, Allison J and Amako K 2003 Geant4—a simulation toolkit *Nucl Instruments Methods Phys Res A* **506** 250–303.

[7] Kawrakow I, Rogers DWO, Tessier F Mainegra-Hing E and Walters B 2016 The EGSnrc Code System : Monte Carlo Simulation of Electron and Photon Transport

[8] Baró J, Sempau J, Fernández-Varea JM and et al 1995 PENELOPE: An algorithm for Monte Carlo simulation of the penetration and energy loss of electrons and positrons in matter. *Nucl Instruments Methods Phys Res Sect B Beam Interact with Mater Atoms* **100** 31–46

[9] Verhaegen F and Seuntjens J 2003 Monte Carlo Modelling of External Radiotherapy Photon Beams. *Phys Med Biol* **48** R107–R164.

[10] Kim J-H, Hill R and Kuncic Z 2012 An evaluation of calculation parameters in the EGSnrc/BEAMnrc Monte Carlo codes and their effect on surface dose calculation. *Phys Med Biol* **57** N267–N278.

[11] Rogers DWO. Variance reduction techniques used in BEAMnrc. Available at: http://indico.ictp.it/event/a06223/session/13/contribution/7/material/0/0.pdf

[12] Abdul Aziz MZ, Yusoff AL, Osman ND, Abdullah R, Rabaie NA and Salikin MS 2015 Monte Carlo dose calculation in dental amalgam phantom. *J Med Phys* **40** 150–5.

[13] Mohammed M, Chakir E, Boukhal H, Saeed M and El Bardouni T 2016 Evaluation of variance reduction techniques in BEAMnrc Monte Carlo simulation to improve the computing efficiency. *J Radiat Res Appl Sci* **9** 424–430

[14] Khan FM 2010 *The physics of radiation therapy*. 4th ed Lippincott Williams & Wilkins

[15] Ma C-M, Jiang SB 1999 Monte Carlo modelling of electron beams from medical accelerators. *Phys Med Biol* **44** R157–R189

## Acknowledgement

The authors would like to extend their highest gratitude to the Department of Higher Education (Malaysia) for providing a grant under Fundamental Research Grants Scheme (FRGS)203/CIPPT/6711341. The author also would like to thank Dr Shazril Imran and Mohd Fahmi for revising this manuscript. We would also like to thank staffs of the Radiotherapy Unit in Advanced Medical and Dental Institute, Universiti Sains Malaysia; the staffs of Radiology and Radiotherapy Departments, School of Medical Sciences, Universiti Sains Malaysia; School of Physics, Universiti Sains Malaysia, and Siemens Medical System U.S.A for their contribution in this work.

# A method for determination of parameters of the initial electron beam hitting the target in linac

**D T Tai[1,3], N D Son[2], T T H Loan[3] and H D Tuan[3]**

[1]Department of Radiation Oncology, Dong Nai General Hospital, Bien Hoa City, Dong Nai Province, Vietnam
[2]Chi Anh Med. Tech. Co.,Ltd., Ho Chi Minh City, Vietnam
[3]Faculty of Physics and Engineering Physics, VNUHCM-University of Science, Ho Chi Minh City, Vietnam

E-mail: thanhtai_phys@yahoo.com

**Abstract**. Detailed characteristics of the electron beam incident on the target of the Linac are almost impossible to be measured. With Monte Carlo technique it is possible to simulate the transport of radiation through the accelerator head and find out the initial beam parameters. These parameters, such as energy spectrum and spatial distribution can be ascertained by matching the simulated dose distributions with the measured dose distributions using trial and error. In this work, a Monte Carlo modeling of the HPD Siemens Primus linear accelerator in 6 MV photon beams at Dong Nai general hospital was performed. The simulated Percent depth doses (PDD) and beam profiles (OCR) were then compared with the measured ones. Excellent agreements were obtained between simulations and measurements with an average difference of 0.7% for PDD less than 2% for OCR. The percentage gamma passing rate is 100% with 1% dose difference and 1 mm distance to agreement as acceptance criteria. The best suited energy and radius of the electron beam incident on the target were found at 6.04 MeV and at FWHM = 1.2 mm respectively.

## 1. Introduction

The EGSnrc (Electron–Gamma–Shower) is a well-known Monte Carlo simulation package widely used in medical physics application. BEAMnrc, built on the EGSnrc, is a user code for modeling radiotherapy sources from linear accelerators [1]. Detailed information (i.e. the position, direction, charge, and energy) of all the particles leaving the accelerator was stored in a phase-space data file. DOSXYZnrc [2] is an EGSnrc user code for calculating radiation dose in Cartesian voxels. In our study, the phase space files obtained with BEAMnrc are used as source inputs to the DOSXYZnrc to simulate dose distributions in a water phantom. Due to difference in structures of linac, beam parameters have to be determined for each linac. This paper introduces a method for determination of the parameters of the electron beam as it leaves the accelerator vacuum and hits the target for Siemens Primus M5497 linac. The simulated and measured percentage depth dose (PDD) and beam profiles (OCR) for a 10x10 $cm^2$ field's size were compared. The energy and radius of the electron were adjusted until the best fit was found.

## 2. Materials and methods

### *2.1. Monte Carlo simulation of the HPD Linac with BEAMnrc*

The BEAMnrc user code was used to simulate a 6 MV photon beams from a Siemens Primus linear (Siemens Medical Solutions USA, Inc.) at Dong Nai general hospital, Vietnam, with a 10x10 $cm^2$ field at SSD=100 cm. This linear accelerator was installed at the department of radiation oncology in 2009. All the physical dimensions and material of the relevant components of the accelerator head were incorporated according to the manufacture's detailed specifications. In BEAMnrc these components were modeled by a number of predefined component modules (CM). The primus accelerator components are shown in figure 1, which consists of 9 component modules: Vacuum envelope (SLABS), the target (SLABS), the flattening filter (FLATFILT), chambers for monitoring the linac output (CHAMBER), mirror made of $SiO_2$ (MIRROR), the jaws X and jaws Y made of tungsten (JAWS), reticle tray of mica (SLABS), and slab air (SLABS).

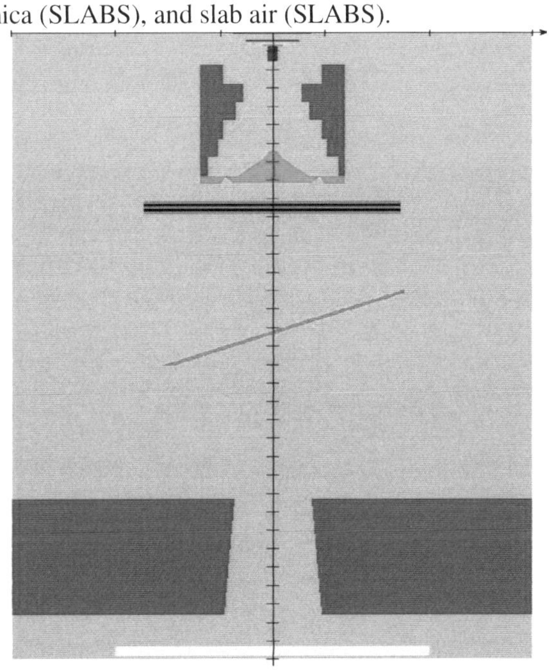

**Figure 1.** Component modules of the accelerator (6 MV)

The global cutoff energy (ECUT) and production cutoff energy (AE, AP) used in simulation were 0.7 MeV for electron and 0.1 MeV for photon. A total of $2\times10^8$ electron histories were simulated. The simulations run on Intel(R) Core i5 2400 processors. The history of particles is ended when its energy is below a predefined energy threshold (ESAVE). In our study, ESAVE was set to 2 MeV in all component modules and directional bremsstrahlung splitting (DBS) was applied [3]. The output of the BEAMnrc simulation is a phase space file, which contains information about all the particles leaving the accelerator such as energy, position, direction, etc. This phase space was scored in a plane perpendicular to the beam axis at 100 cm distance from the target. From this phase space, the BEAMDP was be used to analyze parameters of beam production and plot energy spectrum of photon beam.

### *2.2. Dose calculation with DOSXYZnrc*

The EGSnrc user code DOSXYZnrc was used to calculate dose distribution in a water phantom (50x50x30 $cm^3$ water tank). The voxel size was presented in figure 2. The phantom is divided into 3x5x2 voxels in the x, y, and z directions. In z-axis there were 10 slices of 0.2 cm thickness and 56 slices 0.5 cm of thickness; in the x-axis 2 slices of 24 cm thickness and 1 slice of 2 cm thickness; in the y-axis 2 slices of 17 cm thickness 28 slices of 0.5 cm thickness and 1 slices of 1.5 cm thickness. There were in total 6138 voxels.

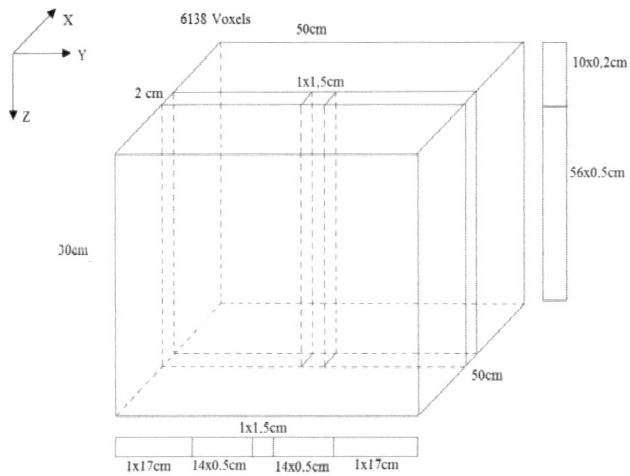


**Figure 2.** Voxels set in DOSXYZnrc simulation

The water phantom was located at source to surface distance (SSD) of 100 cm. The electron and photon cut-off energy (ECUT, PCUT) was set to 0.7 MeV and 0.01 MeV, respectively. $10^9$ histories were simulated. The output file from DOSXYZnrc program was analyzed by an in-house MATLAB code.

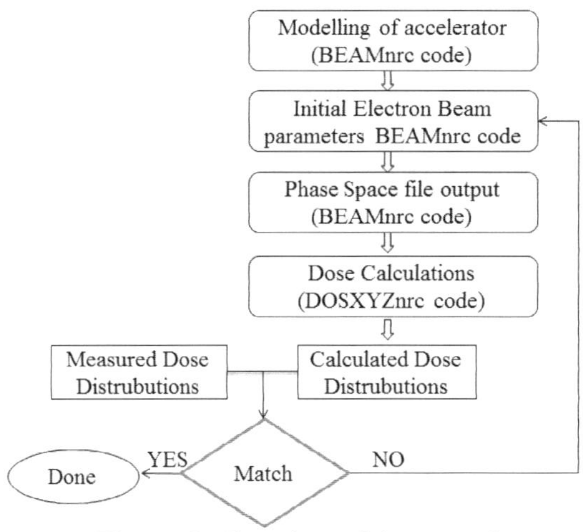


**Figure 3.** Flowchart of the procedure

Before impinging on the x-ray target the electrons were accelerated in the accelerator tube and then bent by magnet system to a net angle of $270^0$. The acceleration and bending process result in a beam of electrons which is heteroenergetic and has spatial distribution [4]. In this work, to simplify the problem, we assumed that the incident electron beam is monoenergetic. The determination of the incident electron beam parameters was an iterative process. Figure 3 shows a flowchart of the process. Initially, Monte Carlo simulations were executed for mono-energetic electron beams ranging from 5.7 to 6.2 MeV. The source routine ISOURC = 0 (Figure 4-a) for a parallel circular beam with a uniform radial intensity distribution, was used. The suitable mean energy was found by varying it in steps of 0.1 MeV until the PDD well matched the measurement [5]. Once the best suited energy was found, it was then applied for the source routine ISOURC = 19 (Figure 4-b), which describes a parallel circular beam with a 2-D Gaussian radial intensity distribution. The suitable FWHM of this beam was found by varying it from 1.0 mm to 1.4 mm and looking for a matching between calculating and measured OCR.

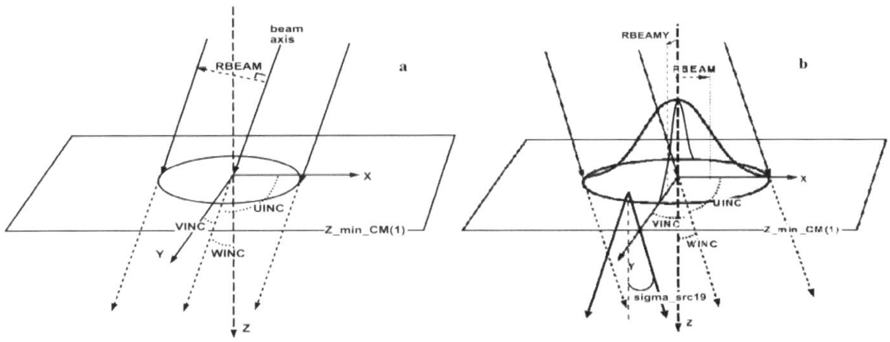


**Figure 4.** The source ISOURC = 0 (a) and the source ISOURC = 19 (b) [1]

### *2.3. Experimental measurements*

In our study, the measurements were performed on Siemens Primus linear accelerator. All the measurements were made using the IBA Dosimetry "blue phantom" scanning system, controlled by Omnipro-Accept 7.4c software. All measurements were done in water using the Scanditronix Wellhofer CC13 cylindrical ionization chamber having an active volume of 0.13 $cm^3$. A reference chamber was placed in the field to correct for beam output variations during scanning. Ranges of measurement, for depth dose curves; it was from a depth of -0.5 mm to a depth of 350 mm and for beam profiles, the off-axis distance was from -125 mm to +125 mm. All measurements were done in continuous scan mode, output step of 1.2 mm, scan speed of 15 mm/s.

### *2.4. Method to compare measured and simulated data*

The agreement between calculations and measurements was usually evaluated by calculating the mean point to point dose error, with flow formula [6]:

$$\Delta D = \frac{1}{N}\sum\left(\frac{|D_c - D_m|}{D_m}\right) \tag{1}$$

ΔD is the mean point-to-point error; N is the number of points, $D_c$ is the simulated dose and $D_m$ is the measured dose.

The evaluation of the result using the point to point errors could give rise to high overall errors in low dose areas and high dose gradient regions. Therefore, calculating the gamma index is an alternative method for this analysis recently [6, 7]. This method has become the gold standard for the comparison between measured and calculated dose distributions. The gamma index is defined for each measurement point $r_m$ as:

$$\gamma(r_m) = \min\{\Gamma(r_m, r_c)\}\forall(r_c) \tag{2}$$

Where

$$\Gamma(r_m, r_c) = \sqrt{\frac{r^2(r_m, r_c)}{\Delta d_M^2} + \frac{\delta^2(r_m, r_c)}{\Delta D_M^2}},\ r^2(r_m, r_c) = |r_m - r_c|, and\ \ \delta(r_m, r_c) = D_c(r_c) - D_m(r_m) \tag{3}$$

$r_m$ is the position at the measurement point and $r_c$ is the position at the calculation point.
$D_c(r_c)$ and $D_m(r_m)$ are calculation and measurement doses.
$\Delta d_M$ (mm) is the distance-to-agreement (DTA) criterion and $\Delta D_M$ (%) is the dose-difference criterion.
We calculated the number of points passing the 1%/1 mm criterion. In order to analysis gamma index, the PDD data from simulation as well as the corresponding beam profiles at various depth was incorporated in a text file which got from eclipse export option. The PDD file from measurement as well as various depths was converted to another text file. This two text files then were compared by ScanDoseMatch software to plot chart gamma index. On the other hand, the percentage passed gamma criterion is calculated by an in house MATLAB code.

## 3. Results and discussion

For the 6 MV photon beam from the Siemens Primus M5497 linac at Dong Nai Hospital, the best agreement between simulation and measured data was found for an electron beam of mono-energy 6.04 MeV with Gaussian radial spread of FWHM 1.2 mm

### *3.1. Percent Depth Dose (PDD)*

Depth dose curves in the 6 MV photon beam is normalized at 1.5 cm where the dose is at maximum. Figure 5 shows that the agreement between the measurement and simulation are excellent. The statistical uncertainties in the calculated depth doses were equal to or less than 0.7% of local value for all voxels at depths from 1.5 to 3 cm.

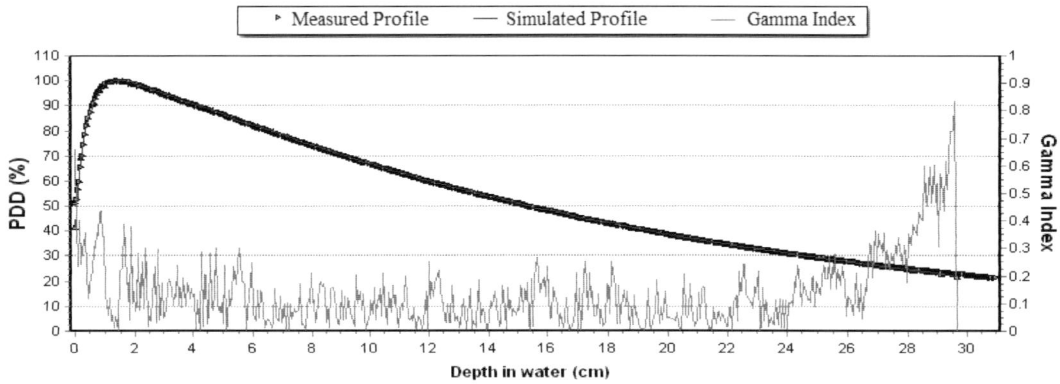


**Figure 5.** Comparisons of simulated with measured PDD

The percentage gamma passing rate (red line in figure 5) is 100% with 1% dose difference and 1 mm distance. In the buildup region and the tail of the PDD curves (at depth from 27 to 30 cm), where the PDD are low, the gamma index is much higher than the one in remaining regions. This observation can be explained based on the general law of statistical uncertainty in radiation counting: the lower the number of events the higher the relative uncertainty of measured and calculated counts, which results in the higher values of gamma index.

### *3.2. Beam profiles (the off-centre ratio-OCR)*

The beam profile at the depth of 10 cm for 10x10 $cm^2$ field is normalized at central axis.

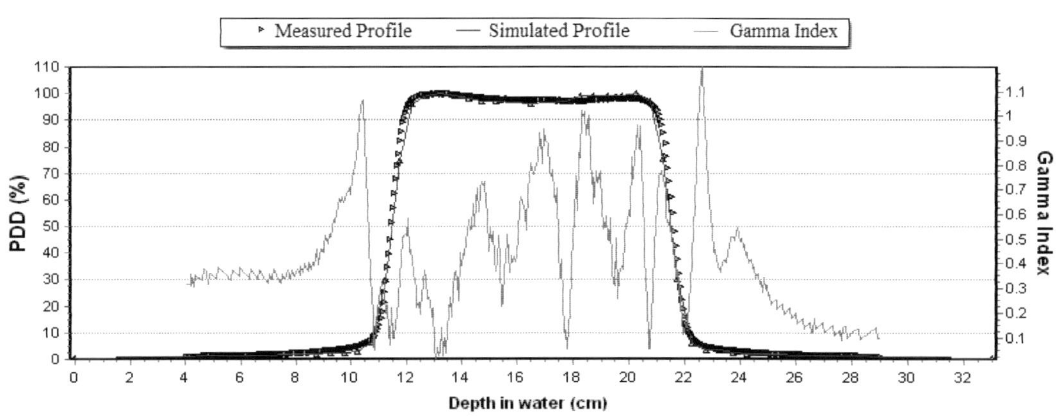


**Figure 6.** The 6MV photon beam profile in water for a 10 x 10 $cm^2$ field at 10 cm depth

Figure 6 shows very good matching between the simulations and the measurements. The average dose of beam profile flatness was less than 2%. The percentage gamma passing rate (red line) is 98% with 2% dose difference and 2 mm distance to agreement as acceptance criteria. In the figure 6 there are some points that gamma index is higher than 1, these points at the edge of penumbra region whose doses are very low. Thus it leads to high percentage difference between calculated and measured dose. The same observation was reported in [8].

## 4. Conclusion

In this work, two parameters of the incident electron beams for the 6 MV photon beam from the Siemens Primus M5497 linac at Dong Nai Hospital have been determined an electron beam with mean energy of 6.04 MeV and an FWHM of radial spread of 1.2 mm was found to give best fit between calculated and measured PDD and OCR.

## 5. References

[1] D. W. O. Rogers, B. Walters and I. Kawrakow 2005, BEAMnrc Users Manual, National Research Council of Canada Report, PIRS –0509a, Ottawa: NRC.
[2] B. Walters, I. Kawrakow and D. W. O. Rogers 2004, DOSXYZnrc User's Manual, National Research Council of Canada Report, PIRS -794, **RevB**, Ottawa: NRC.
[3] D. Sheikh Bagheri, D. W. O. Rogers 2002, Monte Carlo calculation of nine megavoltage photon beam spectra using the BEAM code, Medical Physics, **29(3)**, 391-402.
[4] C. J. Karzmack et al 1993, Medical Electron Accelerators, McGraw-Hill.
[5] K. Aljarrah, C. Greg Sharp, Toni Neicu, and B. Steve Jiang 2005, Determination of the initial beam parameters in Monte Carlo linac simulation, Medical Physics, **32**, 3579.
[6] A. Low Daniel, B. William Harms, Sasa Mutic, and A. James Purdy 1998, A technique for the quantitative evaluation of dose distributions, Medical Physics, **25(5)**.
[7] M. Aljamal and A. Zakaria 2013, Monte Carlo Modeling of a Siemens Primus 6 MV Photon Beam Linear Accelerator , Australian Journal of Basic and Applied Sciences 7(**10**), 340-346.
[8] J. Hrbacek, Tom Depuydt, An Nulens, Ans Swinnen, F. V. Den Heuvel 2007, Quantitative evaluation of a beam-matching procedure using one-dimensional gamma analysis, Medical Physics, **34 (7)**, 2917–2927.

# Contributions of secondary fragmentation by carbon ion beams in water phantom: Monte Carlo simulation

**C K Ying[1], David Bolst[2], Linh T. Tran[2], Susanna Guatelli[2], A. B. Rosenfeld[2], W A Kamil[3]**

[1]Advanced Medical and Dental Institute, AMDI, Universiti Sains Malaysia, Bertam 13200, Kepala Batas, Penang, Malaysia
[2]Centre of Medical Radiation Physics, University of Wollongong, NSW 2522, Australia
[3]School of Medical Sciences, Universiti Sains Malaysia, Kota Bharu 16150, Malaysia

E-mail: ckying7@usm.my

**Abstract**. Heavy-particle therapy such as carbon ion therapy is currently very popular because of its superior conformality in terms of dose distribution and higher Relative Biological Effectiveness (RBE). However, carbon ion beams produce a complex mixed radiation field, which needs to be fully characterised. In this study, the fragmentation of a 290 MeV/u primary carbon ion beam was studied using the Geant4 Monte Carlo Toolkit. When the primary carbon ion beam interacts with water, secondary light charged particles (H, He, Li, Be, B) and fast neutrons are produced, contributing to the dose, especially after the distal edge of the Bragg peak.

## 1. Introduction

Heavy Ion Therapy (HIT) consists of the use of carbon ions to destroy cancer. The main advantage of HIT with respect to proton therapy and X-ray radiotherapy is better conformality to the target, capability to treat deep-seated tumors and a higher Relative Biological Effectiveness (RBE) [1].

When a primary carbon ion is passing through the medium with energy of several hundred of MeV/u, the radiation field is affected by nuclear fragmentation. Nuclear fragments contribute to the dose deposition and are responsible for the dose tail observed after the distal edge of the Bragg Peak [2-3]. Therefore it is crucial to characterise the components of the radiation field produced by the incident carbon ion, in terms of both spatial distribution and LET, as particles with higher LET have a higher probability to kill cancerous cells.

When the primary nuclei are in tissue-like media such as water or PMMA phantom at several hundred MeV, peripheral collisions are the most frequently occurring nuclear reactions. These reactions result in fragmentation of both target and projectile nuclei. This can be described by abrasion-ablation model with two step processes [4-5]. In the first step of peripheral collisions, the incident beam particle may lose one or several nucleons along the penetration depth and the nucleons in the overlapping zone of the incident projectile and target nuclei are abraded, resulting in the formation of a hot zone (fireball). In this process, outer nucleons called "spectators" are only slightly affected.

In the second step (ablation), the remaining projectile and target fragments as well as fireball de-excite with the evaporation of nucleons and light clusters. The produced fragments

IOP Conf. Series: Journal of Physics: Conf. Series **851** (2017) 012033 doi:10.1088/1742-6596/851/1/012033

from the projectile move in the direction of the incident projectile with equal or higher velocity to that of the incident projectile, contribute to the dose deposition until they are completely slowed down. The target fragments have velocities much smaller than that of the projectile velocity, fragments from the fireball have a velocity in the range between target emission and projectile fragments. Nuclear fragmentation reactions lead to an attenuation of the primary beam flux and build-up of low mass fragments which have longer ranges than the primary ions and carry large fractions of the original ion's energy enabling the produced fragments to travel a longer distance of the Bragg peak position. Therefore, the depth dose profile of carbon ion beams shows the characteristic fragment tail beyond the Bragg peak.

### *1.1 Geant4 Simulation toolkit*
GEANT4 (GEometry ANd Tracking) Monte Carlo code [6-7] is a toolkit for the simulation of the passage pf particles through matter. Geant4 is used because of its powerful abilities to simulate all type of particles, greater flexibility, can simulating various geometric variations and different types of physical interactions process and it became the powerful tools for the study of interaction in the fields of medical physics, radiation protection, high energy physics and nuclear physics etc. Many literatures show the Geant4 studies based on Geant4 simulations had good agreement with the experimental results [8-11].

## 2. Materials and Methods

### *2.1. The Geant4 simulation application*
An ad-hoc Geant4 simulation was used to model the radiation field produced by a 290 MeV/u carbon ion beam in a Polymethyl Methacrylate (PMMA) phantom. The beam lateral size was 50 mm. The ion beam line was generated at a distance of 30 cm from the surface of the phantom. The sketch of the beam line is shown in the Figure 1(a). The dose along the phantom depth for each type of nuclear fragment was calculated. The characteristics of the radiation field produced by incident carbon ions were studied as well.

The Geant4 QGSP_BIC_EMY Reference Physics List was adopted. The production of secondary particles was governed by a range cut which was set to 0.1 mm. A total number of $5 \cdot 10^5$ events was simulated in order to obtain an adequate statistics. Figure 1(b) shows a snapshot of the simulation. The incident carbon ions (blue tracks) are incident on the phantom and produce gamma and neutrons (green tracks), positively charged nuclear fragments (blue tracks) and electrons (red tracks). Geant4 version 4.9.6.p01 was used.

IOP Conf. Series: Journal of Physics: Conf. Series **851** (2017) 012033 doi:10.1088/1742-6596/851/1/012033

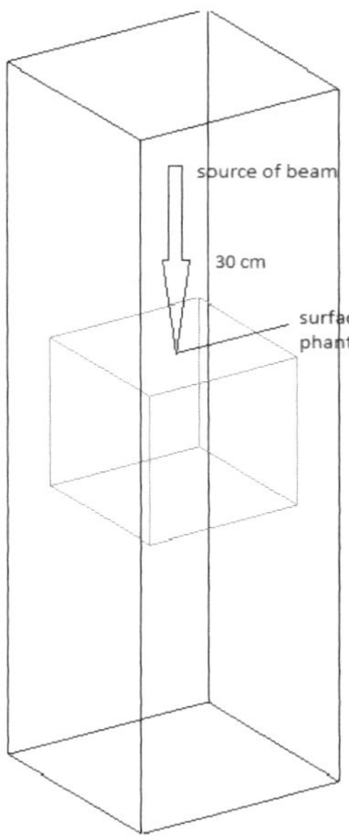


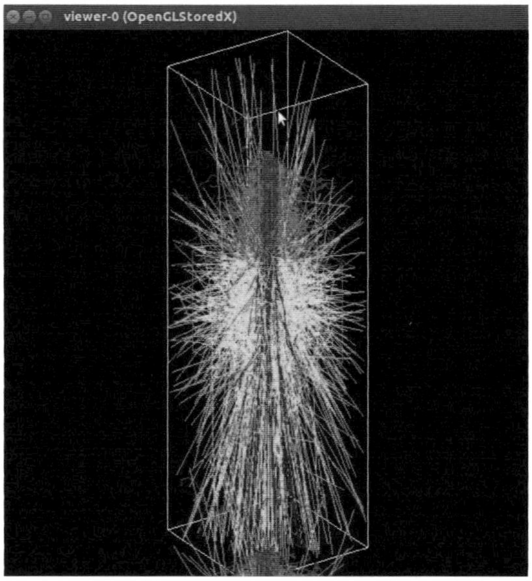


**Figure 1.** (a) Sketch of the simulation experimental set-up and (b) snapshot of the Geant4 simulation. The carbon ion beam (blue tracks) coming from the top interacts with the PMMA target, producing secondary particles. Red and blue tracks are negative and positive charged particles, respectively. Green tracks are neutral particles.

The characteristics of the radiation field produced by the incident carbon ions were studied in the PMMA phantom. The output of the simulation consisted of the energy deposition in the PMMA phantom as well as the position of secondary particles generated within the phantom. The energy deposition derived from the incident beam and from the secondary nuclear fragments was tallied separately. The Bragg peak was calculated along the direction of the incident beam with 0.1 mm spatial resolution. The deposited energy at a given depth and lateral position was stored in the 2D histogram which had 1 $mm^2$ pixels.

## 3. Results and Discussion

### *3.1. Contributions of charged fragments & LET calculations*

Figure 2 (a) shows the Bragg Peak. The energy deposition contributions deriving from incident carbons and secondary fragments, such Hydrogen, Helium, Lithium, Beryllium, Boron, secondary Carbon ions, Nitrogen and Oxygen, are shown as well. The Bragg peak was observed to occur at (126.2 ± 0.1) mm in the PMMA phantom which is in good agreement with the expected results calculated by SRIM [12]. Figure 2 (b) shows the depth dose distribution deriving from secondary fragments only and the Figure is close up of Figure 2 (a).

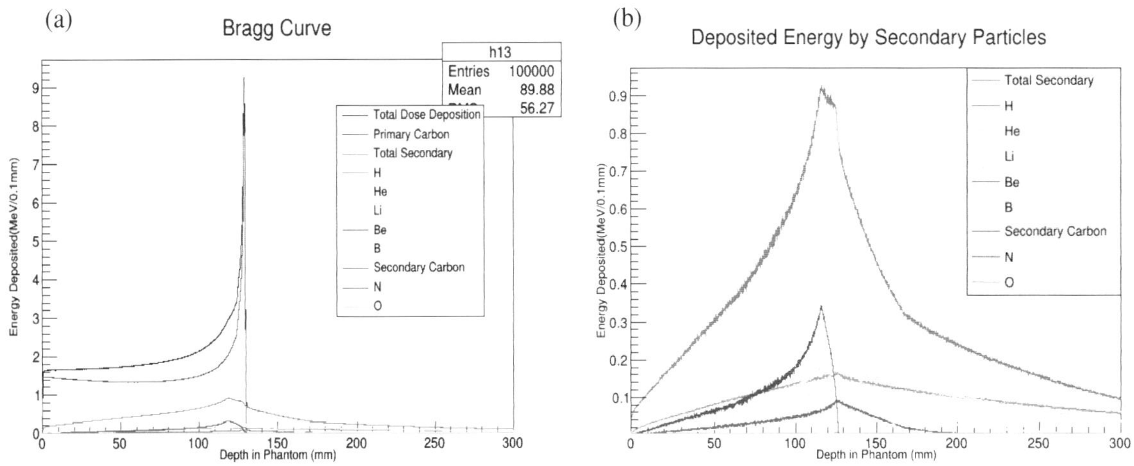


**Figure 2.** (a) Depth dose distribution deriving from the entire radiation field (Bragg curve) in black, and (b) energy deposition deriving from different components of secondary fragments only

Figure 2 (a) clearly shows that the main contribution in the energy deposition is mainly due to the primary carbon ions. Secondary carbon ions result mostly from neutron scattering in PMMA, although the contributions of the total secondary particles become more evident beyond the distal part of the Bragg peak. In the overall plateau region, the total dose of secondary particles contributed less than 1-3% on the total deposited dose registered at the same position and about 0.5% at the peak position. Therefore, the total depth dose deposition (Bragg curve, in black) and primary carbon (in light blue) almost coincide up to the peak position. The contribution of total depth dose deposition just behind the Bragg peak is only due to the secondary particle (100%). This is because the primary carbon beam energy drops to zero drastically.

As shown in Figure 2(b), B fragments are the main contributors to the dose at 0.1 – 10 mm behind Bragg peak, followed by He and H fragments. He and H fragments have the longest range and therefore they contribute to the long energy deposition tail, while heavier fragments contribute to the dose more locally. Secondary neutrons scatter protons in the phantom which again contribute to the dose beyond the Bragg Peak and laterally. The energy deposition peak of the secondary carbon and B ions are evidently larger than the other fragments. Contributions of other fragments, such as N and O ions, are not significant. The lower-Z fragments, Hydrogen (H), Helium (He) and Lithium (Li) have longer ranges than the primary particles. This is the reason of the depth dose profile of heavy ion beams have characteristic fragment tail beyond the Bragg peak.

The fragment tail clearly indicates that the largest fraction of fragmentation reactions result in a small number of nucleons from either the target or projectile nucleus being ejected in "abrasion-ablation" reactions. It is confirmed that this is the origin of the characteristic fragment tail beyond the distal edge of the Bragg peak when heavy ions such as carbon ions are used [13]. This work also shows the similar output with the other literatures for the secondary fragmentation study [14-15].

Figure 3 shows the 2D lateral dose distribution of a set of different fragments. It can be observed that the lighter ions undergo greater lateral scattering and therefore deposit more energy out of the primary radiation field, while the heavier nuclear fragments contribute more substantially to the energy deposition at the distal edge of the Bragg peak.

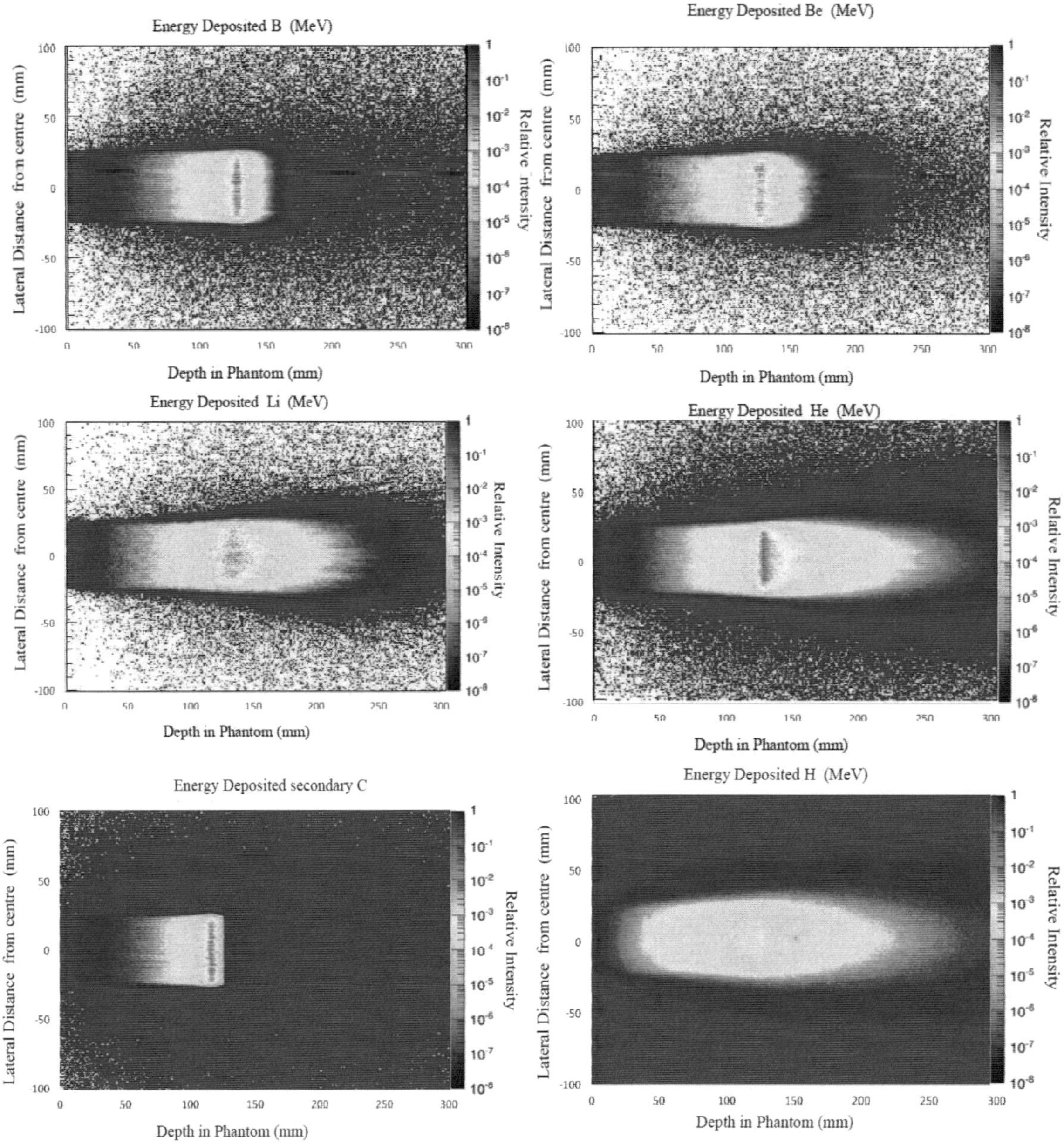


**Figure 3**. 2D histogram of the energy deposited in the phantom different fragments.

## 4. Conclusion

The contribution of total depth dose deposition just behind Bragg peak is 100% due to the secondary particle field. Such energy deposition tail cannot be neglected as organs at risk may be located in this position behind the Bragg Peak. For further study, higher therapeutic beam energies are planned to investigate the secondary fragments especially in the regions outside of the field.

IOP Conf. Series: Journal of Physics: Conf. Series **851** (2017) 012033 doi:10.1088/1742-6596/851/1/012033

## 5. References

[1] IAEA 2008 Relative biological effectiveness in ion beam therapy. (Technical Reports Series) No. 461

[2] Matsufuji N, Kanai T, Kanematsu N, Miyamoto T, Baba M, Kamada T, Kato H, Yamada S, Mizoe J E and Tsujii H 2007 Specification of Carbon Ion Dose at the National Institute of Radiological Sciences (NIRS) *J. Radiat. Res*: **48**; Suppl. A: A81-6.

[3] Matsufuji N, Fukumura A and Komori M *et al.* 2003 Influence of fragment reaction of relativistic heavy charged particles on heavy-ion radiotherapy *Phys. Med. Biol.*, **48**, pp. 1605-1623

[4] Serber R 1947 Nuclear reactions at high energies, Phys. Rev: **72**; 1114–1115.

[5] Gunzert-Marx K, Iwase H, Schardt D, Simon R S 2008 Secondary beam fragments produced by 200 MeVu-1 12C ions in water and their dose contributions in carbon ion radiotherapy, New J. Phys:**10**; 1–21 (075003).

[6] Allison J and Apostolakis J. *et al.* 2006 Geant4 Collaboration. Geant4 Developments and Application *IEEE Transactions on Nuclear Science*: vol. **53** (1) pp. 270-278.

[7] Agostinelli S, Allison J and Amako K *et al.* 2003 Geant4 - a simulation toolkit. *Nucl. Instr. And Meth. in Phys. Res. A*. **506**, 250-30

[8] Apostolakis J, Asai M, Bogdanov A G *et al.* 2009 Geometry and physics of the Geant4 toolkit for high and medium energy applications. Workshop on use of Monte Carlo techniques for design and analysis of radiation detectors. Radiat. Phys. Chem. **78**, 859–873

[9] Lechner A, Ivanchenko V N and Knobloch, J. 2010 Validation of recent Geant4 physics models for application in carbon ion therapy. Nucl. Instrum. Meth. B. **268**, 2343–2354

[10] Toshito T, Bagulya A, Lechner A *et al.* 2011 Validation of New Geant4 Electromagnetic Physics Models for Ion Therapy Applications. Progress in Nuclear Science and Technology, Vol. **2**, pp.918-922

[11] Kameoka S, Amako K, Iwai G *et al.* 2008 Dosimetric evaluation of nuclear interactionmodels in the Geant4 Monte Carlo simulation toolkit for carbon-ion radiotherapy. Radiol Phys Technol
1:183–187

[12] Ziegler J F 2008 Particle Interaction with Matter. SRIM [Online]. Available From World Wide Web: http://www.srim.org

[13] Pshenichnov I, Botvina A, Mishustin I and Greiner W 2010 Nuclear fragmentation reactions in extended media studied with Geant4 toolkit. Nucl. Instr. and Meth. in Phys. Res. B: **268**; 604-615

[14] Gunzert-Marx K, Iwase H, Schardt D, Simon R S 2008 Secondary beam fragments produced by 200 MeVu-1 12C ions in water and their dose contributions in carbon ion radiotherapy, New J. Phys:**10**; 1–21 (075003)

[15] Francis Z, Seif E, Incerti S *et al.* 2014 Carbon ion fragmentation effects on the nanometric level behind the Bragg peak depth. Physics in Medicine and Biology: **59**; 7691-7702

## Acknowledgements

The author would like to thank Dr. Susanna Guatelli and Prof. Anatoly Rosenfeld from Center of Medical Radiation Physics, University of Wollongong for their Geant4 support, constant guidance and support for the study.

# The thermoluminescence response of Ge-doped flat fibre for proton beam measurements: A preliminary study

**M F Hassan[1], W N Wan Abdul Rahman[2], M S Ahmad Fadzil[1], T Tominaga[3], M Geso[4], H Akasaka[5], D A Bradley[6,7] and N Mohd Noor[1]**

[1] Department of Imaging, Faculty of Medicine and Health Sciences, Universiti Putra Malaysia, 43400 Serdang, Selangor, Malaysia
[2] School of Health Sciences, Universiti Sains Malaysia, 16150 Kubang Kerian, Kelantan, Malaysia
[3] Department of Clinical Radiology, Faculty of Health Sciences, Hiroshima International University, Higashi-Hiroshima City, Hiroshima 739-2695, Japan
[4] Discipline of Medical Radiations, School of Medical Sciences, Royal Melbourne Institute of Technology University, Bundoora, Victoria 3083, Australia
[5] Division of Radiation Oncology, Kobe University Hospital, Kobe City, Hyogo 650-0017, Japan
[6] Department of Physics, University of Surrey, Guildford, GU2 7XH, United Kingdom
[7]Sunway University Institute for Health Care Development, Jalan Universiti, 46150 Petaling Jaya, Selangor, Malaysia.

E-mail: noramaliza@upm.edu.my

**Abstract**. The aim of this study was to investigate the thermoluminescence (TL) response of fabricated 2.3 mol% and 6.0 mol% germanium (Ge) doped flat optical fibres to proton irradiation. The fundamental dosimetric characteristics of the fibres have been investigated including dose linearity, reproducibility and fading. The thermoluminescent dosimeters (TLDs) were used as a reference dosimeter to allow the relative response of the fibres. The results show that Ge-doped flat fibres offer excellent dose linearity over the dose range from 1 Gy up to 10 Gy with correlation of determination ($R^2$) of 0.99. The fibres also demonstrated good reproducibility within the standard deviation (SD) of 0.86% to 6.41%. After 96 days post-irradiation, TLD-100 chips gave rise to the least loss in TL signal at around 18% followed by fabricated 2.3 mol% Ge-doped flat fibres about 24%. This preliminary study has demonstrated that the proposed fabricated Ge-doped flat fibre offers a promising potential for use in proton beam measurements.

## 1. Introduction

Protons were first employed in medical treatment in 1946 [1] with the first attempts to treat patients began in 1954 [2]. Since then, proton beam therapy provides an advanced level of radiation treatment and is widely used nowadays in treating cancers. In radiotherapy, the reason of using proton beams instead of photon or electron beam is due to their physical characteristics of the Bragg curve. Proton beams loss their energy rapidly in the last few millimeters of penetration which results in a very precise localized dose peak at a target area, so-called Bragg peak. Changing the Bragg peak depth allows the desired dose to be sharply placed at any point in the patient. However, small changes can make a large difference on accuracy of the dose delivered, particularly in the distal fall-off region [3].

Therefore, verification of dose distribution in proton therapy before the treatment is very important to avoid any radiobiological effects to the patient. To achieve this, there is a need for accurate dosimetry systems which able to verify the radiation dose given to the patient. Over the past few years, a number of research groups have attempted to evaluate the use of doped silica optical fibre in photons [4-7], diagnostic X-rays [8-9], electrons [10], alpha particles [11], fast neutrons [12], and synchrotron radiations [13]. In all such studies, the studied optical fibres have shown considerable potential to be developed as radiation dosimeter in radiotherapy.

Although the dosimetric characteristics of the doped silica optical fibre for various types of irradiations have been reported in the literatures, their investigation in the proton beam has not yet been extensively studied. In the present study, we investigated the dosimetric characteristics of the proposed fabricated germanium (Ge) doped flat optical fibre with respect to dose linearity, reproducibility and signal fading for the proton beam and to compare it with TLD-100 chips.

## 2. Materials and methods

### *2.1. Fabrication of Ge-doped flat optical fibre*

The tailor-made flat optical fibres studied herein were fabricated to produce 2.3 mol% and 6.0 mol% Ge-doped flat fibres with dimension of 643 × 356 $\mu m^2$ and 272 × 69.5 $\mu m^2$ respectively. The fibre preform was fabricated using modified chemical vapour deposition (MCVD) method at MCVD Laboratory in Telekom Research & Development (TM R&D) Sdn. Bhd., Cyberjaya, Malaysia with subsequent fibre pulling process at Flat Fibre Laboratory, Department of Electrical Engineering, Faculty of Engineering in University of Malaya (UM), Malaysia [14]. Figure 1 shows a cross-sectional image of the selected flat fibre and distribution of the germanium at the center of the fibre core.

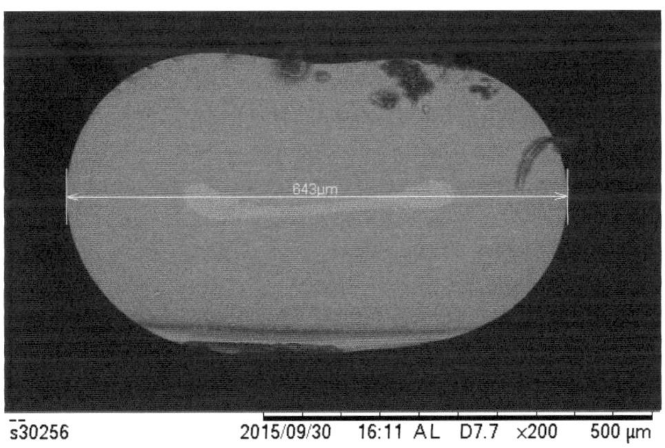


**Figure 1.** Cross-sectional image of a selected fabricated Ge-doped flat optical fibre obtained from a scanning electron microscope (SEM) analysis.

### *2.2. Preparation of samples*

The fabricated flat fibres were cut using a diamond cutter (Thorlabs, USA) into 6.0 ± 1.0 mm length to easily accommodate within the planchet (the heating plate of the TLD reader). A vacuum tweezer (Dymax 5, UK) was used to minimise surface abrasion and deposition of dust or finger oil to the samples during handling. Prior to irradiation, the unscreened fibres were annealed using a furnace (Carbolite, UK) at temperature of 400 °C for one hour to erase any pre-irradiation TL signals. Whilst, TLD-100 chips (Thermo Fisher Scientific, USA) with dimension of 3.2 mm × 3.2 mm × 0.89 mm were annealed using a TLD annealing oven for one hour at 400 °C and subsequently 16 hours at 80 °C [15]. Following annealing procedures, the fibres and TLD-100 chips were retained inside their furnaces to finally cool down to reach room temperature to avoid thermal stress. The samples were kept in a light-tight box at room temperature to minimise exposure to ambient light prior to and following irradiation as it may affect the TL readings.

### *2.3. Irradiation setup*

A total of five units of flat fibres and three units of TLD-100 chips were encased into their respective radiolucent gelatin capsules, used to provide homogeneity in the irradiation. The samples were placed horizontally in between water-equivalent slab phantoms and perpendicularly to the beam axis as shown in figure 2. A slab equivalent in thickness to maximum dose ($d_{max}$) was placed on top of the sample arrangement and another 10 cm of solid water was placed below the sample to provide the build-down for backscatter attenuation. The experiment was carried out based on the clinical treatment source-to-surface distance (SSD) of 100 cm and beam field size of 10 × 10 cm$^2$. Table 1 shows a summary of irradiation settings for proton, gamma, photon and electron beam used in this study.

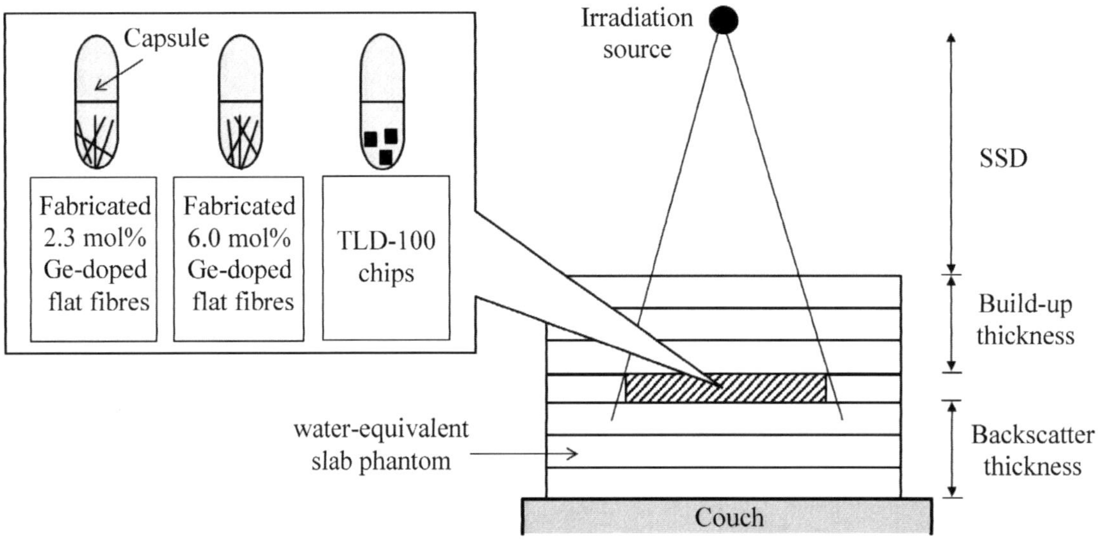


**Figure 2.** Schematic illustration of the experimental setup for sample irradiation.

**Table 1.** Summary of irradiation settings for proton, gamma, photon and electron beam.

| | Proton | Gamma | Photon | Electron |
|---|---|---|---|---|
| Energy | 150 and 210 MeV | 1.25 MeV | 6 MV | 9 MeV |
| Dose | 1-10 Gy | 5 Gy | 5 Gy | 5 Gy |
| Dose rate | 4 Gy/min | 0.0427 Gy/min | 600 MU/min | 600 MU/min |
| Build-up thickness | 14 cm (for 150 MeV) and 25 cm (210 MeV) | Nil | 1.5 cm (bolus) | 2 cm (bolus) |
| Applicator | Nil | Nil | Nil | 10 x 10 cm$^2$ |

*2.3.1. Dose linearity.* The measurement of dose response for proton beam was evaluated over a dose range from 1 up to 10 Gy at a constant energy of 150 MeV.

*2.3.2. Reproducibility.* The samples were irradiated with proton, gamma, photon and electron beam at energy of 210 MeV, 1.25 MeV, 6 MV and 9 MeV respectively at a constant dose of 5 Gy. The same set of samples was reused, subsequent to re-annealing for each irradiation.

*2.3.3. Signal fading.* The samples were irradiated with proton beam at 150 MeV for 5 Gy and readout at day 3, 4, 7, 13, 20, 26, 34, 59 and 96. Throughout the storage period, the samples were stored in a light-tight box at room temperature until readout to avoid thermally stimulated release of trapped electrons.

### *2.4. Readout*

A Harshaw TLD™ Model 3500 (Thermo Fisher Scientific, USA) reader was used to measure the TL signals from the flat fibres and TLD-100 chips, with nitrogen gas atmosphere at 0.5 bar to supress spurious light signals from triboluminescence and minimise surface oxidation. The flat fibres were readout with the time-temperature profile (TTP) at a preheat of 120 °C, acquired at a temperature ramp-rate of 30 °C $s^{-1}$, with an acquisition time of 13 seconds and maximum temperature of 400 °C. Whilst, the TTP for TLD-100 chips was set to preheat at 145 °C for 10 seconds and a temperature ramp-rate of 17 °C $s^{-1}$ up to a maximum temperature of 300 °C for 10 seconds [5].

## 3. Results and discussion

### *3.1. Dose linearity*

In figure 3, the fabricated 2.3 mol% and 6.0 mol% Ge-doped flat fibres were found to show a good linearity with correlation of determination ($r^2$) of 0.9911 and 0.9912 respectively over the investigated dose range. In line with expectation, TLD-100 chips show excellent and relatively flat TL response with $r^2$ of 0.9962 as compared to the flat fibres.

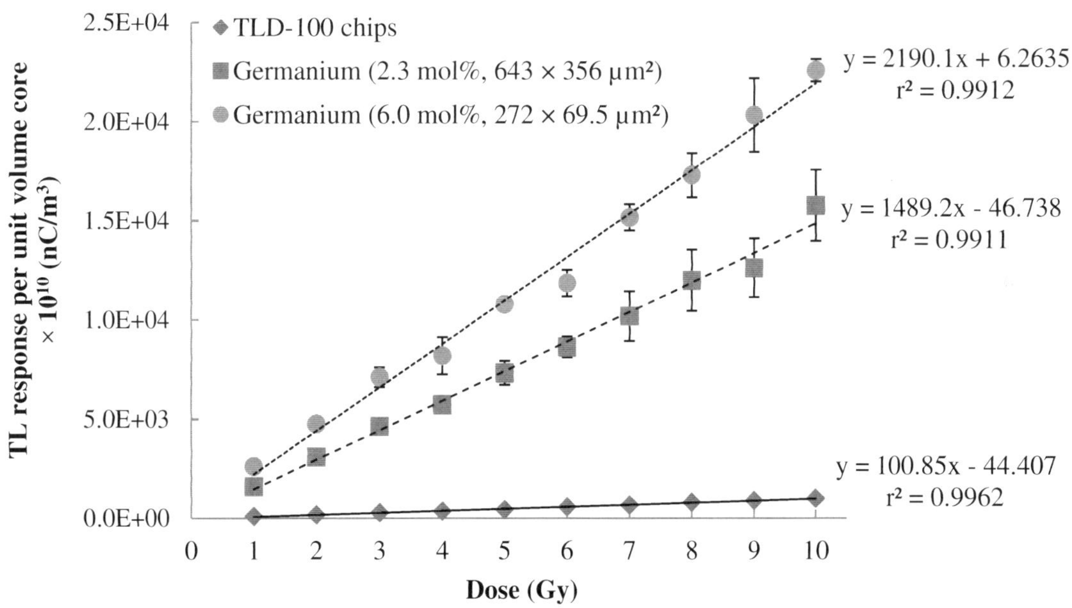


**Figure 3.** Dose linearity of Ge-doped flat fibres compared to that of TLD-100 chips.

### *3.2. Reproducibility*

In table 2, it was apparent that both types of flat fibres show good reproducibility response with the standard deviation (SD) to be within 0.86% to 6.41%, whilst the TLD-100 chips show reproducibility to be always better than 1.06%. The SD value for 6.0 mol% Ge-doped flat fibres (6.41%) obtained in this study is better than the results reported by a previous study which is 7.56% [7].

**Table 2.** Reproducibility response of Ge-doped flat fibres and TLD-100 chips at dose of 5 Gy.

| Sample | Irradiation source | TL response per unit volume × $10^{10}$ (nC/m³) | | |
|---|---|---|---|---|
| | | Mean | SD | SD (%) |
| TLD-100 chips | 1.25 MeV gamma | 379.98 | 4.03 | 1.06 |
| | 9 MeV electron | 376.85 | 3.03 | 0.80 |
| | 210 MeV proton | 420.31 | 2.87 | 0.68 |
| | 6 MV photon | 399.40 | 0.16 | 0.04 |
| 2.3 mol% Ge-doped flat fibre | 1.25 MeV gamma | 89.83 | 2.77 | 3.08 |
| | 9 MeV electron | 119.14 | 1.25 | 1.05 |
| | 210 MeV proton | 125.25 | 1.08 | 0.86 |
| | 6 MV photon | 129.83 | 1.45 | 1.11 |
| 6.0 mol% Ge-doped flat fibre | 1.25 MeV gamma | 106.72 | 6.84 | 6.41 |
| | 9 MeV electron | 151.95 | 7.13 | 4.69 |
| | 210 MeV proton | 126.65 | 6.43 | 5.07 |
| | 6 MV photon | 174.09 | 7.14 | 4.10 |

### *3.3. Signal fading*

Over 96 days post-irradiation, TLD-100 chips gave rise to the least loss in TL signal at around 18% followed by 2.3 mol% Ge-doped flat fibres about 24% as shown in figure 4. The largest signal loss was suffered by 6.0 mol% Ge-doped flat fibres at around 58%.

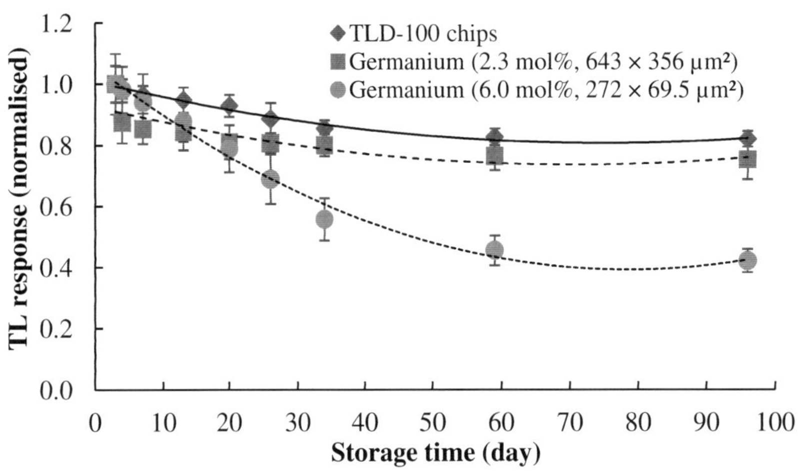


**Figure 4.** The loss of TL signals after normalised to day three post-irradiation.

## 4. Conclusion and future work

Results of the study represented several important key dosimetric characteristics of novel fabricated Ge-doped flat fibres including dose linearity, reproducibility and signal fading. Apparently, flat fibre with 2.3 mol% Ge-doped shows promising dosimetric characteristics as compared to 6.0 mol% Ge-doped. This is mainly due to 2.3 mol% Ge-doped flat fibres have better dose response, good reproducibility with SD better than 3.08%, and least signal loss about 24%. This preliminary study has demonstrated that the proposed fabricated Ge-doped flat fibre offers a promising potential for use in proton beam measurements. In future work, we suggest studying other dosimetric characteristics in more details including measurement of depth-dose distribution of the fabricated fibres in proton beam.

## 5. References

[1] Wilson R R 1946 Radiological use of fast protons *Radiology* **47** 487

[2] Tobias C A, Lawrence J H, Born J L, McCombs R K, Roberts J E, Anger H O, Low-Beer B V A and Huggins C B 1958 Pituitary irradiation with high-energy proton beams: A preliminary report *Cancer Research* **18** 121
[3] Rah J E, Oh D H, Shin D, Kim D H, Ji Y H, Kim J W and Park S Y 2012 Dosimetric evaluation of a glass dosimeter for proton beam measurements *Applied Radiation and Isotopes* **70** 1616
[4] Abdulla Y A, Amin Y M and Bradley D A 2001 The thermoluminescence response of Ge-doped optical fibres subjected to photon irradiation *Radiation Physics and Chemistry* **61** 409
[5] Mohd Noor N, Hussein M, Bradley D A and Nisbet A 2010 The potential of Ge-doped optical fibre TL dosimetry for 3D verification of high energy IMRT photon beams *Nuclear Instruments and Methods in Physics Research* A **619** 157
[6] Mohd Noor N, Bradley D A and Nisbet A 2011 An investigation of the suitability of Ge-doped optical fibres in mailed thermoluminescence dosimetry audits of radiotherapy dose delivery *Radiotherapy and Oncology* **99** S196
[7] Ahmad Fadzil M S, Ramli N N H, Jusoh M A, Kadni T, Bradley D A, Ung N M, Suhairul H and Mohd Noor N 2014 Dosimetric characteristics of fabricated silica fibre for postal radiotherapy dose audits *Journal of Physics: Conference Series* **546** 012010
[8] Ramli N N H, Salleh H, Mahdiraji G A, Zulkifli M I, Hashim S, Bradley D A and Mohd Noor N 2015 Characterization of amorphous thermoluminescence dosimeters for patient dose measurement in X-ray diagnostic procedures *Radiation Physics and Chemistry* **116** 130
[9] Issa F, Latip N A A, Bradley D A and Nisbet A 2011 Ge-doped optical fibres as thermoluminescence dosimeters for kilovoltage X-ray therapy irradiations *Nuclear Instruments and Methods in Physics Research* A **652** 834
[10] Entezam A, Khandaker M U, Amin Y M, Ung N M, Bradley D A, Maah J, Safari M J and Moradi F 2016 Thermoluminescence response of Ge-doped cylindrical-, flat- and photonic crystal silica-fibres to electron and photon radiation *PLoS One* **11** 1
[11] Ramli A T, Bradley D A, Hashim S and Wagiran H 2009 The thermoluminescence response of doped SiO2 optical fibres subjected to alpha-particle irradiation *Applied Radiation and Isotopes* **67** 428
[12] Hashim S, Bradley D A, Saripan M I, Ramli A T and Wagiran H 2010 The thermoluminescence response of doped SiO2 optical fibres subjected to fast neutrons *Applied Radiation and Isotopes* **68** 700
[13] Abdul Rahman A T, Bradley D A, Doran S J, Thierry B, Bräuer-Krisch E and Bravin A 2010 The thermoluminescence response of Ge-doped silica fibres for synchrotron microbeam radiation therapy dosimetry *Nuclear Instruments and Methods in Physics Research* A **619**
[14] Dambul K D, Tamchek N, Sandoghchi S R, Abu Hassan M R, Tee D C and Mahamd Adikan F R 2011 Fabrication and characterization of flat fibers *2011 IEEE 2nd Int. Conf. on Photonics* 17-19 October 2011 Kota Kinabalu, Malaysia
[15] Mohd Noor N, Shukor N A, Hussein M, Nisbet A and Bradley D A 2012 Comparison of the TL fading characteristics of Ge-doped optical fibres and LiF dosimeters *Applied Radiation and Isotopes* **70** 1384

**Acknowledgements**
The authors are highly indebted to Nik Nur Hikmah Ramli who helped in data collection, individuals from MCVD Laboratory at TM R&D Sdn. Bhd., Malaysia as well as Flat Fibre Laboratory in University of Malaya, Malaysia for their technical assistance during fabrication and pulling process of the optical fibres. The authors also would like to express their gratitude to staffs at Hyogo Ion Beam Medical Center, Japan for their assistance for proton irradiations, and Malaysian Nuclear Agency for providing their facilities in this study. This research is supported by Fundamental Research Grant Scheme (FRGS, No. 5524789) from Ministry of Education, Malaysia.

# Feasibility of using the linac real-time log data for VMAT treatment verification

**N S Midi and Hafiz M Zin**

Advanced Medical and Dental Institute, Universiti Sains Malaysia, Bertam, 13200, Kepala Batas, Penang, Malaysia.

E-mail: nshaheeramidi@yahoo.com

**Abstract**. This study investigates the feasibility of using the real-time log data from a linac to verify Volumetric Modulated Arc Therapy (VMAT) treatment. The treatment log data for an Elekta Synergy linac can be recorded at a sampling rate of 4 Hz using the service graphing tool on the linac control computer. A treatment plan that simulates a VMAT treatment was delivered from the linac and all the dynamic treatment parameters including monitor unit (MU), Multileaf Collimator (MLC) position, jaw position, gantry angle and collimator angle were recorded in real-time using the service graphing tool. The recorded raw data were extracted and analysed using algorithms written in Matlab (MathWorks, Natick, MA). The actual treatment parameters logged using the service graphing tool was compared to the prescription and the deviations were analysed. The MLC position errors travelling at the speed range from -3.25 to 5.92 cm/s were between -1.7 mm to 2.5 mm, well within the 3.5 mm tolerance value (AAPM TG-142). The discrepancies of other delivery parameters were also within the tolerance. The real-time linac parameters logged using the service graphing tool can be used as a supplementary data for patient specific VMAT pre-treatment quality assurance.

## 1. Introduction

Advanced radiotherapy treatment such as Intensity Modulated Radiation Therapy (IMRT) and Volumetric Modulated Arc Therapy (VMAT) allows conformity of radiation beam to the tumour using intensity modulated beams. The conformal beam is delivered using simultaneous movement of the dynamic multileaf collimator (MLC), the linac gantry and the collimator angle while the dose rate varied during irradiation [1]. The complexity of the MLC delivery makes IMRT and VMAT prone to delivery errors and requires patient specific treatment plan verification. Treatment verification using phantoms and clinical detectors requires difficult calibrations and is prone to errors if performed incorrectly. Alternatively, the MLC positions and other machine parameters are logged by the linac during treatment delivery may be used for verification. The adaptation of the advanced delivery technique is relatively new in Malaysia. Without appropriate training and quality assurance protocol, the quality of the treatment may be jeopardised. The study may provide an efficient safety net to detect any erroneous delivery of IMRT and VMAT for new centres implementing the advanced radiotherapy technique.

Varian linac (Varian Medical System, Palo Alto, CA), records the MLC positions based on the motor current feedback that provides the surrogate information at every 0.05 s in a text file. The file also contains delivery parameters such as the beam state, the gantry angle and the dose fraction that can be utilised for dose reconstruction [2–4] and mechanical error analysis [5]. However, the surrogate MLC positions data is not able to detect error due to the MLC mechanical faults such as loose t-nut in the MLC system [6]. The study focuses on Elekta linac (Elekta, Crawley, UK) real-time MLC data measured by the propriety optical tracking system in the linac. The tracking system records the MLC position and other treatment parameters at every 0.25 s using the service graphing (SG) tool on the linac control computer. Our study investigates the feasibility of using the logged data for VMAT treatment verification.

## 2. Material and Methods

### *2.1. Treatment delivery*

The study was performed on Elekta Synergy linac (Elekta, Crawley, UK) equipped with Agility multileaf collimator (MLC). Agility consists of 80 MLC pairs of 5 mm leaf width. The MLC positions are tracked by the proprietary optical tracking system.

A dynamic motion of the MLC (dMLC) was first studied to investigate the structure of the data logged using the SG tool. All the MLCs in the right bank moves synchronously for 10 control points of different MLC positions. The MLC positions were 10.0 cm, 7.5 cm, 10.0 cm, 5 cm, 10.0 cm, 2.5 cm, 10.0 cm, 0 cm and 10.0 cm in the order of the control points. This back and forth of MLC motion produces opening and closing of a rectangular field. The other treatment parameters such as gantry angle and collimator angle remain constant.

A treatment prescription simulating a VMAT treatment delivery was delivered with 6 MV beam. The range of the prescribed treatment parameters are shown in Table 1. The MLC leaves move dynamically but asynchronously between 0.25 cm to 11.50 cm (the dynamic MLC shape will be explained in Section 3) and the pair of parallel collimator jaws position move from 5.0 cm to 11.5 cm. The gantry moves from 0° to 40° as the MLCs were moving. Total monitor unit (MU) was 50 MU.

**Table 1**. Prescription range of delivery parameters

| Parameters | Range of the prescription |
|---|---|
| MLC position | 0.25 to 11.50 cm |
| Upper jaw | 5 to 11.5 cm |
| Lower jaw | 5 to 11.5 cm |
| Collimator rotation (°) | 0 to 50 |
| Gantry (°) | 0 to 40 |
| Monitor unit | 50 MU |

The delivery parameters for the dMLC and the VMAT treatment were recorded using the SG tool on the linac control computer in the service mode. In this feasibility study, the prescribed MLC position logged during the delivery was only for the first MLC, i.e., MLC 1 and the middle MLC, i.e., MLC 40, while the MLC positions during treatment delivery were logged for all leaves. The limited number of the MLC positions recorded in the feasibility study is to minimise the complexity of the raw data recorded in the log file. Trigger function was set up to ensure the SG tool begins the data logging when the beam delivery started. The logged data were saved in an extensive mark-up language (XML) file after each treatment delivery.

### *2.2. Data analysis*

The log file recorded by the SG tool was retrieved from the linac control system. It contains two-structured line in which the first line is the file header. The second line is the attributes containing the logged data recorded during delivery. Data extraction and analysis were performed using algorithm developed in Matlab R2013b (MathWorks, Natick, MA).

The MLC speed and gantry speed during treatment delivery were calculated using equation 1 and equation 2 respectively. The numerator calculates the difference between the two consecutive values recorded at every 0.25 s, i.e., the MLC positions in equation 1 and gantry angle in equation 2. The denominator describes the time between the two consecutive values. The delivered parameters were compared to the VMAT treatment prescription. Deviation between the prescribed and the delivered position for MLC 1 and 40 were also calculated. In addition to that, the influence of MLC speed, gantry angle and gantry speed to the MLC position errors of the middle leaf (MLC 40) were studied.

$$\text{MLC Speed (cm/s)} = \frac{\text{MLC Position (i)} - \text{MLC Position (i+1)}}{(0.25)} \quad (1)$$

$$\text{Gantry speed (°/s)} = \frac{\text{Gantry angle (i)} - \text{Gantry angle (i+1)}}{(0.25)} \quad (2)$$

## 3. Results

The range and the mean of deviation between the prescribed and the delivered value of MLC 1, MLC 40 and other delivery parameters are shown in Table 2. In the dMLC study, the range of the MLC error is between -1.7 mm to 2.5 mm. No deviation was seen in the delivered MU and jaw position. As the collimator and gantry were at a static position, no deviation was observed for these parameters. For the VMAT study, the range of MLC position error is between -0.6 mm to 1.6 mm. Figure 1 shows the actual MLC position during VMAT treatment from the logged data. Figure 1(a) shows the minimum field aperture at 0.25 s and Figure 1(b) shows the maximum field aperture at 9.25 s. Both sides of the jaws deviated to a maximum of 0.3 mm from prescription value. Collimator rotation and gantry angle position deviates to a maximum of 0.1° to 0.2° respectively.

**Table 2**. Deviation range and mean of delivery parameters in dMLC and VMAT

| Parameters | dMLC<br>range (mean) | VMAT<br>range (mean) |
|---|---|---|
| MLC Leaf 1 (mm) | -1.7 to 2.5 (0.02) | -0.6 to 0.5 (0.003) |
| MLC Leaf 40 (mm) | -1.7 to 2.5 (0.002) | -1.5 to 1.6 (0.003) |
| Monitor unit (MU) | 0 | 0 |
| Upper jaw (mm) | - | 0-0.3 (0.03) |
| Lower jaw (mm) | - | 0-0.3 (0.03) |
| Collimator rotation (°) | - | 0-0.1 (0.03) |
| Gantry (°) | - | 0-0.1 (0.09) |

MLC speed was calculated to investigate whether it influences the MLC position error. MLC 40 in the dMLC study travels at a speed range from -3.24 cm/s to 2.64 cm/s. In VMAT the MLC travels at a higher speed from -5.88 cm/s to 5.92 cm/s. Figure 2 shows the position error for MLC 40 of the right MLC bank at different MLC speed for the dMLC study (Figure 2a) and the VMAT study (Figure 2b). Transient MLC position errors (spikes in the plot indicating higher error values) were seen when the leaves were changing its direction.

In the VMAT study, as the MLC speeds were modulating, gantry speed also varies throughout delivery. The range of the gantry speed during the treatment is between -10°/s to 8.8°/s as shown in Figure 3. Figure 4 shows the MLC position error as a function of gantry angle (Figure 4a) and gantry speed (Figure 4b). 97 % of the MLC 40 position errors are between -0.5 mm to 0.5 mm at most gantry speed and gantry angle.

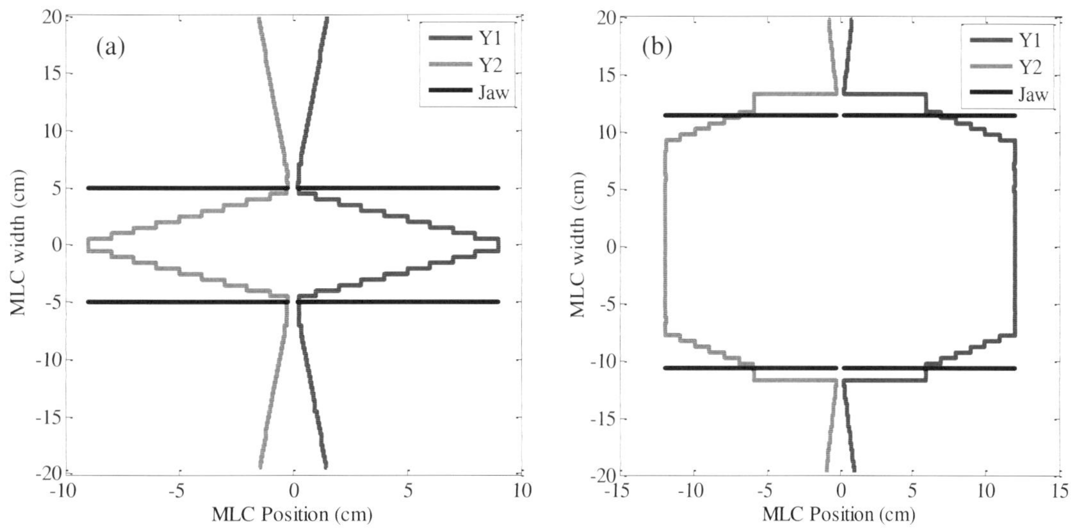


**Figure 1.** Plot of the MLC positions for the minimum (a) and the maximum (b) field size of VMAT study.

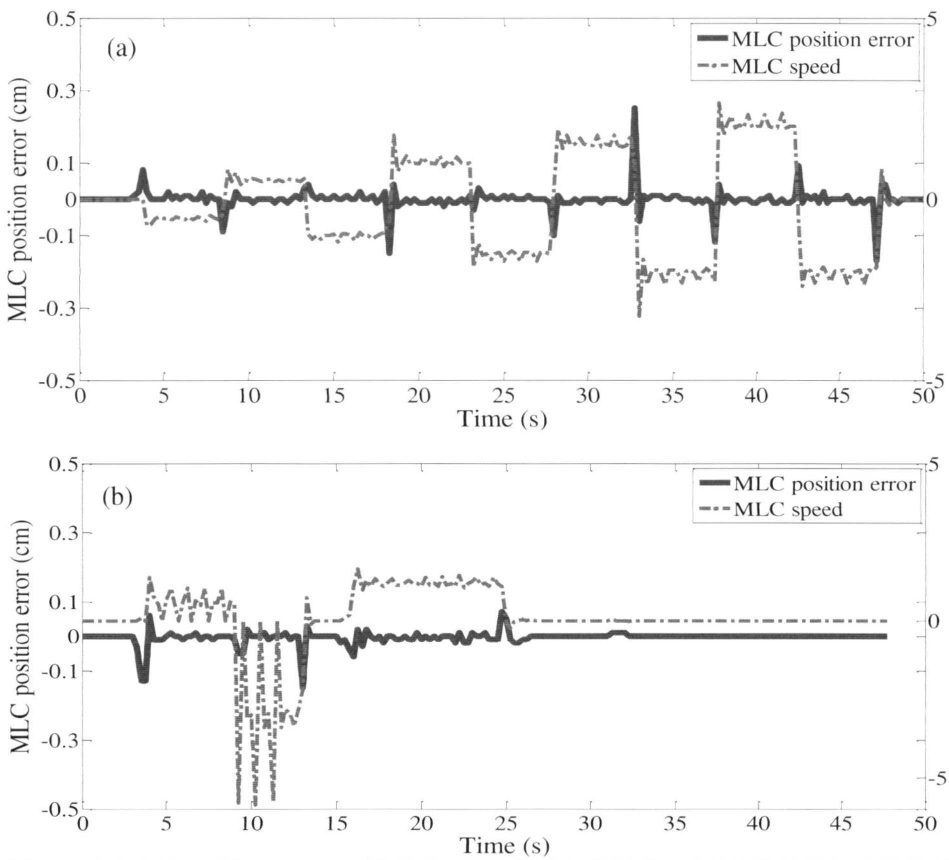


**Figure 2**. MLC position error and MLC speed for (a) dMLC and (b) VMAT for MLC 40.

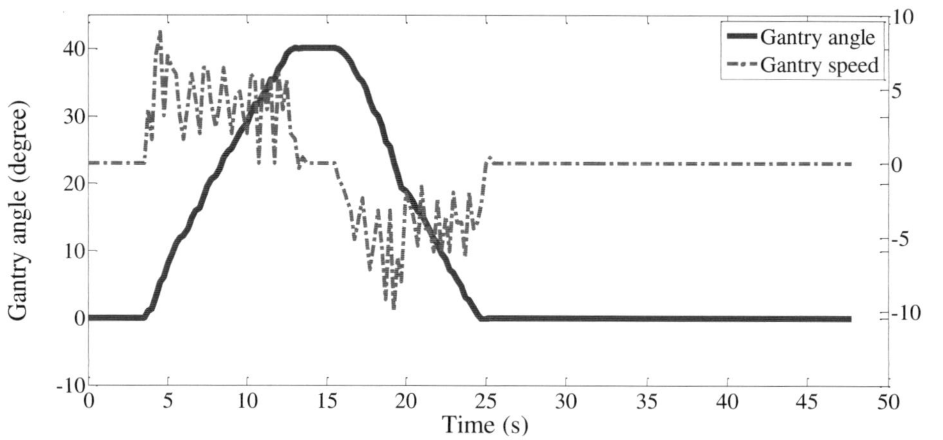


**Figure 3.** Gantry angle and gantry speed during VMAT delivery.

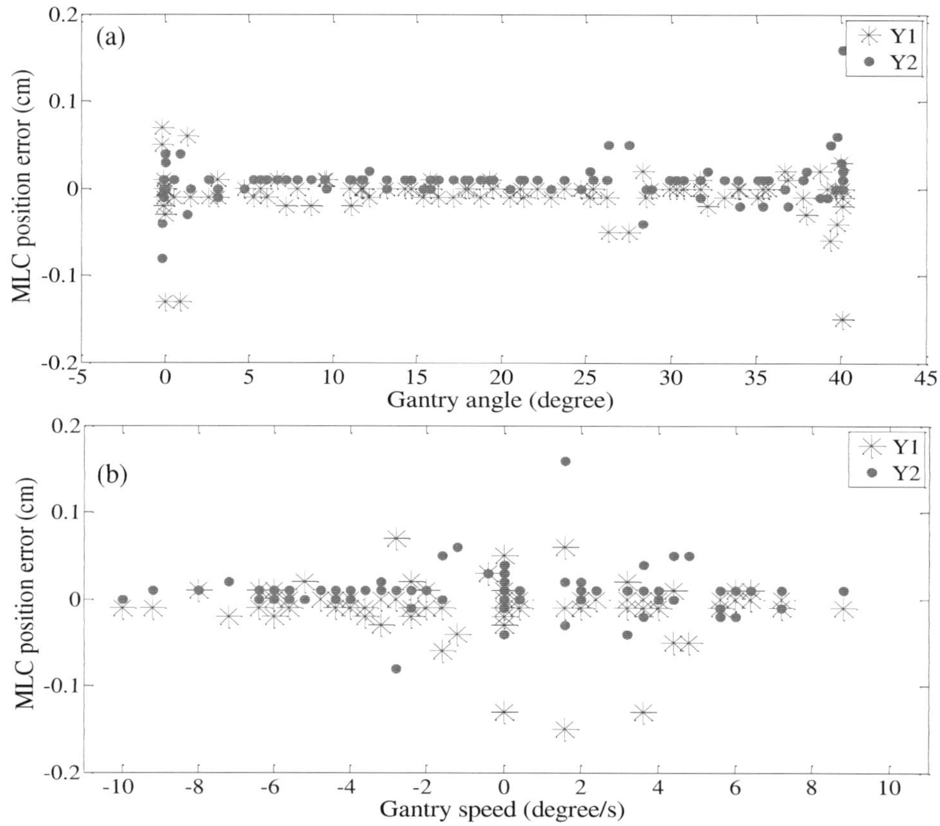


**Figure 4.** MLC position error at different (a) gantry angle and (b) gantry speed for VMAT.

## 4. Discussions

We have utilised the use of real-time log data of delivery parameters such as MLC position, monitor unit, jaw position, collimator rotation and gantry angle for VMAT treatment verification. Analysis of the deviation of the delivered MLC position from the prescribed MLC position shows errors ranges from -0.6 mm to 1.6 mm, well within the 3.5 mm tolerance value stipulated in AAPM TG-142 [7]. The dMLC study shows higher MLC position error than the VMAT although it records slower MLC speed. Leaf position error happened due to potential recoil when leaves changes direction [8] which explain why higher MLC position error happened at time where the MLC had to change its direction in the dMLC study. It also explains higher MLC position error at 0° and 40° gantry angles in VMAT. It is at the initial and final angle of each of the arc in which the MLC are required to reverse their movement direction. The errors were not affected by gravity. Clarke et al study the correlation of gravity effect on the stability of MLC using electronic portal device at different gantry angle and collimator rotation. Their results shows MLC position errors were consistent for all gantry angles except the initial and final showing that MLC error does not affected by gravity [9]. 97% of the MLC 40 error were within -0.05 cm to 0.05 cm at different gantry speed. Gantry angle and collimator rotation error in VMAT are within the machine tolerance value.

SG tool provide real-time data of treatment delivery parameters at sampling time of 0.25 s. The parameters are MLC position, beam MU, gantry angle, collimator angle and jaw position. The logged data includes prescribe and delivered value of parameters. Range of MLC position error in this study is between -1.7 mm to 2.5 mm. Pasler *et al* also reported that SG tool detect MLC position error as much as 6.5 mm when the dynamic parameter are pushed to maximum range [9].

## 5. Conclusion

The treatment parameters logged using the SG tool can be used as a supplementary verification method for IMRT and VMAT delivery. This study was performed on a simulated VMAT treatment fields. Verification of clinical treatment plans will be performed using the SG tool and the result will be compared to the measurements from clinical detectors.

## 6. References

[1] Masi L, Doro R, Favuzza V, Cipressi S and Livi L 2013 Impact of plan parameters on the dosimetric accuracy of volumetric modulated arc therapy Med.Phys. **40** 071718.

[2] Dinesh Kumar M, Thirumavalavan N, Venugopal Krishna D, and Babaiah M 2006 QA of intensity-modulated beams using dynamic MLC log files. J Med Phys. *J Med Phys*. **31** 36.

[3] Calvo-ortega JF, Teke T, Moragues S, and Pozo M 2014 A Varian DynaLog file-based procedure for patient dose-volume histogram – based IMRT QA *Journal of Applied Clinical Medical Physics* **15.2** 5–7.

[4] Tae-Suk S, Jeong-Woo L, Jeong-Hoon P, Jin-Beom C, Ji-Yeon P, Bo-Young C, Doo-Hyun L, Semie H, Min-Young K and Kyoung-Sik C 2009 Inverse Verification of Dose Distribution for Intensity Modulated Radiation Therapy Patient-specific Quality Assurance Using dynamic MLC Log Files *J Korean Phys Soc*. **55** 1649.

[5] Kerns JR, Childress N, and Kry SF 2014 A multi-institution evaluation of MLC log files and performance in IMRT delivery *Radiat Oncol* **9** 176.

[6] Agnew A. Agnew C, Grattan M, Hounsell A and McGarry C 2014 Monitoring daily MLC positional errors using trajectory log files and EPID measurements for IMRT and VMAT deliveries *Phys.Med.Biol.* **59** N49-N63.

[7] Klein EE, Hanley J, Bayouth J, Yin FF, Simon W, Dresser S, Serago C, Aguirre F, Ma L, Arjomandy B, Liu C, Sandin C and Holmes T 2009 Task Group 142 report: quality assurance of medical accelerators *Med.Phys* **36** 4197–212.

[8] Mu G, and Xia P 2007 Impact of MLC leaf position errors on simple and complex IMRT plans for head and neck cancer *Phys.Med. Biol.* **53** 77-88.

[9] Clarke M, and Budgell G 2007 Use of an amorphous silicon EPID for measuring MLC calibration at varying gantry angle *Phys.Med.Biol.* **53** 473- 485.

**Acknowledgement**

This research is funded by Fundamental Research Grant Scheme, Ministry of Education Malaysia, 203/CIPPT/6771383. Thanks to the linac service engineer, Kho K.L. (ABEX Medical System) for the help with the service graphing tool in the initial stage of the work.

# Fabrication and characterisation of phantom material made of Tannin-added *Rhizophora spp*. particleboards for photon and electron beams

**M F Mohd Yusof[1,3], P N K Abd Hamid[1], A A Tajuddin[1], R Hashim[2], S Bauk[1], N Mohd Isa[4] and M J Md Isa[4]**

[1] School of Physics, Universiti Sains Malaysia, 11800 Penang, Malaysia
[2] School of Industrial Technology, Universiti Sains Malaysia, 11800 Penang, Malaysia
[3] School of Health Sciences, Universiti Sains Malaysia, 16150 Kelantan, Malaysia
[4] Health Physics Group, Malaysian Nuclear Agency, 43000 Selangor, Malaysia

E-mail: mfahmi@usm.my

**Abstract.** Particleboards made of *Rhizophora spp*. with addition of tannin adhesive were fabricated at target density of 1.0 g/cm$^3$. The physical and mechanical properties of the particleboards including internal bond strength (IB) and modulus of rupture (MOR) were measured based on Japanese Industrial Standards (JIS A-5908). The characterisation of the particleboards including the effective atomic number, CT number and relative electron density were determined and compared to water. The mass attenuation coefficient of the particleboards were measured and compared to the calculated value of water using photon cross-section database (XCOM). The results showed that the physical and mechanical properties of the particleboards complied with Type 13 and 18 of JIS A-5908. The values of effective atomic number, CT number and relative electron density were also close to the value of water. The value of mass attenuation coefficients of the particleboards showed good agreement with water (XCOM) at low and high energy photon indicated by the $\chi^2$ values.

## 1. Introduction

Phantom material is defined as materials that simulated the absorption and scattering properties of human tissues towards ionising radiation (Khan, 2010). Phantom materials are used mainly in quality assurance and dosimetry works involving ionising radiation. Water is commonly used as phantom materials it has mas density near to human soft tissues at 1.0 g/cm$^3$. The use of water however is not always convenience due to its liquidity. Several solid-state phantom materials are introduced and made water-equivalent to replace water such as acrylic and perspex. These water equivalent materials however do not always give attenuation properties similar to water as they still differs to water in term of effective atomic number and electron density. These phantom materials also may develop electron contaminations that could alter the dose measurement.

Earlier studies have suggested that *Rhizophora spp*. as potential phantom material as it has mass density and attenuation coefficient close to water (Bradley at al., 1991; Tajuddin et al., 1996). However the use of *Rhizophora spp*. solid woods as phantom materials has several limitations including tendency to crack over period of times. The size of *Rhizophora spp*. trunk is also limited to construct a full size phantom besides its density inhomogeneity across the trunk. The binderless particleboards made of *Rhizophora spp*. was introduced with advantages includes better density uniformity and able to be fabricated at various size and shapes without compromising its attenuation properties (Marashdeh et al., 2011). However, the strength of the binderless particleboards made of *Rhizophora spp*. becoming a concern due to rigidity and heavy workload as phantom materials. To date, studies had been carried out to improve the strength of the particleboards made of *Rhizophora spp*. Addition of adhesives or binder becoming major option to increase the strength of the particleboards with at the same time retaining its water-equivalent properties.

The synthetic-based adhesives such as urea-formaldehyde (UF) commonly used in industrial particleboards manufacturing failed to retain the attenuation properties of *Rhizophora spp*. particleboards in comparison to water (Surani, 2008, Ngu et al., 2015). Tannin has been used as alternative binder for particleboards and plywood to replace commonly used formaldehyde-based adhesives (Pizzi and Merlin, (1981); Pizzi and Scharfetter, (1989); Oo, et al, (2008). Tannin is a biological-based adhesive that can be extracted mainly from barks including *Rhizophora spp*. (Mohd Yusoff et al., 1988). Previous study had suggested the suitability of tannin adhesive for *Rhizophora spp*. particleboards as phantom at low energy photons (Safian, 2012). However, the tannin-based particleboards with low treatment level of tannin (5%) still failed to satisfy the industrial standards for physical and mechanical properties. Previous studies had suggested that the suitable percentage of biological-based adhesives for *Rhizophora spp*. is within 10% without changing their attenuation properties (Ghasemi, 2012; Tousi et al., 2014; Abu Arra et al., 2014). This study focused on the fabrication and characterisations of *Rhizophora spp* particleboards with 10% percentage of tannin adhesive as phantom materials in application of photon and electron beams.

## 2. Methodology

### *2.1. Preparation of Sample*

The *Rhizophora spp*. trunk was obtained from a charcoal factory in Kuala Sepetang, Perak, Malaysia. Only trunk from the middle part of the tree was chosen according to the study by Shakhreet et al., (2013). The trunk was cleaned and dried before planed in order to reduce it into smaller wood chips. The wood chips were then ground into smaller wood particle and screened using horizontal screen machine to obtained only wood particle with particle size of 104 μm according to the study by Marashdeh et al., (2011). Particleboards were fabricated using hot press machine with target density of 1.0 g/cm$^3$. The target density of the particleboards was determined based on the calculation of the mass wood particle required to fabricate the particleboards at density of 1.0 g/cm$^3$. And amount of 10% tannin is added into the wood particle based on the dried mass of wood particle. The actual density of the fabricated particleboards were measured based on the external dimension and mass of the particleboards given by the equation of

$$\rho = \frac{m}{V} \quad (1)$$

With *m* is the measured mass of the particleboards and *V* is the volume of the particleboards measured based on its external dimension given by the equation of

$$V = length \times width \times thickness \quad (2)$$

The particleboards were cut into smaller piece sample with size 5.0 cm x 5.0 cm as shown in Figure 1(a) for mechanical and characterisation tests. A replication of particleboard was sawn circularly as shown in Figure 1(b) to construct density plug phantoms that will be discussed in chapter 2.4. The internal bond (IB) strength of the particleboards were measured based on the Japanese Industrial Standard (JIS A 5908: 2003). The IB strength was calculated using the equation

$$Internal\ Bond = \frac{P_{max}}{2wl} \quad (3)$$

With $P_{max}$ is the maximum load at failure of the particleboards and *w* and *l* is the width and length of the particleboard sample respectively.

**Figure 1.** The test piece for mechanical test (a) and circularly sawn

*2.2. Elemental Composition and Effective Atomic Number*

The percentages of elemental composition of the particleboards were determined using energy dispersive x-ray analysis (EDXA) from the scanning electron microscopy (SEM). The percentages of elemental composition were used to calculate the effective atomic number, $Z_{eff}$ of the particleboards. The attenuation properties of medium are related to the effective atomic number, therefore the effective atomic number value is used to determine the water equivalent and attenuation properties of the particleboard in comparison to water. The effective atomic number is calculated according to the study by Duvauchelle et al., (1999) using the equation

$$Z_{eff} = \left[\sum_{i=1}^{n}(\alpha_i z_i^{m})\right]^{\left(\frac{1}{m}\right)} \quad (4)$$

where $a_i$ and $z_i$ are electron fraction and atomic number of $i^{th}$ element in the sample respectively. $m$ is the experimental coefficient for biological materials and water with value of 3.4. The electron fraction of the $i^{th}$ element can be calculated by the equation

$$\alpha_i = \frac{wi\left(\frac{z_i}{A_i}\right)}{\Sigma w_i\left(\frac{z_i}{A_i}\right)} \quad (5)$$

where $w_i$ and $A_i$ are fractional weight and atomic mass of the $i^{th}$ element respectively.

*2.3. Measurement of CT Number and Relative Electron Density*

A study by Saito, (2012) had suggested that the electron density of a medium is very much related to it CT number. Thus the electron density of a medium can be derived directly from it CT number. A density plug phantom was constructed from the particleboard made compatible with computed tomography (CT) electron density phantom model CIRS 062M. The particleboard was sawn circularly and stacked together using wood adhesive to obtain a cylindrical-shaped phantom with approximate diameter of 3.0 cm. The density plug phantoms were scanned using the CT scanner at 120 kVp CT energy along with the water, liver muscle, and fat equivalent density plug phantoms. The average CT numbers of samples were measured and the relative electron density curves were plotted based on the recommendation by the manufacturer as shown in Figure 2. The relative electron density of tannin-added *Rhizophora spp.* particleboards was determined from the electron density curve.

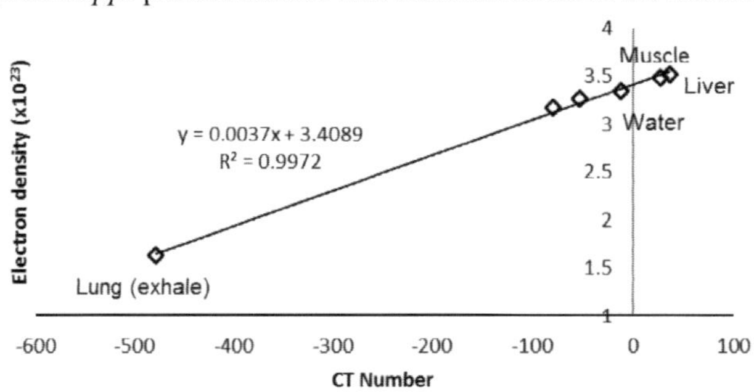


**Figure 2.** The relative electron density curve of various tissue equivalent density plug phantoms based on CT number at 120 kVp CT X-ray energy.

The relative electron density of tannin-added *Rhizophora spp.* particleboard can be calculated using equation (6)

$$RED = 0.0037x + 3.4089 \quad (6)$$

with $x$ is the Hounsfield unit or the CT number of the sample

*2.4. Measurement of Mass Attenuation Coefficient*

The mass attenuation coefficients of the particleboards were measured at low and high energy x-ray. The mass attenuation coefficient at low energy photons were carried out according to the study by Maarashdeh et al., (2012) based using the x-ray fluorescent (XRF) set up as shown in Figure 3 (a). A $^{241}$Am source were used in conjunction with metal plates of Niobium, Palladium, Molybdenum and silver were used to obtain $K_{\alpha 1}$ x-ray between 16.59 and 25.26 keV. The mass attenuation coefficient at high energy photon were measured using Ladlum set up in as shown in Figure 3 (b). The Ladlum set up was used conjunction with $^{137}$Cs and $^{60}$Co gamma sources that provided the gamma energy of 0.662 and 1.2 MeV respectively.

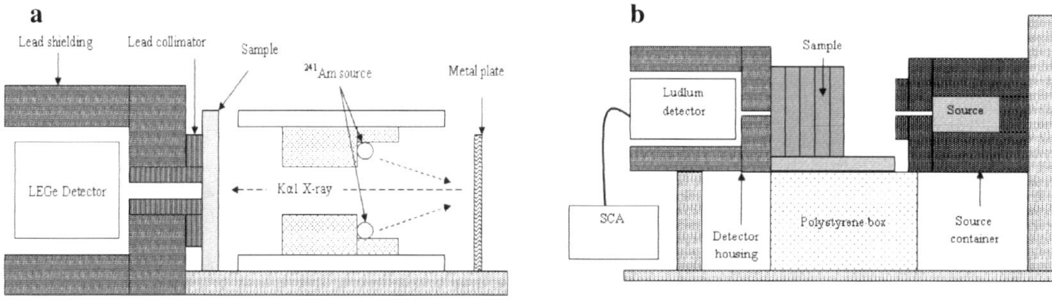


**Figure 3.** The experimental set up of X-ray fluorescent set up (a) and Ladlum set up (b) for measurement of mass attenuation coefficient.

The linear attenuation coefficient of the particleboards were measured using the transmitted photon through the sample according to the Beer-Lambert equation of

$$I = I_o e^{-\mu x} \tag{7}$$

with $I_o$ is initial intensity of photons, and μ is linear attenuation coefficient of sample in. The value of linear attenuation coefficient is calculated using the equation

$$\mu = \frac{1}{x} \ln\left(\frac{I_o}{I}\right) \tag{8}$$

The value of mass attenuation coefficient is obtained by dividing the value of linear attenuation coefficient with the gravimetric calculation of sample density. The measured mass attenuation cofficient is compared to the calculated value of water using photon cross-section database (XCOM) according to the study by Berger and Hubbell, (1987).

## 3. Results

The average mass density of tannin-added *Rhizophora spp.* particleboards measured using gravimetric method is presented in Table 1. The results showed that the particleboards were able to be fabricated at target density of 1.0 g/cm$^3$. The particleboards were also showed good density uniformity shown by the standard deviation (SD) of measured density.

**Table 1.** The average density of tannin-added *Rhizophora spp.* particleboards in comparison to binderless *Rhizophora spp.* and binderless *Rhizophora spp.* particleboards.

| Sample | Measured Density (g/cm$^3$) | | | Standard |
|---|---|---|---|---|
| | Average | Max. | Min. | Deviation |
| *R. spp*. raw wood | 0.991 | 1.004 | 0.973 | 0.011 |
| Binderless *R. spp*. | 1.037 | 1.095 | 0.955 | 0.061 |
| Tannin-added *R. spp*. | 1.004 | 1.040 | 0.980 | 0.023 |

The average IB strength of the particleboards measured based on JIS A 5908: 2003 were illustrated in Figure 4. The results showed that the tannin-added *Rhizophora spp.* in this study satisfied the JIS A 5908: 2003 for particleboard Type 18 in comparison to the binderless particleboard from the previous study by Marashdeh et al., (2012). A comparison to the tannin-based *Rhizophora spp.* particleboards with lower percentage of tannin showed that the tannin-based particleboards from the previous study by Safian (2012) did not satisfy any of the industrial standards.

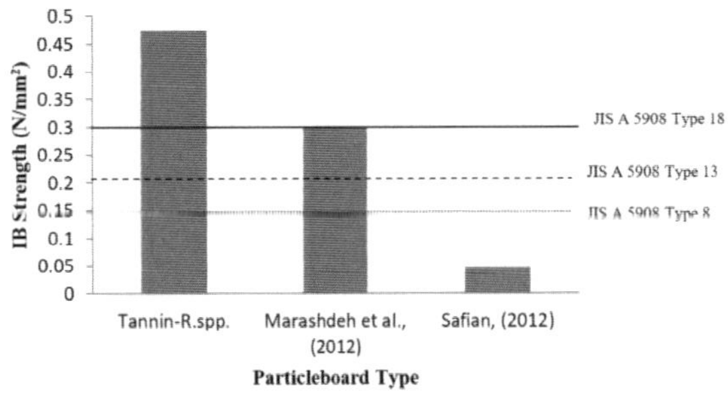


**Figure 4.** Average IB strength of the particleboards based on JIS A 5908: 2003.

### *3.1. Elemental Compositions and Effective Atomic Number*

The percentage of elemental compositions and calculated effective atomic number of samples are presented in Table 2. The results showed that the particleboards made of *Rhizophora spp.* consists of mainly carbon and oxygen similar to that in human soft tissues. Addition of tannin did not give significant change of elemental compositions to the particleboards indicating the similarities of elemental compositions of tannin to the wood particles. The effective atomic number can be used to determine the attenuation coefficient of a medium (Duvauchelle et al., 1999). One medium can be said to have water equivalent properties if the effective atomic numbers are equal. The calculated effective atomic number of samples as shown in Figure 2 indicated that the value of tannin-added *Rhizophora spp.* were close to the value of water. This had indicated the addition of tannin had improved the effective atomic number and water equivalent property of tannin-added *Rhizophora spp.*

**Table 2.** The percentage of elemental composition and calculated effective atomic number of particleboards made of *Rhizophora spp.* in comparison to water.

| Sample | Elemental composition (%) | | | | | $Z_{eff}$ |
|---|---|---|---|---|---|---|
| | H | C | O | N | F | |
| Water[a] | 11.11 | | 88.89 | | | 7.50[a] |
| *R. spp* raw wood[b] | 5.41 | 40.16 | 54.40 | 0.03 | | 5.85 |
| Binderless *Rhizophora. spp.*[c] | | 32.93 | 38.98 | 28.08 | | 7.18[c] |
| Tannin-added *Rhizophora spp.*[c] | | 51.25 | 43.11 | | 5.64 | 7.22[c] |

[a]AAPM-21 (1983) [b]Che Wan Sudin (1991) [c]Current study

### *3.2. CT Number and Relative Electron Density*

The average CT number and relative electron density of *Rhizophora spp.* particleboards in comparison to various tissue equivalent density plug phantoms is presented in Table 3. The results showed that the average CT number of tannin-added *Rhizophora spp.* particleboards was close to the value of water equivalent density plug phantoms.

**Table 3.** The average CT number and relative electron density of tannin-added *Rhizophora spp.* particleboards and various tissue equivalent density plug phantoms.

| Sample | Average CT number (HU) | Relative electron density (x $10^{23}$ electron/cm$^3$) |
|---|---|---|
| Lung (exhale) | -477.74 | 1.632 |
| Adipose tissue | -78.75 | 3.171 |
| Breast (50/50) | -52.09 | 3.261 |
| *R. spp.* raw wood | -34.07 | 3.193 |
| Tannin-added *R. spp.* | -13.69 | 3.341 |
| Water | -11.61 | 3.340 |
| Muscle | 27.36 | 3.483 |
| Liver | 36.45 | 3.516 |

The calculation of relative electron density also showed close value of tannin-added *Rhizophora spp.* to water. This had indicated good agreement of attenuation coefficient of tannin-added *Rhizophora spp.* to water as the CT number is very much related to the attenuation coefficient of medium to the x-ray. A study by Kurudirek, (2014) suggested that the electron density of tissue is very much related to the attenuation coefficient of the tissue. Therefore, the results also indicated the similarity of attenuation coefficient of tannin-added *Rhizophora spp.* particleboards to water in based on their close value of relative electron density.

### *3.3. Mass Attenuation Coefficient*

The mass attenuation coefficient of tannin-added *Rhizophora spp.* particleboards in comparison to water (XCOM) and previous studies at low energy photon measured using XRF set up is illustrated in Figure 5. The detailed value of linear and mass attenuation coefficient of the samples is presented in Table 4. The results showed that the linear and mass attenuation coefficient of tannin-added *Rhizophora spp.* particleboards was in good agreement to the value of water (XCOM). This had indicated the water equivalent property and similar attenuation properties of tannin-added *Rhizophora spp.* particleboards. The results also showed that the mass attenuation coefficient of tannin-added *Rhizophora spp.* particleboards was in good agreement with the binderless *Rhizophora spp.* particleboards from the study by Marashdeh et al., (2012). The addition of tannin was found to retain the attenuation properties of the particleboards made of *Rhizophora spp.*

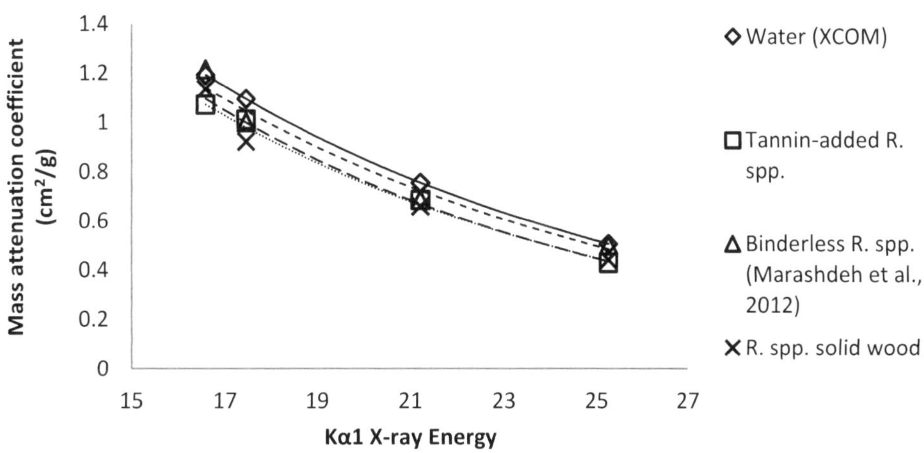


**Figure 5.** Mass attenuation coefficient of tannin-added *Rhizophora spp.* particleboards in comparison to water and previous studies.

**Table 4.** The linear and mass attenuation coefficient of samples at low energy photons measured using XRF set up.

| Kα1 Energy (keV) | | Tannin-added *R. spp.* | *R.spp.* raw wood | Water (XCOM) |
|---|---|---|---|---|
| 16.59 | $\mu$ | 1.142 | 1.136 | 1.193 |
| | $\mu/\rho$ | 1.072 | 1.135 | 1.193 |
| | $\sigma_{(\mu/\rho)}$ ±% | 2.23 | 2.18 | - |
| 17.46 | $\mu$ | 1.072 | 0.921 | 1.095 |
| | $\mu/\rho$ | 1.011 | 0.921 | 1.095 |
| | $\sigma_{(\mu/\rho)}$ ±% | 2.69 | 2.02 | - |
| 21.21 | $\mu$ | 0.696 | 0.659 | 0.755 |
| | $\mu/\rho$ | 0.685 | 0.659 | 0.755 |
| | $\sigma_{(\mu/\rho)}$ ±% | 2.24 | 2.34 | |
| 25.26 | $\mu$ | 0.436 | 0.440 | 0.506 |
| | $\mu/\rho$ | 0.429 | 0.440 | 0.506 |
| | $\sigma_{(\mu/\rho)}$ ±% | 1.57 | 1.40 | - |

The mass attenuation coefficient at high energy photon measured using Ladlum set up is presented in Table 5. The results showed that the mass attenuation coefficient of tannin-added *Rhizophora spp.* particleboards were close to the value of water (XCOM). The mass attenuation coefficient at $^{137}$Cs gamma energy was slightly lower to the value of water with percentage of discrepancies of 17% to that in water. The mass attenuation coefficient at $^{60}$Co gamma energy however was in good agreement to water with lower percentage of discrepancies of 11%. The results had indicated that tannin-added *Rhizophora spp.* having attenuation coefficient close to to the value of water at both low and high energy photons.

**Table 5.** The linear and mass attenuation coefficient of samples at high energy photons measured using Ladlum set up.

| Gamma energy (MeV) | | Tannin-added *R. spp.* | *R.spp.* raw wood | Water (XCOM) |
|---|---|---|---|---|
| 0.622 | $\mu$ | 0.071 | 0.067 | - |
| | $\mu/\rho$ | 0.071 | 0.067 | 0.086 |
| | $\sigma_{(\mu/\rho)}$ ±% | 0.017 | 0.030 | - |
| 1.20 | $\mu$ | 0.052 | 0.057 | - |
| | $\mu/\rho$ | 0.052 | 0.048 | 0.059 |
| | $\sigma_{(\mu/\rho)}$ ±% | 0.012 | 0.009 | - |

## 4. Conclusion

The addition of tannin had increased the mechanical property of particleboards made of *Rhizophora spp.* shown by the average IB strength. The calculation of effective atomic number based on the percentage of elemental composition showed close value of tannin-added *Rhizophora spp.* particleboards to the value of water by AAPM-21 (1983). The measurement of CT number and relative electron density also showed value of tannin-added *Rhizophora spp.* particleboards to water indicating close attenuation coefficient to water. The measurement of mass attenuation coefficient showed good agreement of tannin-added *Rhizophora spp.* particleboards to the calculated value of water using XCOM at both low and high energy photons. The results had indicated the suitability of tannin-added *Rhizophora spp.* particleboards as phantom materials in applications of low and high energy photons.

## 5. References

[1] AAPM-21, 1983 Protocol for the determination of absorbed dose from high energy photon and electron beams; Task Group 21 *Med. Phys.* 10, 741-771

[2] Abu Arra A Hashim R Bauk S Kandaiya S and Tousi E T 2014 Fabrication and characterization of gum Arabic bonded *Rhizophora spp.* particleboards, *Materials and Designs*, 60, 108-115

[3] Banjade D P Tajuddin A A and Shukri A 2000 A study of *Rhizophora spp.* wood phantom for dosimetric purposes using high-energy photon and electron beams, *Appl. Radiat. Iso.*55 297-302

[4] Berger M J and Hubbell J H 1987 XCOM: Photon cross-sections on a personal computer. NBSIR 87-3597 Washington

[5] Bradley D A Tajuddin A A Che Wan Sudin C W A and Bauk, S 1991 Photon attenuation studies on tropical hardwoods. *International Journal of Radiation Applications and Instrumentation. Part A. Appl. Radiat. Isotopes* 42 771-773

[6] Brown S Bailey D L Willowson K and Baldock C 2008 Investigation of the linear relationship between linear attenuation coefficients and CT Hounsfield units using radionuclides for SPECT, *Appl. Radiat. Isotopes* 66 1206-1212

[7] Duvauchelle P Peix G and Babot D 1999 Effective atomic number in the Rayleigh to Compton scattering ratio. *Nucl. Instrum. Methods Section B: Beam Interactions with Materials and Atoms,* 155 221-228

[8] Ghasemi L (2013). Fabrication and Characterization of Bio Plastic *Rhizophora spp.* Particleboard and Determination of Their Mass Attenuation Coefficient. M.Sc. Thesis, Universiti Sains

Malaysia
[9] Japanese Industrial Standard Particleboards (JIS A 5908) 2003 Standardization Promotion Department, Japanese Standard Association, Tokyo
[10] Khan M F 2010 The physics of radiation therapy, Lippincott William & Wilkins, Philadelphia
[11] Kurudirek M 2014 Effective atomic numbers and electron densities of some human tissues and dosimetric materials for mean energies of various radiation sources relevant to radiotherapy and medical applications *Radiation Phys. Chem.*102 139-146
[12] Marashdeh M W Hashim R Tajuddin A A Bauk S and Sulaiman O 2011 Effect of particle size on the characterization of binderless particleboard made from *Rhizophora spp.* Mangrove wood for use as phantom material *Bioresource Technol.* 6 4028-4044
[13] Marashdeh M W Bauk S Tajuddin A A Hashim R 2012 Measurement of mass attenuation coefficients of *Rhizophora spp.* binderless particleboards in the 16.59–25.26 keV photon energy range and their density profile using x-ray computed tomography *Appl. Radiat. Isotopes* 70 656-662
[14] Mohd Yusoff M N Chew L T Mohd Ali A R Abdul Kadir A 1988 Mangrove tannin as an adhesive for wood-based panels. JTFS. 1, 97-102
[15] Ngu K T Bauk S Hashim R Tajuddin A A and Shukri A 2015 Fabrication of formaldehyde-based particleboards and their mass attenuation coefficients at 15.77, 17.48, 21.18 and 25.27 keV photon energies. JPS 26 (1), 27-33
[16] Oo C Pizzi A Pasch H and Kassim M 2008 Study on the structure of mangrove polyflavonoid tannins with MALDI – TOF mass spectrometry. *J. Appl. Polym. Sci* 109 963-967
[17] Pizzi A and Scharfetter H 1989 Adhesives and techniques open new possibilities for the wood processing industry. Part1: Experience with tannin based adhesives. *Eur. J. Wood Wood Prod.* 39 85–89
[18] Pizzi A and Merlin M 1981 A new class of tannins adhesives for exterior particleboard. *Int. J. Ahhes. Adhes.* 1(5) 261–264
[19] Safian A 2012 Tannin-based *Rhizophora spp.* Particleboard Breast Phantom at Mammographic Energy. M.Sc. Thesis, Universiti Sains Malaysia
[20] Saito M 2012 Potential of dual-energy subtraction for converting CT numbers to electron density based on a single linear relationship, *Med Phys* 39(4) 2021-2030
[21] Shakhreet B Z Bauk S Tajuddin A A and Shukri A 2009 Mass attenuation coefficnet on natural *Rhizophora spp.* woods for X-rays in the 15.77 to 25.27 keV range *Radiat. Prot. Dosim* 135 4-53
[22] Surani B T 2008 The Suitability of PF, UF and PRF Resins in Term of Structure and Attenuation Properties to be Used in *Rhizophora spp.* Particleboard Phantom. M.Sc. Thesis, Universiti Sains Malaysia
[23] Tajuddin A A Che Wan Sudin C W A and Bradley D A 1996 Radiographic and scattering investigation on the suitability of *Rhizophora spp.* as tissue equivalent medium for dosimetric study. *Radiat. Phys. Chem.* 47 739-740
[24] Tousi E T Bauk S Hashim R Jaafar M S Abuarra A 2014 Measurement of mass attenuation coefficients of *Eremurus-Rhizophora spp.* particleboards for X-ray in the 16.63-25.30 keV energy range. *Radiat. Phys. and Chem.* 103, 119-125
[25] Tursucu A Demir D and Onder P 2013 Effective atomic number determination of rare earth oxides with scattering intensity ratio, *Sci. Tech. Nucl. Inst. 2013* (738978)

**Acknowledgement**
The authors would like to acknowledge the financial support by Short Term Grant no. 304/PPSK/61313068. The authors also would like to acknowledge Radiology Department, Hospital Universiti Sains Malaysia (HUSM) for the use of CT scanner.

IOP Conf. Series: Journal of Physics: Conf. Series **851** (2017) 012037 doi:10.1088/1742-6596/851/1/012037

# Sequence-specific inhibition of microRNA-130a gene by CRISPR/Cas9 system in breast cancer cell line

**Nur Ainina Abdollah[1], Kumitaa Theva Das[2], Narazah Mohd Yusoff[3], Siti Razila Abdul Razak[1]**

[1]Oncology and Radiological Sciences Cluster, Advanced Medical and Dental Institute, Universiti Sains Malaysia, Bertam, 13200 Kepala Batas, Pulau Pinang, Malaysia. [2]Infectomics Cluster, Advanced Medical and Dental Institute, Universiti Sains Malaysia, Bertam, 13200 Kepala Batas, Pulau Pinang, Malaysia. [3]Regenerative Medicine Cluster, Advanced Medical and Dental Institute, Universiti Sains Malaysia, Bertam, 13200 Kepala Batas, Pulau Pinang, Malaysia.

E-mail: sitirazila@usm.my and aininaabdollah@gmail.com

**Abstract**. MicroRNAs (miRNAs) are short stranded noncoding RNA that play important roles in apoptosis, cell survival, development and cell proliferation. However, gene expression control via small regulatory RNA, particularly miRNA in breast cancer is still less explored. Therefore, this project aims to develop an approach to target microRNA-130a using the Clustered Regularly Interspaced Short Palindromic Repeat (CRISPR)/Cas9 system in MCF7, breast cancer cell line. The 20 bp sequences target at stem loop, 3' and 5' end of miR130a were cloned into pSpCas9(BB)-2A-GFP (PX458) plasmid, and the positive clones were confirmed by sequencing. A total of 5 μg of PX458-miR130a was transfected to MCF7 using Lipofectamine® 3000 according to manufacturer's protocol. The transfected cells were maintained in the incubator at 37 $^{0}$C under humidified 5% CO2. After 48 hours, cells were harvested and total RNA was extracted using miRNeasy Mini Kit (Qiagen). cDNAs were synthesised specific to miR-130a using TaqMan MicroRNA Reverse Transcription Kit (Applied Biosystems). Then, qRT-PCR was carried out using TaqMan Universal Master Mix (Applied Biosystems) to quantify the knockdown level of mature miRNAs in the cells. Result showed that miR-130a-5p was significantly downregulated in MCF7 cell line. However, no significant changes were observed for sequences targeting miR-130a-3p and stem loop. Thus, this study showed that the expression of miR-130a-5p was successfully down-regulated using CRISPR silencing system. This technique may be useful to manipulate the level of miRNA in various cell types to answer clinical questions at the molecular level.

## 1. Introduction

Breast cancer is the most common cancer among women worldwide. The mortality rates have significantly reduced in the past 20 years due to the implementation of screening programs and the development of new therapy [1]. Even though there are many findings in understanding the development and progression of breast cancer, only part of the molecular mechanisms underlying breast cancer pathogenesis is fully understood.

MicroRNAs (miRNAs) are short stranded noncoding RNA with 18-25 nucleotides in length, highly conserved in evolution and play important roles in almost all biological pathways [2]. miRNAs post-transcriptionally regulate the expression of protein-coding genes by binding to the specific binding sites in their 3'UTR and causing either mRNA degradation or translational termination [2]. Briefly, primary-miRNA (pri-miRNA) is transcribed by RNA polymerase II [3]. In nuclease, enzymes Drosha and Pasha will recognize and cleaves pri-miRNA leaving two nucleotides overhangs in the 3'-end (3p) [4]. The precursor-miRNA (pre-miRNA, Figure 1) exit the nuclease and in the cytoplasm, Dicer will cleaves off the stem loop (SL) leaving 3'-end (3p) and 5'-end (5p) [5, 6]. One of the strand, usually 5p, will assemble into AGO to form RISC complex which will then cause translational repression [7]. The other strand will degrade rapidly [7].

In cancer, it has been demonstrated that dysregulation of miRNA can alter the synthesis of either oncogenic protein or tumor suppressor protein, which then leads to tumour formation and progression [8, 9]. miR-130a has been reported to be overexpressed or under-expressed in different types of cancer including colon cancer, non-small lung cancer, chronic myeloid leukaemia and hepatocellular carcinoma indicating the important roles of miR-130a in cancer development, progression and metastasis [8-11]. A study reported that miR130a is an important tumor suppressor in breast cancer by targeting RAB5A [12]. However, at present, there is a limited evidence of the role of miR-130a in oncogenesis of breast cancer.

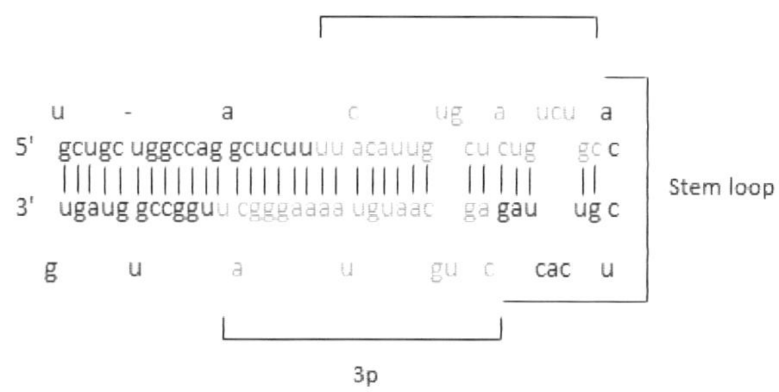


**Figure 1.** Figure shows the structure of pre-miR130a consisting of 5' (5p) end, stem loop and 3' (3p) end.

Thus, in order to investigate functional activity of miRNAs, selection of precise and efficient miRNA silencing technique is important. One of the genome editing technology, Clustered Regularly Interspaced Short Palindromic Repeats (CRISPRs) system is increasingly being employed to silence protein-coding genes in several model organisms [13]. CRISPRs are DNA loci with short repetitions of base sequences [14]. Short segments of "spacer DNA" are present in between each repetition from previous exposures to a virus [14]. Cas9 is a nuclease guided by small RNAs which promotes genome editing via double strand break at a targeted genomic locus [13]. Cas9 decreases the off-target effect as it is highly specific, efficient and well-suited for high throughput and multiple gene editing for different cell types and organisms [13, 15]. This double-nicking strategy using CRISPR-Cas9 approach has successfully generated stable knockout cell line models in human [16]. Thus, CRISPR system was used as a tool to silence miR130a in this study. This may then assist in studying the role of miR130a at molecular level in breast cancer.

In this study, the expression of miR130a-3p and miR130a-5p in different types of cell lines (human breast cancer cell line (MCF7), acute myeloid leukaemia cell lines (Kasumi-1 and SKNO-1), human normal breast cell line (MCF10A), human lung cancer cell line (A549) were compared adjacent to non-cancerous cell line (HEK293T) by real-time quantitative PCR. CRISPRs system was used to silence the expression of mature miR130a-3p and miR130a-5p in MCF7 cell line. The suppression activity was then confirmed by real-time quantitative PCR.

## 2. Materials and Methods

### *2.1 Cell and culture conditions*

MCF7 was cultured in RPMI 1640 (Gibco, Thermo Fisher Inc, Grand Island, USA) medium supplemented with 10% heat inactivated fetal bovine serum (FBS) and 1% penicillin-streptomycin. MCF10A was cultured in DMEM (Gibco, Thermo Fisher Inc, Grand Island, USA) medium supplemented with 5% (v/v) horse serum, 0.2% EGF (20 mg/mL), 0.5 mg/mL hydrocortisone, 10 μg/mL insulin and 1% penicillin-

IOP Conf. Series: Journal of Physics: Conf. Series **851** (2017) 012037 doi:10.1088/1742-6596/851/1/012037

streptomycin. HEK293T and A549 were cultured in DMEM (Gibco, Thermo Fisher Inc, Grand Island, USA) medium supplemented with 10% heat inactivated FBS and 1% penicillin-streptomycin. Cultures were incubated in humidified atmosphere with 5% $CO_2$ at 37 $^0$C.

*2.2 Construction of plasmid miR130a-CRISPR*

The oligonucleotides for site-specific chromatin cleavage of miR-130a target regions (Table 1) were designed using a bioinformatics filtering tool (http://crispr.mit.edu/). Oligonucleotides used to target site-specific chromatin cleavage were: miR130a-3p (3p) forward 5'-CACCGTGCAATGTTAAAA GGGCAT-3', reverse 5'-AAACATGCCCTTTTAACATTGCAC-5'; miR130a-5p (5p) forward 5'-CAC CGCACAATGTGAAAAGAGCTC-3', reverse 5'-AAACGAGCTCTTTTCACATTGTGC-3'; and miR130a-stem loop (SL) forward 5'-CACCGTGTAACACGATGACAGACG-3', reverse 5'-AAACCGTCTGTCATCGTGTTACAC-3'.

Each pair of oligonucleotide was phosphorylated and annealed at 37 °C (30 min) followed by 95 °C (5 min and then ramp down to 25 °C at 5 °C /min) in a thermal cycler (Bio-Rad MyCycler Thermal Cycler PCR). The annealed oligonucleotides were diluted at 1:200 with RNase-free water and cloned into the *BpiI* sites of pSpCas9(BB)-2A-GFP (Cp458) plasmid [13]. The plasmid was a gift from Feng Zhang (Addgene plasmid # 48138). The miR130a-CRISPR plasmids were then transformed into stable competent *E.coli* (NEB C040H) according to manufacturer's protocol. The transformed cells were spread on LB agar with ampicillin (100 μg/mL) and incubated overnight at 37 °C.

*2.3 Screening oligonucleotide targets*

A trace of two to three single colonies were picked and screened for oligonucleotide targets. PCR was carried out in a thermal cycler (Bio-Rad MyCycler Thermal Cycler PCR) using the following primers: forward 5'-GAGGGCCTATTTCCCATG-3'; reverse 5'-AAGGTCATGTACTGGGCACA-3'. The size (~600 bp) of each target was confirmed by running on 1% 1× TAE agarose gel according to the established protocol. The positive clones were then sequenced (1$^{st}$ BASE Sequencing, Selangor, Malaysia) to further confirm the targets for each clones.

*2.4 Transfection*

A total of 5 μg plasmids were transfected into 0.18 × $10^6$ MCF7 cells by single transfection of 3p (miR130a-3p-Cp458), 5p (miR130a-5p-Cp458) or SL (miR130a-SL-Cp458) using Lipofectamine® 3000 (Life Technologies, Carlsbad, USA) according to manufacturer's protocol. Empty plasmid (Cp458) was used as a control. The transfected cells were maintained in the incubator at 37 $^0$C under humidified 5% $CO_2$.

*2.5 RNA isolation, reverse transcription (RT) and quantitative real-time PCR (RT-qPCR)*

After 48 hours, cells were harvested and total RNA was extracted using miRNeasy Mini Kit (Qiagen, Germany) according to manufacturer's protocol. Reverse transcription and quantitative real-time PCR of mature miR130a-3p, miR130a-5p and small nucleolar U6 (RNU6B) normalisation control were performed as described [17, 18] with some minor changes. Briefly, 500 ng of total RNA was reverse transcribed by Taqman® microRNA reverse transcription kit (Applied Biosystems, Foster City, USA). Quantitative PCR (qPCR) was performed by Taqman® Universal Master Mix *II*, no UNG (Applied Biosystems, USA) using ABI StepOne™Plus Real-Time PCR Systems. The level of mature miR130a-3p (hsa-miR-130a-3p, Taqman® microRNA Assays) and miR130a-5p (has-miR-130a-5p, Taqman® microRNA Assays) expression was normalised to those of internal control (RNU6B, Taqman® microRNA Control Assays). The relative expression levels were evaluated using the $2^{-\Delta\Delta Ct}$.

*2.6 Statistical analysis*

Data was analysed using Microsoft Excel 2013. Statistical differences were determined by one tailed *t* test, with values of $p < 0.05$ considered statistically significant. Results were expressed as mean ± S.E.M. from minimum of three independent experiments.

## 3. Results

*3.1 Screening on the expression miR130a-3p and miR130a-5p in different cell lines*

The expression of miR130a-5p and miR130a-3p of HEK293T, MCF7 and MCF10A, were determined by RT-qPCR. Expression of miR130a-5p was significantly up-regulated in MCF7 ($p < 0.01$) when compared to HEK293T (Figure 2 A). Expression of miR130-3p was significantly up-regulated in MCF7 ($p < 0.05$) compared to MCF10A (Figure 2 B). HEK293T and MCF10A cell line were used as control for comparative CT method ($2^{-\Delta\Delta CT}$) expression analysis as they are considered as a non- or low-tumorigenic cell line. The results were presented as means ± S.E.M. from three independent experiments.

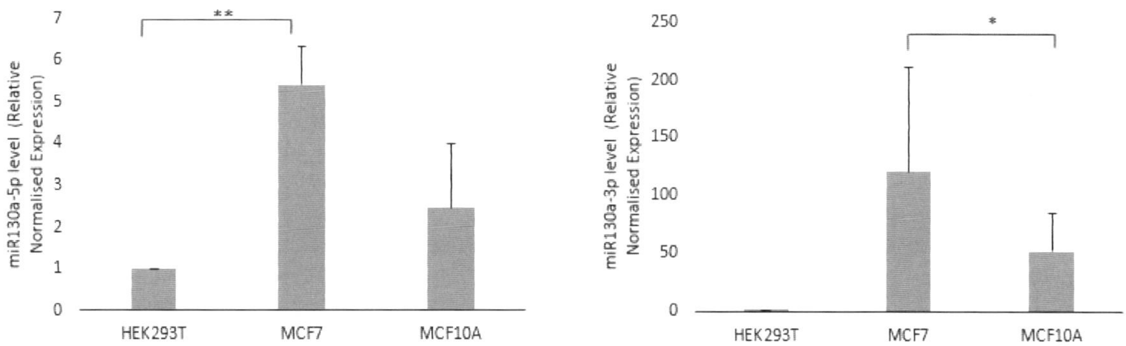


**Figure 2.** HEK293T, MCF7 and MCF10A expression levels of miR130a-5p (A) and miR130a-3p (B) detected by RT-qPCR (*$p < 0.05$; ** $p < 0.01$ (one tail, paired *t* test)).

*3.2 Analysis of single transfection*

5p single transfection resulted in significant down-regulation of miR130a-5p by 2-fold compared to control. Both SL and 3p did not cause down-regulation of miR130a-5p expression (Figure 3 A). In contrast, single transfection of plasmid stem loop, 3p and 5p did not down-regulates the expression of miR130a-3p when compared to control (Figure 3 B). Transfection of empty Cp458 plasmid into MCF7 was used as a control for comparative CT method ($2^{-\Delta\Delta CT}$) expression analysis. Results were obtained from three independent biological replicates.

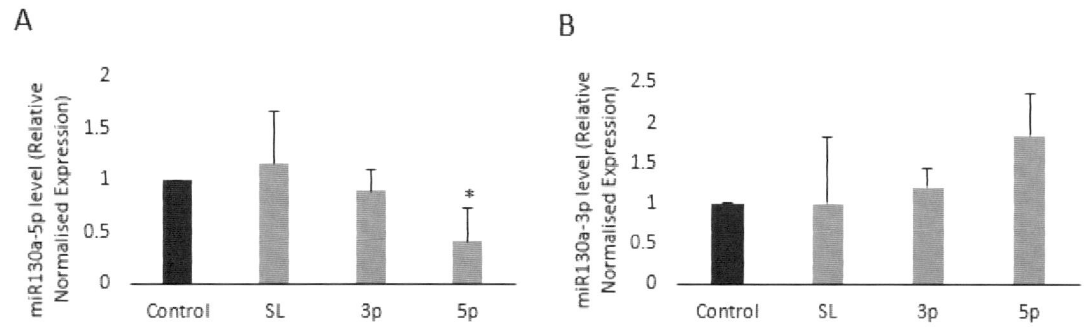


**Figure 3.** Down-regulation of the miR130a-5p (A) and miR130a-3p (B) expression by plasmid targeting 3p (miR130a-3p-Cp458), 5p (miR130a-5p-Cp458) or stem loop (miR130a-SL-Cp458) (*$p < 0.05$ (one tail, paired *t* test)).

## 4. Discussion

miRNAs play important roles in the regulation of regulatory factors involved with the pathogenesis of cancer [19]. The changes in their expression is a common character of many cancers, including breast cancer [20]. In this study, miR130a expression analysis showed that the miR130a-5p was observed at highest level in MCF7 compared to HEK293 ($p < 0.01$). Expression of mir130a-3p was significantly higher in MCF7 compared to MCF10A ($p < 0.05$). The overall expression of miR130a-3p was higher compared to miR130a-5p in all cell lines. The reason behind this is that 3p remains as mature miRNA and therefore is preferentially integrated into the RISC [19]. As the expression of miR130a-5p and miR130a-3p was significantly high in MCF7, MCF7 was used as a model to down-regulate the expression of miR130a-3p and miR130a-5p by CRISPR/Cas9 system.

CRISPR/Cas9 systems has been shown to increase the efficacy of editing the genome differentiated in human cells [15, 21]. Thus, this system was used to down-regulate miR130a expression. The CRISPR/Cas9 system silenced either 3p, 5p or stem loop of the pre-miR130a in the cytoplasm. In the cytoplasm, pre-miR130a will be processed by the RNAse III enzyme Dicer producing mature miRNAs that are incorporated into a multi-protein complex [22]. Result shows that the expression of miR-130a-5p was significantly down-regulated with 5p single transfection. On the other hand, all single transfection does not cause significant change of miR130a-3p. This suggest that our CRISPR system works to suppress the mature level of miR130a-5p but not the miR130a-3p.

## 5. Conclusion

The expression of miR130a-5p was successfully down-regulated by single 5p transfection. This technique may be useful in answering certain clinical question at molecular level, such as role of miR130a in breast cancer cells.

## 6. References

[1] Youlden DR, Cramb SM, Dunn NA, Muller JM, Pyke CM and Baade PD 2012 The descriptive epidemiology of female breast cancer: an international comparison of screening, incidence, survival and mortality *Cancer Epidemiology* 36(3)237-48

[2] Bartel DP 2014 MicroRNAs: genomics, biogenesis, mechanism, and function *Cell* 116(2)281-97

[3] Ha M, Kim VN 2014 Regulation of microRNA biogenesis *Nature Reviews Molecular Cell Biology* 15(8)509-24

[4] Krol J, Loedige I and Filipowicz W 2010 The widespread regulation of microRNA biogenesis, function and decay *Nature Reviews Genetics* 11(9)597-610

[5] Hutvágner G, McLachlan J, Pasquinelli AE, Bálint É, Tuschl T and Zamore PD 2001 A cellular function for the RNA-interference enzyme Dicer in the maturation of the let-7 small temporal RNA. *Science* 293(5531)834-8

[6] Ketting RF, Fischer SE, Bernstein E, Sijen T, Hannon GJ and Plasterk RH 2001 Dicer functions in RNA interference and in synthesis of small RNA involved in developmental timing in C. elegans *Genes & Development* 15(20)2654-9

[7] Schwarz DS, Hutvágner G, Du T, Xu Z, Aronin N and Zamore PD 2003 Asymmetry in the assembly of the RNAi enzyme complex *Cell* 115(2)199-208

[8] Wang X-C, Tian L-L, Wu H-L, Jiang X-Y, Du L-Q, Zhang H et al 2010 Expression of miRNA-130a in nonsmall cell lung cancer *The American Journal of the Medical Sciences* 340(5)385-8

[9] Zhou P, Jiang W, Wu L, Chang R, Wu K and Wang Z 2012 miR-301a is a candidate oncogene that targets the homeobox gene Gax in human hepatocellular carcinoma *Digestive Diseases and Sciences* 57(5)1171-80

[10] Chen Q, Ye Z, Lin S-c and Lin B 2012 Recent patents and advances in genomic biomarker discovery for colorectal cancers. *Recent Patents on DNA & Gene Sequences* 4(2)86-93

[11] Suresh S, McCallum L, Lu W, Lazar N, Perbal B and Irvine AE 2011 MicroRNAs 130a/b are regulated by BCR-ABL and downregulate expression of CCN3 in CML *Journal of Cell Communication and Signaling* 5(3)183-91
[12] Pan Y, Wang R, Zhang F, Chen Y, Lv Q, Long G et al 2015 MicroRNA-130a inhibits cell proliferation, invasion and migration in human breast cancer by targeting the RAB5A *International Journal of Clinical and Experimental Pathology* 8(1)384-93
[13] Ran FA, Hsu PD, Wright J, Agarwala V, Scott DA and Zhang F 2013 Genome engineering using the CRISPR-Cas9 system *Nature Protocols* 8 (11)2281-308
[14] Marraffini LA and Sontheimer EJ 2010 CRISPR interference: RNA-directed adaptive immunity in bacteria and archaea *Nature Reviews Genetics* 11(3)181-90
[15] Mali P, Yang L, Esvelt KM, Aach J, Guell M, DiCarlo JE et al 2013 RNA-guided human genome engineering via Cas9. *Science* 339(6121)823-6
[16] Zhao Y, Dai Z, Liang Y, Yin M, Ma K, He M et al 2014 Sequence-specific inhibition of microRNA via CRISPR/CRISPRi system. *Scientific Reports* 4 doi:10.1038/srep03943
[17] Raymond CK, Roberts BS, Garrett-Engele P, Lim LP and Johnson JM 2005 Simple, quantitative primer-extension PCR assay for direct monitoring of microRNAs and short-interfering RNAs. *RNA* 11(11)1737-44
[18] Fulci V, Chiaretti S, Goldoni M, Azzalin G, Carucci N, Tavolaro S et al 2007 Quantitative technologies establish a novel microRNA profile of chronic lymphocytic leukemia *Blood* 109(11)4944-51
[19] Di Leva G, Garofalo M and Croce CM 2014 MicroRNAs in cancer. *Annual Review of Pathology* 9:287
[20] Esquela-Kerscher A and Slack FJ 2006 Oncomirs—microRNAs with a role in cancer *Nature Reviews Cancer* 6(4)259-69
[21] Pépin NL, Chapdelaine P, Rodriguez Y, Tremblay J-P and Fortier MA 2014 Generation of human endometrial knockout cell lines with the CRISPR-Cas9 system confirms the prostaglandin F2α synthase activity of aldo-ketoreductase 1B1 *Molecular Human Reproduction* 20(7)650-63
[22] Hurst DR, Edmonds MD and Welch DR 2009 Metastamir: the field of metastasis-regulatory microRNA is spreading *Cancer Research* 69(19)7495-8

**Acknowledgement**
This project is supported by USM Short Term Grant (304/CIPPT/6313203).

IOP Conf. Series: Journal of Physics: Conf. Series **851** (2017) 012038 doi:10.1088/1742-6596/851/1/012038

# Study on IL-2 and CA 15-3 level as combined biomarkers in monitoring chemotherapeutic response among invasive breast cancer patients

**Ahmed Muthanna Abdul Hameed, Auni Fatin Abdul Hamid , Nurul Shahfiza Noor, Gokula Kumar Appalanaido and Shahrul Bariyah Sahul Hamid**

Advanced Medical and Dental Institute (AMDI), Universiti Sains Malaysia, Kepala Batas, Penang, 13200, Malaysia.

E-mail: ahmad_al_tuhafi@yahoo.com and shahrulbariyah@usm.my

**Abstract.** In Malaysia, breast cancer is the most frequent type of disease among women. This study was designed to determine the clinical usefulness of carbohydrate antigen (CA 15-3) and interleukin 2 (IL-2) levels as combined biomarkers in monitoring breast cancer patient's response to chemotherapy. Ethical approval was obtained to recruit patients with histologically confirmed invasive ductal carcinoma (IDC) attending Oncology Clinic at Advanced Medical and Dental Institute. Whole blood was collected from 10 IDC breast cancer patients' pre and post primary chemotherapy. Plasma was separated from the whole blood to determine the CA 15-3 level and IL-2 level using enzyme-linked immunosorbent assay (ELISA) pre and post-treatment. In addition, the histological findings, tumour stage and other patients' data were obtained from the medical record. Findings showed that IL-2 had borderline significant changes between pre- and post-chemotherapy ($p$ = 0.074) whereas for CA 15-3, there was insignificant differences of CA 15-3 level between pre and post-chemotherapy ($p > 0.05$). It was noted that only CA 15-3 level had significant correlation with tumour size. This study demonstrates that IL-2 level requires further investigation in a larger sample size to correlate its potential use as combined biomarker with CA 15-3 in monitoring response to chemotherapy.

## 1. Introduction

The *MUC1* gene was first recognized in human milk, where it is shed from the milk secreting epithelial cells [1]. Further, it is now known to be expressed on different types of hematopoietic cells as well [2, 3]. However, this polymorphic transmembrane glycoprotein is frequently overexpressed on malignant glandular cell surfaces, resulting in increased levels shed into the blood of some patients with carcinomas [4, 5]. Numerous circulating mucinous markers, including carbohydrate antigen 15-3 and 27.29 (CA 15-3 and CA 27.29) are secreted products of the polymorphic *MUC1* gene, and can be used as biomarkers in patients with breast cancer [2]. Although high levels of the protein CA 15-3 is produced by tumour tissue, it is not usually elevated in serum until the cancer has reached an advanced stage. A high CA15-3 level of about >32 U/mL typically indicates advanced breast cancer and a larger tumour burden [6]. Both CA 15-3 and CA 27.29 antigen are approved by the FDA as a biomarker for breast cancer in 1997 for the detection of recurrent breast cancer, and also in the monitoring of the response to therapy for metastatic breast cancer [6, 7]. However, they have to be used in conjunction with diagnostic imaging, clinical history, and physical examination [8, 9].

Interleukin-2 (IL-2) is a multi-potential cytokine secreted by T helper 1 cells (Th1) which is involved in the growth and differentiation of a wide range of immunologic functions, particularly in T

IOP Conf. Series: Journal of Physics: Conf. Series **851** (2017) 012038 doi:10.1088/1742-6596/851/1/012038

and B cells. Additionally, it enhances the cytolytic functions of natural killer (NK) cells [10-13]. It is also known to have some function in the proliferation of several non-lymphoid cells [11]. Interleukin-2 gene expression in hepatocarcinoma patient with interleukin-2 positive tumours was reported to show favourable prognosis [14]. A previous study also suggested that reduced plasma IL-2 level concentration in patients with breast cancer was a possible risk factor for relapse [15]. In fact, this study is the first to assess IL-2 level as a potential biomarker in monitoring breast cancer patient's response to chemotherapy. Early detection of breast cancer is a challenge due to its heterogeneity and complexity. Thus, a profile test with multiple biomarkers should be conducted to characterize the full spectrum of this heterogeneous disease rather than a single analyte [16]. In this study, CA 15-3 and IL-2 levels were measured pre and post-chemotherapy treatment to assess their usefulness in monitoring patient's response to chemotherapy.

## 2. Materials and Method

The aim of this study was to monitor response to chemotherapy based on both CA-153 and IL-2 level instead of a single biomarker. This was achieved by measuring the difference in both biomarkers (CA-153 and IL-2) level pre- and post-chemotherapy among invasive breast cancer patients.

### *2.1. Sample size, patients selection and ethical approval*

Sample size was calculated using PS software version 3.0.43 (Dupont & Plummer, 1997) based on comparing 2 means. To detect the difference of 7.5 pg/µl in IL-2 with 80% power and $\alpha = 0.05$, 9 subjects are required (SD was estimated 7.19 pg/µl). By anticipating a 10% attrition rate, a total of 10 subjects were recruited in this study. A total of 10 patients attending Oncology Clinic at Clinical Trial Complex (CTC), Advanced Medical and Dental Institute (AMDI) were recruited in this study after obtaining the ethical approval. Inclusion criteria were based on histologically confirmed invasive breast cancer patients of ductal origin and women who were post pubertal. Meanwhile, exclusion criteria include tumours of non-ductal origin, male breast carcinoma, and women who were pre pubertal.

### *2.2. Biochemical analysis*

From each patient, (1-2 ml) venous whole blood sample was collected pre and post primary chemotherapy in EDTA anticoagulant tubes. The samples were centrifuged at 4000 rpm for 5 minutes to separate the plasma for clinical chemistry analysis. The plasma was transferred into Eppendorf tubes and stored at -80°C until analysis. Clinical chemistry analyses were done to determine the levels of the tumour markers i.e. CA 15-3 and IL-2 in the patients' blood by ELISA method (enzyme-linked immunosorbent assay). CanAg CA15-3 EIA kit (Fujirebio Diagnostics, Japan) was used for measuring the level of CA 15-3 in the patient's plasma as recommended by the manufacturer's instructions.

### *2.3. Statistical analysis*

Data was analysed using IBM SPSS Statistics v.20 software (SPSS Inc., Chicago, IL, USA). Distribution normality tests were performed using histogram (SPSS). The quantitative variables were summarized as median and Interquartile Range (IQR) with 95% Confidence interval (CI) as they were not normally distributed. Accordingly, Wilcoxon Signed Ranks non-parametric two-tailed test was used to compare the median of the tumour marker levels before and after chemotherapy. The correlation between the levels of each tumour marker and tumour size and lymph node metastasis was determined by using Spearman's Rank Correlation Coefficient test and $p$-values less than 0.05 were regarded as statistically significant.

## 3. Results

### *3.1. Patients' demographic characteristics and histological findings*

All of the 10 patients were females from Malaysian population with histologically confirmed IDC (invasive ductal carcinoma), and who were post pubertal receiving primary chemotherapy. The histological findings, tumour stage and other patients' data were obtained from the patients' medical record in Oncology Clinic at AMDI. The mean age of the patients was 48.4 years (range 40-67 years). Tumour size was classified as T2 (tumour size between 2 and 5 cm in greatest dimension) in 7/10 patients (70%) and T4 (tumour extends to chest wall) in 3/10 (30%) of cases according to the TNM classification. The percentage of patients with negative lymph nodes and positive nodes was 50 % for each. Whole blood samples were collected pre and post different cycles of primary chemotherapy. The number of samples collected in cycle 1 was 3/10 (30%) and in cycle 2 was 6/10 (60%) whereas in cycle 4 was 1/10 (10%).

### *3.2. Biochemical findings*

The level of CA 15-3 for each sample before and after treatment is shown in Figure 1. There was no significant difference of CA 15-3 levels in the samples of the patients pre and post-chemotherapy as shown in Table 1. Correlation between the levels of CA 15-3, tumour size and lymph node status was determined by using Spearman's Rank Correlation Coefficient test. There was a significant moderate positive correlation between the CA 15-3 level and the tumour size $r = 67\%$ ($p < 0.05$) (Figure 2). Meanwhile, the correlation between CA 15-3 level and lymph node status was moderate positive $r = 56\%$ ($p > 0.05$) even though the $p$-value did not show any significant results. The level of IL-2 for each sample before and after treatment is illustrated in Figure 3. No significant difference of IL-2 levels was found in the samples of the patient's pre and post-chemotherapy as shown in Table 2. Correlation between the IL-2 level, tumor size and lymph node status was determined by using Spearman's Rank Correlation Coefficient test. No significant correlation was observed between IL-2 level and the tumour size $r = 25\%$ ($p > 0.05$) or between IL-2 level and lymph node status in patients $r = 22\%$ ($p > 0.05$) as well.

**Table 1:** The patients' demographic data

| Study participant No. | Age (year) | Height (cm) | Weight (kg) | Cycle of collection | Chemotherapy x no. of cycles | Tumor stage (TNM) system | Tumor size (cm) | Size (cm) in max. dimen. |
|---|---|---|---|---|---|---|---|---|
| 1 | 50 | 138 | 35.0 | C4 | TC x4 | T4NxMx | 15x15x5 | 15 |
| 2 | 40 | 154 | 56.5 | C2 | AC-T x4 | T4N2Mx | 12x12 | 12 |
| 3 | 43 | 156 | 63.6 | C1 | AC x4 | T2N0Mx | 4.5x3.5x3.4 | 4.5 |
| 4 | 41 | 160 | 62.0 | C2 | AC-T x4 | T2N2M0 | 4x3x2.5 | 4 |
| 5 | 45 | 162 | 73.8 | C2 | AC-T x4 | T2N1M0 | 2.8x2.5x3.8 | 3.8 |
| 6 | 42 | 155 | 56.0 | C2 | FEC x6 | T2N0Mx | 3.5x2.5x2.5 | 3.5 |
| 7 | 60 | 142 | 79.0 | C2 | AC-T x4 | T2N2Mx | 3.5x2.5x2 | 3.5 |
| 8 | 49 | 158 | 53.2 | C2 | AC x4 | T2N0Mx | 1x1x3 | 3 |
| 9 | 67 | 149 | 67.4 | C1 | AC-T x4 | T4NxM1 | 17x13 | 17 |
| 10 | 47 | 142 | 51.7 | C1 | AC-T x4 | T2N1Mx | 5x3x2 | 5 |

A, Adriamycin; C, Cyclophosphamide; T, Docetaxel; F, Fluorescent; E, Epirubicin

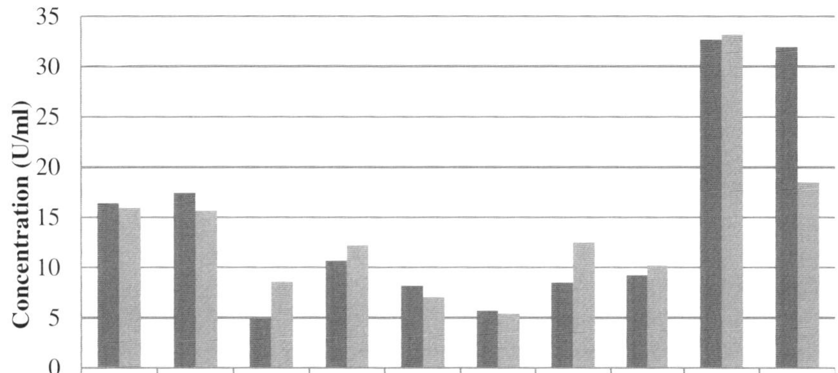


| | 1 | 2 | 3 | 4 | 5 | 6 | 7 | 8 | 9 | 10 |
|---|---|---|---|---|---|---|---|---|---|---|
| ■ CA 15-3 pre-treatment | 16.43 | 17.41 | 4.93 | 10.65 | 8.18 | 5.65 | 8.5 | 9.22 | 32.7 | 31.98 |
| ■ CA 15-3 post-treatment | 15.91 | 15.65 | 8.57 | 12.21 | 7.01 | 5.39 | 12.47 | 10.19 | 33.17 | 18.51 |

**Figure 1:** The levels of CA 15-3 in the samples pre and post-chemotherapy

**Table 2**: Statistical analysis on difference of CA 15-3 concentration pre and post-treatment

| Variable | No. of subjects | Pre-treatment Median (IQR) | Post-treatment Median (IQR) | *Z* statistic[a] | *P* value[a] |
|---|---|---|---|---|---|
| **CA15-3 (all C)** | $n$ =10 | 9.9 (13.51) | 12.3 (8.38) | - 0.153 | 0.878 |
| **CA 15-3 (C1)** | $n$ =3 | 31.9 (-) | 18.5 (-) | 0.0 | 1.0 |
| **CA 15-3 (C2)** | $n$ =6 | 8.8 (4.79) | 11.2 (6.66) | - 0.314 | 0.73 |

[a] Wilcoxon Signed Ranks Test, IQR = Interquartile Range, C = Cycle of chemotherapy

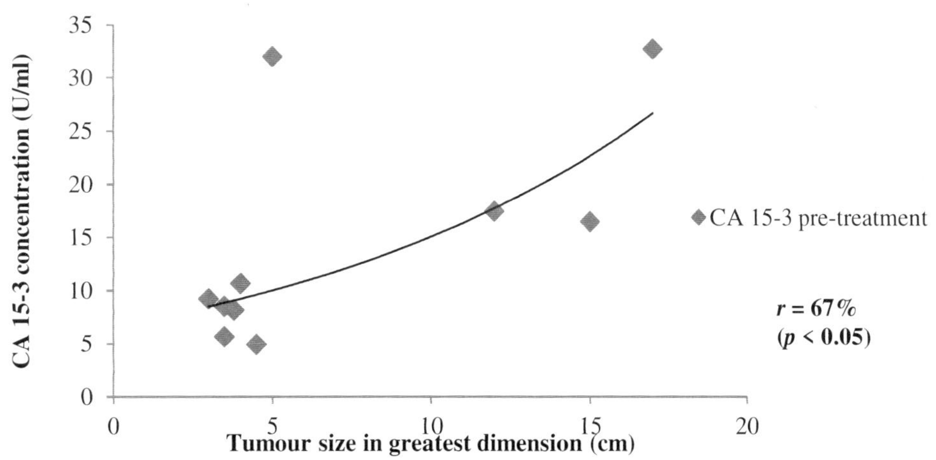


******r*: correlation coefficient

**Figure 2:** The correlation between the level of CA 15-3 in the samples of the patients (x-axis) and tumour size in the greatest dimension (y-axis)

IOP Conf. Series: Journal of Physics: Conf. Series **851** (2017) 012038 doi:10.1088/1742-6596/851/1/012038

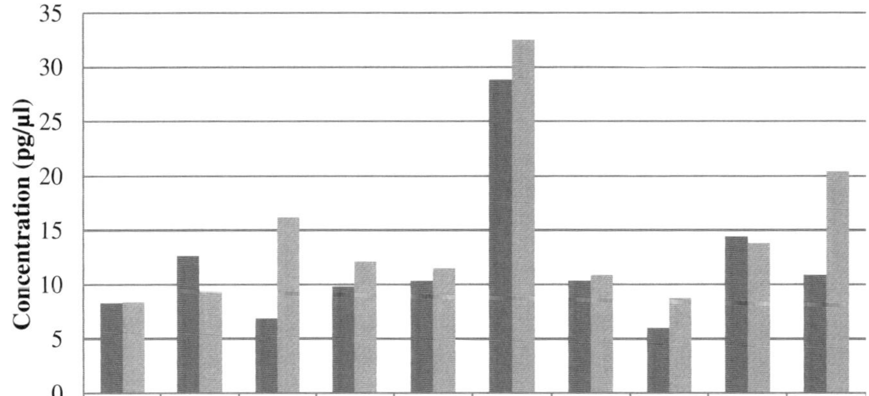


| | 1 | 2 | 3 | 4 | 5 | 6 | 7 | 8 | 9 | 10 |
|---|---|---|---|---|---|---|---|---|---|---|
| ■ IL-2 pre-treatment | 8.26 | 12.61 | 6.85 | 9.8 | 10.35 | 28.82 | 10.35 | 5.96 | 14.39 | 10.88 |
| ■ IL-2 post-treatment | 8.36 | 9.27 | 16.2 | 12.13 | 11.52 | 32.51 | 10.88 | 8.76 | 13.78 | 20.41 |

**Figure 3:** The levels of IL-2 in the samples of the patients' pre and post-chemotherapy

**Table 3**: Statistical analysis on difference of IL-2 concentration pre and post-treatment

| Variable | No. of subjects | Pre-treatment Median (IQR) | Post-treatment Median (IQR) | *Z* statistic[a] | *P* value[a] |
|---|---|---|---|---|---|
| **IL-2 (all C)** | *n* = 10 | 10.3 (5.15) | 11.8 (8.11) | - 1.784 | 0.074 |
| **IL-2 (C1)** | *n* = 3 | 10.8 (-) | 16.2 (-) | -1.069 | 0.285 |
| **IL-2 (C2)** | *n* = 6 | 10.3 (7.28) | 11.2 (8.08) | - 1.153 | 0.249 |

[a] Wilcoxon Signed Ranks Test, IQR = Interquartile Range, C = Cycle of chemotherapy

## 4. Discussion

In this study, the results revealed that only 20 % of the patients had CA 15-3 level higher than the cutoff (> 30 U/ml) and there was insignificant differences of CA 15-3 level between pre and post-chemotherapy ($p > 0.05$). Additionally, no significant correlation ($p > 0.05$) was found between CA 15-3 and lymph node status, whilst there was a significant correlation ($p < 0.05$) between the level of this biomarker and the tumour size, suggesting that tumour tissue itself may be the source of serum CA 15-3 [16]. Previously higher CA 15-3 level was reported to correlate with larger tumour size and more advanced disease [17-19]. A study in which 818 patients with histologically verified diagnosis of invasive breast cancer had their CA 15-3 serum level determined pre-operatively. Finding showed that 15.2% of patients with clinically non-metastatic breast cancer had high serum CA 15-3 levels (>30 U/ml) [20]. However, according to the present literature, CA 15-3 is not specific or sensitive enough to detect early breast cancer and the elevation of CA15-3 is not a reliable indicator in the diagnosis of breast cancer since normal serum levels can also be found in women with breast cancer [18].

Further analysis on IL-2 level was done to correlate its level with CA 15-3. We noted 80 % of the patients showed higher levels of IL-2 in post-chemotherapy than pre-chemotherapy and this result was in line with studies done by Fang et al., (2006) and Orditura et al., (2000). The IL-2 level showed borderline significant changes between pre- and post-chemotherapy ($p = 0.074$). Additionally, no significant correlation was observed between IL-2 level and the tumour size ($p > 0.05$) or between IL-2 level and lymph node status in patients ($p > 0.05$). IL-2 is a cytokine released from T helper lymphocytes. It promotes the generation, proliferation, differentiation of T lymphocytes, enhances the activity of natural killer cells, induces the generation of lymphokine-activated killer (LAK) cells, and promotes the production of antibodies by B lymphocytes. Through these mechanisms, it plays a vital role in antitumor immune responses [10, 21]. The increased IL-2 level in this study demonstrated that anti-tumour immune response of the patients was increased after receiving chemotherapy. Similar result was obtained from another study on IL-2 serum level among 60 advanced non-small-cell lung

(NSCL) cancer patients whereby a significant increase in IL-2 level was observed following chemotherapy [22].

## 5. Conclusion
This study revealed that IL-2 level was borderline significant when compared between pre- and post-chemotherapy, whereas CA15-3 serum level had significant correlation with the tumour size. It suggests the use of IL-2 in monitoring response to chemotherapy but requires further investigation in a larger sample size to correlate its potential as combined biomarker with CA 15-3.

## 6. References
[1] Taylor-Papadimitriou J, Burchell J, Miles D W and Dalziel M 1999. MUC1 and cancer. Biochimica et Biophysica Acta (BBA) - Molecular Basis of Disease, 1455, 301-313.
[2] Begum M, Karim S, Malik A, Khurshid R, Asif M, Salim A, Nagra S A, Zaheer A, Iqbal Z, Abuzenadah A M, Alqahtani M H and Rasool M 2012. CA 15-3 (Mucin-1) and Physiological Characteristics of Breast Cancer from Lahore, Pakistan. Asian Pacific Journal of Cancer Prevention, **13**, 5257-5261.
[3] Engelmann K, Shen H and Finn O J 2008. MCF7 side population cells with characteristics of cancer stem/progenitor cells express the tumor antigen MUC1. Cancer Res, **68**, 2419-26.
[4] Handy B 2009. The Clinical Utility of Tumor Markers. Laboratory Medicine, **40**, 99-103.
[5] Zaretsky J, Barnea I, Aylon Y, Gorivodsky M, Wreschner D and Keydar I 2006. MUC1 gene overexpressed in breast cancer: structure and transcriptional activity of the MUC1 promoter and role of estrogen receptor alpha (ERα) in regulation of the MUC1 gene expression. Molecular Cancer, **5**, 1-14.
[6] Joy J E, Penhoet E E and Petitti D B 2005. Saving Women's Lives: Strategies for Improving Breast Cancer Detection and Diagnosis, Washington (DC), National Academies Press (US).
[7] Misek D E and Kim E H 2011. Protein Biomarkers for the Early Detection of Breast Cancer. International Journal of Proteomics, 2011, 1-9.
[8] Harris L, Fritsche H, Mennel R, Norton L, Ravdin P, Taube S, Somerfield M R, Hayes D F, Bast R C Jr and American Society of Clinical O 2007. American Society of Clinical Oncology 2007 update of recommendations for the use of tumor markers in breast cancer. J Clin Oncol, **25**, 5287-312.
[9] Khatcheressian J L, Hurley P, Bantug E, Esserman L J, Grunfeld E, Halberg F, Hantel A, Henry N L, Muss H B, Smith T J, Vogel V G, Wolff A C, Somerfield M R, Davidson N E and American Society of Clinical O 2013. Breast cancer follow-up and management after primary treatment: American Society of Clinical Oncology clinical practice guideline update. J Clin Oncol, **31**, 961-5.
[10] Hu X B, Ouyang L Z and Tang L L 2013. Interleukin-2 gene polymorphisms and prognosis of breast cancer. Genet Test Mol Biomarkers, **17**, 453-7.
[11] Garcia-Tunon I, Ricote M, Ruiz A, Fraile B, Paniagua R and Royuela M 2004. Interleukin-2 and its receptor complex (alpha, beta and gamma chains) in in situ and infiltrative human breast cancer: an immunohistochemical comparative study. Breast Cancer Res, **6**, R1-7.
[12] Saglam S, Suzme R and Gurdol F 2009. Serum tumor necrosis factor-alpha and interleukin-2 concentrations in newly diagnosed ERBB2 (HER2/neu) positive breast cancer patients.
[13] Romagnani S 1999. Th1/Th2 cells. Inflamm Bowel Dis, **5**, 285-94.
[14] Ikeguchi M and Hirooka Y 2005. Interleukin-2 gene expression is a new biological prognostic marker in hepatocellular carcinomas. Onkologie, **28**, 255-9.
[15] Arduino, S., et al., Reduced IL-2 level concentration in patients with breast cancer as a possible risk factor for relapse. Eur J Gynaecol Oncol, 1996. 17(6): p. 535-7.
[16] Sauter E R and Daly M B 2010. Breast Cancer Risk Reduction and Early Detection, New York, USA, Springer.
[17] Hewala T I, Abd El-Monaim N A, Anwar M and Ebied S A 2012. The clinical significance of serum soluble Fas and p53 protein in breast cancer patients: comparison with serum CA 15-3. Pathol Oncol Res, **18**, 841-8.

[18] Daniele A, Divella R, Trerotoli P, Caringella M E, Paradiso A, Casamassima P, Abbate I, Quaranta M and Mazzocca A 2013. Clinical Usefulness of Cancer Antigen 15-3 in Breast Cancer Patients Before and After Surgery. Surgery, 472, 66.1.

[19] Lumachi F, Basso S M, Brandes A A, Pagano D and Ermani M 2004. Relationship between tumor markers CEA and CA 15-3, TNM staging, estrogen receptor rate and MIB-1 index in patients with pT1-2 breast cancer. Anticancer Res, **24**, 3221-4.

[20] Leboeuf R, Langlois M, Martin M, Ahnadi C and Fink G 2005. Hook effect in calcitonin immunoradiometric assay in patients with metastatic medullary thyroid carcinoma: Case report and review of literature. J Clin Endocrin Metab, **91**, 361 - 364.

[21] Fang N, Li Y, Xu Y S, Ma D, Fu P, Gao H Q, Gao F T, Yang H S and Yang Z J 2006. Serum concentrations of IL-2 and TNF-alpha in patients with painful bone metastases: correlation with responses to 89SrCl2 therapy. J Nucl Med, **47**, 242-6.

[22] Orditura M, Romano C, De Vita F, Galizia G, Lieto E, Infusino S, De Cataldis G and Catalano G 2000. Behaviour of interleukin-2 serum levels in advanced non-small-cell lung cancer patients: relationship with response to therapy and survival. Cancer Immunol Immunother, **49**, 530-6.

**Acknowledgement**
This study was funded by AMDI Student Research Fund, USM.

# Estimating glomerular filtration rate in oncology patients receiving Cisplatin chemotherapy: Predicted creatinine clearance against $^{99m}$Tc-DTPA methods

**Sharifah Khaidah Syed Sahab[1,2], Mahayuddin Manap[1,2], Fadzilah Hamzah[2]**

[1] Advanced Medical and Dental Institute, Universiti Sains Malaysia, 13200 Kepala Batas, Pulau Pinang, Malaysia
[2]Nuclear Medicine Department, Penang Hospital, 10450 Georgetown, Pulau Pinang, Malaysia

E-mail: skss12_snm004@student.usm.my

**Abstract**. The therapeutic potential of cisplatin as the best anticancer treatment for solid tumor is limited by its potential nephrotoxicity. This study analyses the incidence of cisplatin induced nephrotoxicity in oncology patients through GFR estimation using $^{99m}$Tc-DTPA plasma sampling (reference method) and to compare with predicted creatinine clearance and Tc-99m renal scintigraphy. A prospective study of 33 oncology patients referred for GFR estimation in Penang Hospital. The incidence of cisplatin induced nephrotoxicity was analysed via radionuclide and creatinine based method. Of 33 samples, only 21 selected for the study. The dose of cisplatin given was 75 mg/m$^2$ for each cycle. The mean difference of GFR pre and post chemotherapy (PSC 2) was 13.38 (-4.60, 31.36) ml/min/1.73m2 ($p$ 0.136). Of 21 patients, 3 developed severe nephrotoxicity (GFR $<$ 50ml/min/1.73 m$^2$) contributing 14.3% of incidence. Bland-Altman plot showed only PSC 1 is in agreement with PSC 2 technique. Intraclass Correlation Coefficients (ICC) also showed that PSC 1 has high degree of reliability in comparison to PSC 2 ($p < 0.001$). The other methods do not show reliability and agreement in comparison to PSC 2 ($p < 0.05$). 3 of 21 patients (14.3%) developed severe nephrotoxicity post cisplatin chemotherapy. This percentage is much less than the reported 20 – 25% of cases from other studies, probably due to small sample size and biased study population due to strict exclusion criteria. Radionuclide method for evaluating GFR is the most sensitive method for the detection of cisplatin induced nephrotoxicity by showing 3 of 21 patients developing severe nephrotoxicity. PSC 1 was found to be a reliable substitute of PSC 2. The other methods are not reliable for detection of early nephrotoxicity. We will recommend the use of single plasma sampling method (PSC 1) for GFR estimation in monitoring post cisplatin chemotherapy patients.

## 1. Introduction

Cisplatin remains as one of the best anticancer agent for the treatment of solid tumour over the last 30 years [1]. Despite its well-known desirable effect on cancer treatment per se, its full therapeutic potential has been limited by its potential toxicity. Many articles reported that the incidence of nephrotoxicity following high dose cisplatin chemotherapy happens in the range of 20 to 25% [2, 3]. Therefore, current clinical practice requires close monitoring of the kidney function pre and post chemotherapy to anticipate any significant decline of renal function. Therefore, estimating the glomerular filtration rate (GFR) has been accepted at large as a parameter to represent the functional status of the kidney [4]. There are various techniques which has been proposed to estimate GFR using endogenous or exogenous markers, but the most important aspect in choosing which method to use will have to take into account the simplicity, cost and availability without compromising the accuracy of the result.

## 2. Materials and Methods

This is a prospective study involving Nuclear Medicine Department and Radiotherapy and Oncology Department in Hospital Pulau Pinang. The sample is derived using PS Power and Sample Size Ver. 3.0.10 using paired t-test. 21 subjects with various solid tumors referred for GFR evaluation (16 males and 5 females, mean age 55.1) were enrolled in this study. The inclusion criteria include cases for high dose cisplatin. Those with comorbidities or GFR $\leq$ 60 ml/min/1.73m$^2$ were excluded from the study. GFR was calculated for each subject by plasma two samples clearance method (PSC 2), plasma one sample clearance method (PSC 1), Gates method, predicted creatinine clearance by modified diet in renal disease (MDRD) method and Cockroft-Gault (CG) method at prechemotherapy (3 days prior to start of cisplatin) and within 3 days to 1 week at post completion of chemotherapy cycles.

### *2.1. Gates and PSC Methods*

Tc-99m DTPA is prepared using a commercially available freeze-dried kit (Eczacibasi Monrol Nükleer Ürünler San, Turkey). Two sets of radiopharmaceuticals are prepared in the range of 185MBq to 250MBq. One of the radiopharmaceutical is set aside as the standard, while the other one is used for patient dose. The injection dose and the standard dose are calibrated in a dose calibrator. The percent difference between the dose and the standard should not exceed 5% under any circumstances. The whole Tc-99m DTPA injection dose is injected into the patient at the start of the renal scintigraphy study and the time is recorded. After injection, the syringe is assayed and any residue will be recorded. If the residue in the syringe post injection is >3% of the dose, an appropriate correction in the calculation should be made. The patient then undergoes renal scintigraphy using the gamma camera. The GFR (GFR Gates) is automatically estimated by a commercially available computer software (E.Soft 2.0 GHz Siemens Medical Solutions USA Inc.) according to the Gates' algorithm.

In two samples method, 10 ml of blood sample is taken at 60 and 180 minutes after the radiopharmaceutical injection, and in one sample method, only single sample drawn at 180 minutes from the opposite arm into an EDTA collection tube and the time is recorded. The blood should be mixed well with anticoagulant. Blood samples are centrifuged at 1000 grams for 10 minutes to separate the red blood cells from the plasma. 100 microliter aliquot of plasma will be pipetted into a plain tube labeled as plasma sample. The counts should be taken from the 3 tubes using gamma well counter, which consists of plasma sample tube, standard solution (1:10000 dilution) and blank test tube containing 100 microliter of distilled water for the purpose of background correction. Counts were obtained for 1 minute from each tube in an automatic gamma well counter (Biodex Atomlab 950 Uptake Stand). The activity of each tube will be recorded automatically and GFR value will be calculated automatically as well by the computer system.

### *2.2 Predicted creatinine clearance*

The GFR ($GFR_{MDRD}$) was predicted from the serum creatinine (SCr) level using MDRD study using equation (1) from [5]. The $GFR_{MDRD}$ is multiplied with a factor of 0.742 if female and 1.212 if African American.

$$GFR_{MDRD} = 32788\times (S_{cr}(umol/L)^{-1.154}\times(Age)^{-0.203} \quad (1)$$

The GFR ($GFR_{CG}$) was predicted from the serum creatinine (SCr) level using Cockroft-Gault's equation (2) where Constant is 1.23 for men and 1.04 for women [6].

$$GFR_{CG} = ((140 - \text{Age}) \times \text{Mass (kg)} \times \text{Constant}) / \text{Serum Creatinine (umol/L)} \quad (2)$$

The GFR (GFRCKD-EPI) was predicted from the serum creatinine (SCr) level using CKD-EPI equation (3) from [7]. GFRCKD- EPI is multiplied with a factor of 1.018 if female and 1.159 if African American.

$$\text{GFRCKD-EPI} = 141 \times \min(S_{cr}/\kappa,\ 1)^{\alpha} \times \max(S_{cr}/\kappa,\ 1)^{-1.209} \times 0.993^{\text{Age}} \quad (3)$$

## 3. Data Analysis and Statistics

The sample is derived using PS Power and Sample Size Ver. 3.0.10 using paired t-test. The required size to achieve a power of 80% and confidence interval of 95% is 33 patients. The demographic, clinical, GFR from serum creatinine and the 99mTc-DTPA methods are recorded in the data collection sheets and analysed using Statistical Package for Social Science (SPSS) software version 21.0.0. GFR values obtained from pre and post chemotherapy were normalised according to body surface area. Mean GFR calculated for each method (CG, MDRD, CKD-EPI, Gates, PSC 1 and PSC 2). Paired t test was performed to obtain mean difference of GFR pre and post chemo. Test was significant at p value of $< 0.05$. Scatter plot was used to correlate between each different method. Intraclass Correlation Coefficients was used to assess reliability of all methods in comparison to PSC 2.

## 4. Results

The patients are given high dose cisplatin (75 mg/m$^2$) for every cycle of chemotherapy. Out of 21 patients, 3 developed severe nephrotoxicity (GFR $<$ 50ml/min/1.73m$^2$). PSC 2 (reference method) estimated the mean fall in GFR of 13.38 (-4.60, 31.36) ml/min/1.73m$^2$ (p 0.136). The other methods showed mean fall in GFR as 7.03 (2.43, 11.63), 6.70 (0.02, 13.38), 0.86 (-5.65, 7.36), 7.55 (0.75, 14.34) and 20.30 (6.45, 34.15) ml/min/1.73m$^2$ as estimated by CG, MDRD, CKD-EPI, Gates and PSC 1 respectively (Figure 1). We observed that the GFR declined post chemotherapy by all methods.

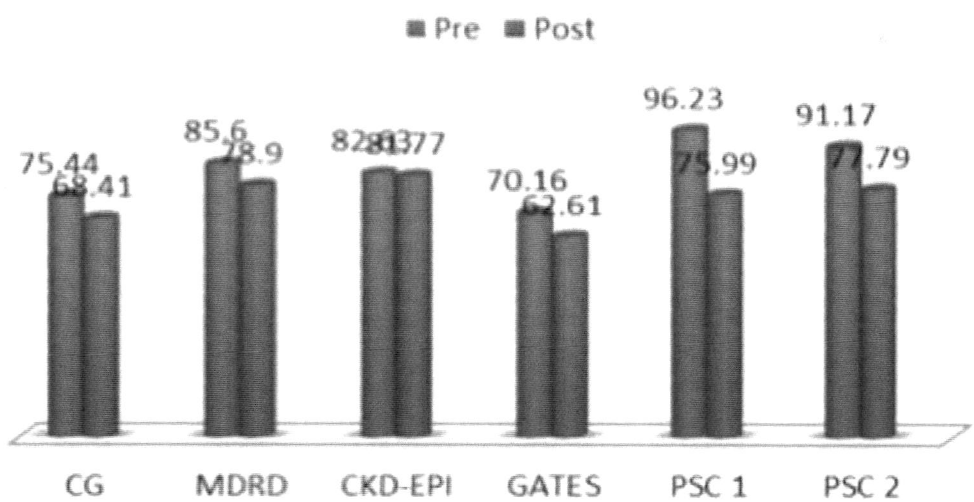


**Figure 1.** Difference in GFR value according to different methods pre and post cisplatin chemotherapy.

**Table 1.** Detection of nephrotoxicity post chemotherapy by all methods.

| Methods | No. of Cases Detected | Percentage (%) of patients developing nephrotoxicity (n = 21) |
|---|---|---|
| CG | 1 | 4.8 |
| MDRD | 0 | 0.0 |
| CKD-EPI | 0 | 0.0 |
| Gates | 4 | 19.0 |
| PSC 1 | 3 | 14.3 |
| PSC 2 | 3 | 14.3 |

The table above shows the number of cases developing nephrotoxicity post completion of chemotherapy. In this study, Gates method showed the highest percentage of case detection (19%) as compared to other methods. This is followed by PSC 1 and PSC 2 whereby both methods' detection rate at 14.3% followed by the CG method (4.8%). The MDRD and CKD-EPI methods are unable to detect any evidence of nephrotoxicity in this study. Thus, from this observation we may conclude that MDRD and CKD-EPI are not sensitive in the measurement of changes in GFR and the detection of nephrotoxicity post chemotherapy.

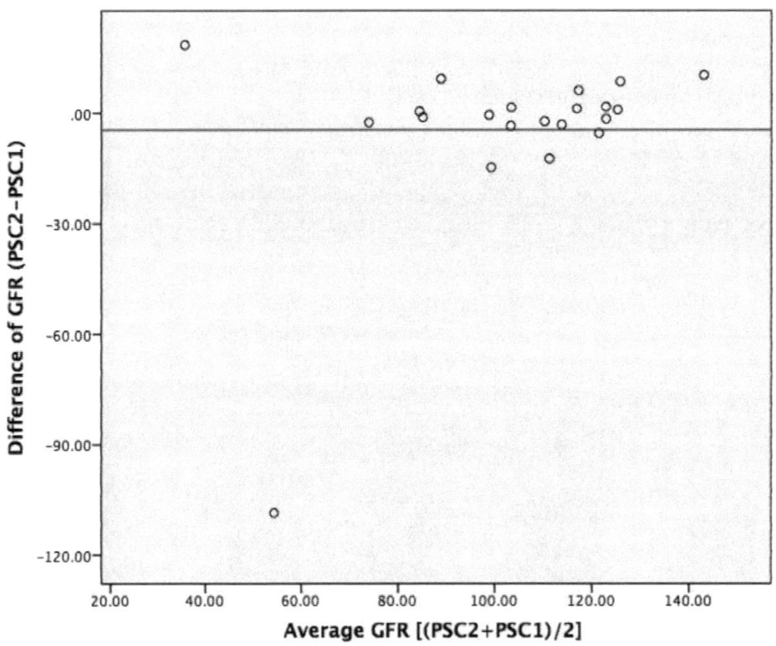


**Figure 2.** The Bland-Altman plot of 99mTc-DTPA for GFR (n=21) comparing between PSC 2 and PSC 1 methods.

The Bland-Altman analysis indicates that the 95% limits of agreement between PSC2 and PSC 1 methods is narrow ranging from -15.9 to 6.6. The scatter plots is also more consistent and close to the line across the graph. The rest of the methods showed inconsistent plot and away from the line across the graph. Hence, we may conclude that PSC 1 and PSC 2 are in agreement with each other (Figure 2).

**Table 2.** Intraclass Correlation Coefficients (ICC)

| | Intra-class Correlation Coefficients | **(95% CI)** | **P value** |
|---|---|---|---|
| PSC2 – Cockroft-Gault | 0.085 | (-0.350, 0.490) | 0.353 |
| PSC2 - MDRD | -0.063 | ( -0.473, 0.369) | 0.610 |
| PSC2 – CKD Epi | 0.101 | ( -0.336, 0.502) | 0.328 |
| PSC2 - Gates | 0.171 | ( -0.271, 0.554) | 0.223 |
| PSC2 – PSC1 | 0.696 | ( 0.388, 0.864) | <0.001 |

From the ICC, a high degree of reliability in measuring GFR was found between PSC 1 in comparison to PSC 2 method which shows a significant $p$ value of less than 0.001 and a 95% confidence interval from 0.338 to 0.864 ($p < 0.05$, CI 0.338, 0.864). The rest of the methods showed insignificant result and do not show reliability in measuring GFR in comparison to PSC 2 method.

## 5. Discussion

Our mean age of respondents is 55.1 years (10.80). This represents the majority of our patients are within the middle-aged group. Increasing age physiologically will lead to reducing GFR level [8]. Since cancer risk is also increasing with increasing age, therefore renal resistance towards cisplatin will also be reduced as age increases [9].

Upon completion of 3 cycles of chemotherapy, the average decline in GFR was 13.38 (-4.60, 31.36) ml/min/1.73m$^2$. This shows 14% reduction from baseline GFR and is much less than the reported average declines in GFR post completion of chemotherapy by Fatima N. [10], which showed 43.86 ± 16.10 ml/min/1.73m$^2$ reduction. This factor can be explained by the difference in the completion of chemotherapy cycle itself as we only conducted up to 3 cycles of high dose cisplatin as compared to 6 cycles of cisplatin in the former study with higher average cumulative dose. The concept of cisplatin induced nephrotoxicity is based on repetitive exposure and the effect is cumulative and dose dependent [3]. Therefore we may conclude that the more cycle of cisplatin and the higher is the cumulative dose, the more likelihood for cisplatin induced nephrotoxicity to occur.

The creatinine based methods as shown by many studies do not correlate with the radionuclide methods. CG underestimated GFR as much as 14% whereas MDRD underestimated GFR by 10% in comparison to PSC 2 [10]. The reasons are probably attributable to inaccuracies of formulae secondary to intra-assay and intra-individual variability from serum creatinine and GFR values, lack of calibration of serum creatinine result in laboratories as well as measurement of other variables in predicted equations affecting the precision and accuracy of renal clearance prediction equations [11]. From our result, Gates method detected the highest number of cases for cisplatin induced nephrotoxicity by giving the highest percentage of 19.0%. The next lower percentage of patients developing nephrotoxicity is followed by PSC 1 and PSC 2 method by giving 14.3% each. CG method showed only 1 case by giving percentage of 4.8%. MDRD and CKD-EPI showed the worst detection rate and was the least sensitive by not having any case detection at all. Gates appears to have the highest case detection than the reference method itself. Since Gates was found to give the highest underestimation of GFR (23% in this study), this maybe a false positive result leading to falsely high percentage of case detection. The other creatinine based methods do not show good detection of cases and is thus least sensitive of all methods conducted here. This maybe attributable to the fact that serum creatinine tends to underestimate GFR and creatinine will not be raised until the kidneys lost 60% of its total function [12].

In conclusion, utilisation of radionuclide methods has shown better detection in GFR changes compared to creatinine based techniques. Finally, we will recommend the use of one plasma sampling method (PSC 1) for GFR estimation in monitoring post cisplatin chemotherapy patients. In comparison to other study, the findings of this study show similar trend with all methods; radionuclide methods are superior in detecting early GFR changes than creatinine based techniques [10].

## 6. References

[1] Stathopoulos G. Cisplatin: process and future. J BUON. 2013;18(3):564-9.
[2] Gonzalez R, Borrego A, Zamora Z, Romay C, Hernandez F, Menendez S, et al. Reversion by ozone treatment of acute nephrotoxicity induced by cisplatin in rats. Mediators of inflammation. 2004;13(5-6):307-12.
[3] Yao X, Panichpisal K, Kurtzman N, Nugent K. Cisplatin nephrotoxicity: a review. The American journal of the medical sciences. 2007;334(2):115-24.
[4] Itoh K. Determination of glomerular filtration rate by means of newly developed plastic scintillation counter both with and without dilution procedures. Journal of nuclear medicine : official publication, Society of Nuclear Medicine. 2001;42(10):1484-8.
[5] Levey, A., et al. (2000). "A simplified equation to predict glomerular filtration rate from serum creatinine [Abstract]: J Am Soc." Nephrol 11.
[6] Cockcroft, D. W. and M. H. Gault (1976). "Prediction of Creatinine Clearance from Serum Creatinine." Nephron 16(1): 31-41.
[7] Levey, A. S., et al. (2009). "A new equation to estimate glomerular filtration rate." Annals of internal medicine 150(9): 604-612)
[8] Lindeman RD, Tobin J, Shock NW. Longitudinal studies on the rate of decline in renal function with age. Journal of the American Geriatrics Society. 1985;33(4):278-85.
[9] Mathe C, Bohacs A, Duffek L, Lukacsovits J, Komlosi ZI, Szondy K, et al. Cisplatin nephrotoxicity aggravated by cardiovascular disease and diabetes in lung cancer patients. The European respiratory journal. 2011;37(4):888-94.
[10] Nosheen F, Maseehuz Z, Shahid K, Abid H. Detection of early cisplatin induced nephrotoxicity by serial estimation of glomerular filtration rate: comparison of various methods. Nephro-Urology Monthly. 2010;2(03):422-30.
[11] Mittal B, Kher V, Gulati S, Maini A, Arora P, Kumar U, et al. Measurement of GFR using 99m Tc DTPA-A comparative analysis of Gates and plasma clearance technique. Indian J Nephrol. 1996;6:4-8.
[12] Kuster N, Cristol JP, Cavalier E, Bargnoux AS, Halimi JM, Froissart M, et al. Enzymatic creatinine assays allow estimation of glomerular filtration rate in stages 1 and 2 chronic kidney disease using CKD-EPI equation. Clinica chimica acta; international journal of clinical chemistry. 2014;428:89-95.

# Pre and post PET-CT impact on oesophageal cancer management: a retrospective analysis.

**NA Azmi[1, 2], HRA Razak[1], S Vinjamuri[3]**

[1] Department of Medical Imaging, Faculty of Health Sciences, Universiti Teknologi MARA (UiTM) Selangor, Puncak Alam Campus, 42300 Puncak Alam, Selangor, Malaysia.
[2] Diagnostic Imaging & Radiotherapy Programme, Faculty of Health Sciences, Universiti Kebangsaan Malaysia, Jalan Raja Muda Abdul Aziz, 50300 Kuala Lumpur, Malaysia.
[3] Department of Nuclear Medicine, Royal Liverpool University Hospital, Liverpool, L7 8XP, United Kingdom.

E-mail: hairil@puncakalam.uitm.edu.my

**Abstract**. Assessment of the retrospective cancer incidence, prevalence and crude survival rates of oesophageal cancer to allow comparison between pre and post PET-CT introduction are part of 4 phase cost effectiveness research. It will provide baseline data for to assess PET or PET-CT cost effective potential for staging. A total of 849 patient's data received from NWCIS databases with various stages of oesophageal cancer between 2001 and 2008. The fundamental activities are retrospective analysis of patient data. In most cases where appropriate, results are presented with 95 percent confidence intervals (CI). Variances between patient groups and variables are assessed using chi-square test. In cases where it deems vital, multiple logistic regression are used to modify for potential confounder such as age and sex. All p- values are two-sided and any value lower than 0.05 were considered to suggest a statistically significant result. Retrospective analysis were categorised into two categories, patients from 2001-2003 considered as pre PET and post PET for 2004-2008. This categorisation allows better comparison of patients' survival trend to be made between both groups. Rates are presented in percentages and being grouped by tumour characteristics and other variables associated with demographic profile, diagnosis, staging and treatment. Results allowed comparison of oesophageal cancer trends between the pre and post PET-CT introduction such as changes in incidence rate or changes in survival. These data were used to normalise the decision tree model so that cost- effectiveness analysis can be performed across the whole population.

## 1. Introduction

Oesophageal cancer, although a malignancy with a relatively low incidence, has a very high mortality, related among other factors, to delayed diagnosis and inaccurate pre-therapy assessment on the extent of the disease (1, 2). Evolution of times brings together a continuous rapid advancement in diagnosis and treatment of oesophageal cancer that not only can improve chances of survival but also make early diagnosis for its late symptom diseases (3, 4). Despite of this advancement, oesophageal cancer continues to be a major health problem worldwide.

As the treatment for oesophageal cancer remains in evolution, the fundamental principle of cancer management is patient's evaluation for surgical treatment. The domain prognostic factor for survival is whether the patient can be completely resected. As oesophageal cancer comes with significant morbidity, mortality and cost associated with the surgical treatment, it is important to identify and to exclude patient who will not benefit from the surgical treatment. Improved staging techniques may decrease the number of surgically treated patients and result in improving the survival figures for the more selected group. It is a common practice to evaluate patients with potentially operable disease and ensure that all detectable metastases are included.

Oesophageal cancer has high fluorodeoxyglucose (FDG) avidity (5-7). 18-FDG PET is increasingly used for the staging of oesophageal cancer, but provides limited anatomic information. Hybrid PET-CT imaging improve characterisation of FDG activity in the vicinity of highly tracer-avid primary tumours before surgery, in close proximity to organs with high physiological uptake and in the presence of postsurgical distorted anatomy. Integrated PET-CT enables optimal anatomic delineation of PET findings and identification of FDG-negative lesions on computed tomography (CT) images and results in patient management improvement.

In recent study of patients with oesophageal cancer (initial staging, evaluation of neo-adjuvant chemotherapy, or postsurgical follow-up), PET-CT appeared superior to PET or CT alone and had an effect on further management in 22% of patients (8). Management changes were related to better localising PET abnormalities, retrospectively detecting true abnormalities on concurrent contrast CT, guiding endoscopy to the site of suspicious lesions, and eliminating the need for further work-up that would have been necessary with CT or PET findings alone.

This study purpose is to assess the retrospective cancer incidence, prevalence and treatment pathway of oesophageal cancer between pre and post PET-CT. This allowed comparison to be made on oesophageal cancer trends between the pre and post PET-CT introduction such as changes in incidence rate or changes in survival. The prognosis of any patient with disseminated disease is very poor with expected 5 year survival rate less than 5%. For patients with no disseminated disease the survival is also poor but may extend to 10% - 30% (9-12). This is part of a large study on cost effectiveness analysis of PET-CT for oesophageal cancer.

## 2. Materials & Methods

Ethics clearance has been obtained for a retrospective data collection from Merseyside and Cheshire Cancer Registry databases. It provides information on the use of PET and PET-CT in 849 with various stages (0 to 4) of oesophageal cancer from 2001-2008.

All patients were categorised into two categories, in which patients diagnosed from 2001-2003 considered as pre PET and post PET for 2004-2008. This is due to the fact PET only introduced in RLBUHT in 2004 and PET/CT in 2007 (mobile).

Data analysis was done with Microsoft Excel spreadsheet, Version 2007 (Microsoft Corporation, Redmond, Washington, United States) and SPSS PASW Statistics for Windows Version 23 (IBM Company, New York, United States).

## 3. Results

Demographic profiles on 849 patients evaluated in this phase are as described in Table 1. Rates are presented in percentages and being grouped by tumour characteristics, treatment and other variables associated with demographic profile, diagnosis, staging and treatment information.

In most cases where appropriate, results are presented with 95 percent confidence intervals (CI). Variances between patient groups and variables are assessed using chi-square test. In cases where it deems vital, multiple logistic regression are used to modify for potential confounder such as age and sex. All p- values are two-sided and any value lower than 0.05 were considered to suggest a statistically significant result.

**Table 1.** Trend Rate Survey Analysis on demographic and clinical profile

| N = 849 | Pre PET (2001-2003) | Post PET (2004-2008) | Pre Percentage (%) | Post Percentage (%) |
|---|---|---|---|---|
| ***Gender*** | | | | |
| **Male** | **271** | **142** | **63.5** | **60.7** |
| **Female** | **156** | **92** | **36.5** | **39.3** |
| ***Age median ( range)*** | | 65 (25-91 years) | | |
| ***Grade*** | | | | |
| **Well differentiated** | 25 | 31 | 5.9 | 13.2 |
| **Moderately differentiated** | 151 | 83 | 35.4 | 35.5 |
| **Poorly differentiated** | 121 | 68 | 28.3 | 29.1 |
| **Undifferentiated** | 7 | 4 | 1.6 | 1.7 |
| **Not appropriate/not assessable** | - | 1 | 0 | 0.4 |
| **Not known** | 103 | 45 | 24.1 | 19.2 |
| ***Treatment Plan*** | | | | |
| **Surgery** | 245 | 150 | 57.4 | 64.1 |
| **Palliative treatment** | - | 1 | 0 | 0.4 |
| **Chemotherapy** | 87 | 24 | 20.4 | 10.3 |
| **Radiotherapy** | 92 | 58 | 21.5 | 24.8 |
| **Unknown Drugs** | 3 | 7 | 0.7 | 0.4 |

In North West; between 2001-2003 and before the introduction of PET-CT, the proportion of patients diagnosed with oesophageal cancer was 30% higher in comparison to post PET-CT (35%). Mean patient age was 65 years, and most patients were male (63%). Inclusion criteria specified that patients have histologically confirmed cancer.

Over the period of 2001-2008, the number of patients for all grades of cancer has decreased from approximately 407 to 232 during post PET-CT period. This study also reported that post PET-CT has shown decreased in treatment plans from 64% to 36%.

### *3.1. Pre and post PET-CT comparison on oesophageal cancer grades.*

This study provides evidence that post PET-CT introduction leads to changes in the oesophageal cancer grades. Major changes can be seen in moderately differentiated grade (from 151-83) as shown in Figure 1.

This study provides evidence that post PET-CT introduction leads to changes in the treatment plans from 64% to 36% of all patients. The most frequent major change was surgery from 245 to 150 cases. Changes can also be seen for chemotherapy (from 87 to 24) and radiotherapy (92 to 58) as shown in Figure 2.

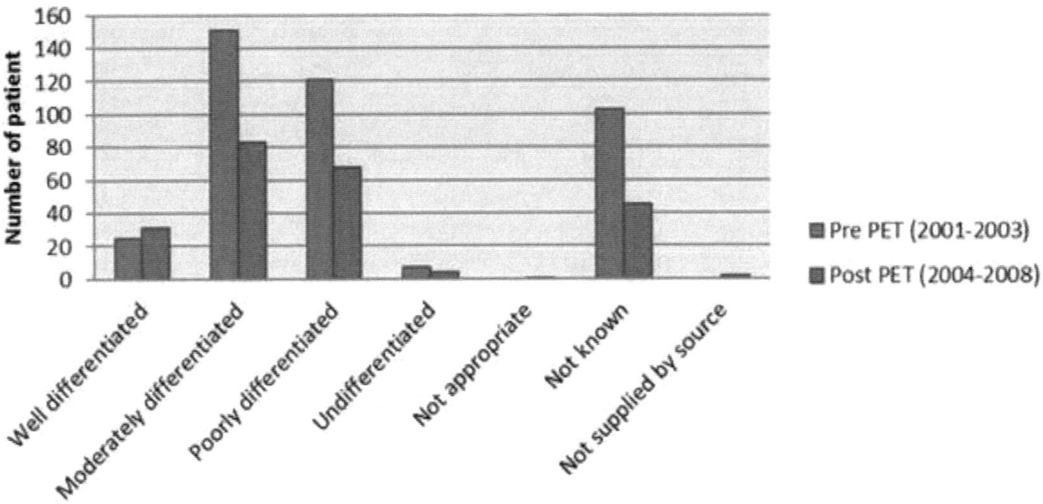


**Figure 1.** Pre and post PET-CT comparison on oesophageal cancer grades

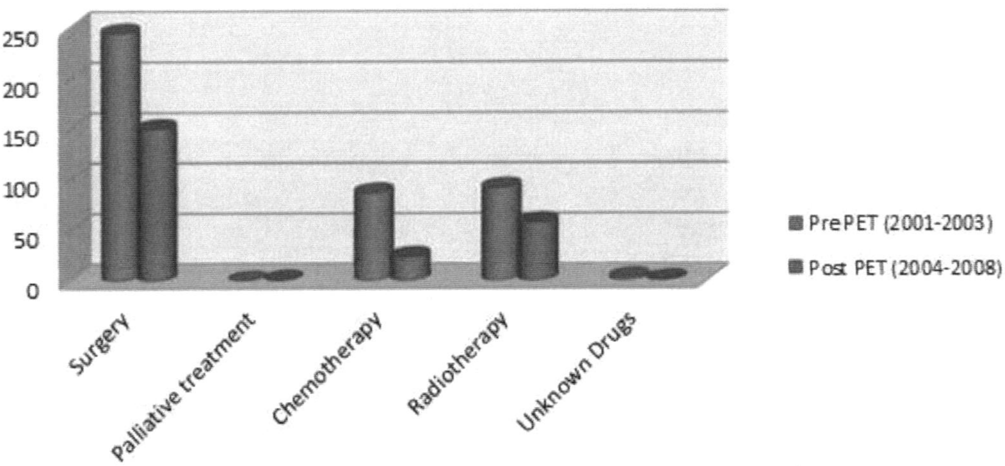


**Figure 2.** Pre and post PET-CT comparison on treatment plan

## 4. Discussion and Conclusion

This study allowed comparison of oesophageal cancer trends between the pre and post PET-CT introduction such as changes in incidence rate or changes in grades and treatment plan. This accumulating data including the current study that was part of cost- effectiveness analysis study, were then used to normalise the decision tree model so that cost- effectiveness analysis can be performed across the whole population.

Retrospective comparison between pre and post PET-CT introduction shows that the cancer incidence and prevalence of oesophageal cancer in North West have decreased by almost half (46%) post PET-CT introduction.

The main role of PET-CT for pre-treatment staging of primary oesophageal cancer patients is to identify other disease that would preclude definitive curable treatment options. Introduction of PET-CT improves pre-treatment staging accuracy and grading, thus improving patient outcomes and results in more appropriate treatment management.

There are some limitations with this current study. The comparative study of definitive versus palliative therapies were not identified. Therefore, a case linked for increasing patient outcomes post PET-CT introduction for these indications cannot be made.

This study suggests a potential evidence that PET-CT introduction in primary oesophageal cancer patients leads to change in grading and treatment plans in substantial portion of patients. PET-CT itself is a costly technology. However, the potential role in diagnosing oesophageal cancer and directing therapeutic strategies to improve patient outcome may outweigh the cost of scanning and lower the financial individual health-care cost. Thus, the results from this study will provide sequential data for larger ongoing cost-effectiveness analysis study to highlight the role of PET-CT in oesophageal cancer management. The complete research will be beneficial to Malaysia cancer research in term of improving the selection of PET/CT to be made widely available in cancer management based on the clinical values and cost effectiveness.

## 5. References

[1] Dehdashti F, Siegel BA. Neoplasms of the esophagus and stomach. Seminars in Nuclear Medicine. 2004;34(3):198-208.
[2] Dehdashti F, Siegel BA. Neoplasms of the esophagus and stomach. Semin Nucl Med. 2004;34(3):198-208.
[3] Eslick GD. Epidemiology of esophageal cancer. Gastroenterol Clin North Am. 2009;38(1):17-25, vii.
[4] Eslick GD. Esophageal Cancer: A Historical Perspective. Gastroenterology Clinics of North America. 2009;38(1):1-15.
[5] Torrance AD, Almond LM, Fry J, Wadley MS, Lyburn ID. Has integrated 18F FDG PET/CT improved staging, reduced early recurrence or increased survival in oesophageal cancer? Surgeon. 2015;13(1):19-33.
[6] Noble F, Bailey D, Panel SUGT, Tung K, Byrne JP. Impact of integrated PET/CT in the staging of oesophageal cancer: a UK population-based cohort study. Clin Radiol. 2009;64(7):699-705.
[7] Wong WL, Chambers RJ. Role of PET/PET CT in the staging and restaging of thoracic oesophageal cancer and gastro-oesophageal cancer: a literature review. Abdom Imaging. 2008;33(2):183-90.
[8] Bar-Shalom R, Guralnik L, Tsalic M, Leiderman M, Frenkel A, Gaitini D, et al. The additional value of PET/CT over PET in FDG imaging of oesophageal cancer. European journal of nuclear medicine and molecular imaging. 2005;32(8):918-24.
[9] Greene FL, American Joint Committee on C. AJCC cancer staging handbook : TNM classification of malignant tumors. New York; London: Springer; 2002.
[10] Headrick JR, Nichols FC, 3rd, Miller DL, Allen MS, Trastek VF, Deschamps C, et al. High-grade esophageal dysplasia: long-term survival and quality of life after esophagectomy. Ann Thorac Surg. 2002;73(6):1697-702; discussion 702-3.
[11] Pera M, Trastek VF, Carpenter HA, Allen MS, Deschamps C, Pairolero PC. Barrett's esophagus with high-grade dysplasia: an indication for esophagectomy? Ann Thorac Surg. 1992;54(2):199-204.
[12] Reed CE. Surgical Management of Esophageal Carcinoma. 1999. p. 95-105.

**Acknowledgement**
Ministry of Higher Education, Malaysia, National University of Malaysia, Malaysia, University of Technology MARA (600-IRMI/MYRA/5/3/LESTARI (0017/2016) Grant), Malaysia and Royal Liverpool & Broadgreen University Hospital Trust, United Kingdom.

# Definitive radiotherapy for locally advanced squamous cell carcinoma of the vulva and technical issues: a case report

**Gokula Kumar[1], Norhafizah I[1], Shazril I[1], Nursyatina AR[1], Abdul Aziz MZ[1], Hafiz M Zin[1], Zakir MK[1], Norjayadi[1], Norliza AS[1], Ismail A[2] and Khairun N[1]**

[1]Advanced Medical & Dental Institute (AMDI), Universiti Sains Malaysia (USM), Bertam, 13200 Kepala Batas, Penang, Malaysia
[2]O&G Department, Hospital Sultanah Bahiyah, Alor Setar, Kedah, Malaysia

E-mail: gokula@usm.my

**Abstract**. This case report describes a complex radical 3D-Conformal Radiotherapy treatment planning, dosimetric issues and outcome of definitive treatment of un-resectable carcinoma of the vulvar in a 42-year old lady. The patient presented with large fungating mass of the vulva which was biopsy confirmed as Keratinizing Squamous Cell Carcinoma. Further staging investigation revealed locally advanced disease (T4), with bilateral inguinal lymph nodes involvement. There is no systemic metastasis or intra-pelvic nodes. The patient was seen by Gynae-Oncology team and the disease was deemed un-resectable without significant morbidity. She was treated to a total dose of 64.8Gy in 36 fractions over 7 weeks with concurrent weekly Cisplatinum in 2 phases. 3D-Conformal radiotherapy technique using the modified segmental boost technique (MSBT, large PA and small AP photon fields with inguinal electron matching) was used. TLD chips were used for in-vivo dose verification in phase 1 and 2 of the treatment. At completion of planned radiotherapy, patient had a complete clinical response, grade 2-3 skin toxicity, grade 2 rectal toxicity, and grade 2 dysuria Vulval Squamous Cell Carcinomas are very radiosensitive tumours and the skills of the treating Radiation Oncologist, Dosimetrists, Physicist, Radiation Therapist and also nurses is of foremost importance is ensuring good clinical outcomes.

## 1. Introduction

Squamous Cell Carcinoma (SCC) of the vulva is an uncommon cancer mostly associated with HPV viral infection. Two common histologies are described – the HPV 16, 18 and 33 associated basaloid or condylomatous SCC and non-HPV associated keratinizing SCC's. In nearly 60% of vulval cancers, HPV infection is likely the causative factor.

This case report will discuss the successful treatment of a Moderately Differentiated Keratinizing SCC of the vulva with definitive chemo-radiotherapy using the modified segmental boost technique (MSBT), large PA and small AP photon fields with inguinal electron matching. This technique is published in the literature for lower pelvic malignancies and Moran M *et al* in his paper described the use of single prescription point in MSBT. In Advanced Medical & Dental Institute (AMDI), we further modified this technique using dual prescription points in the phase I followed by further tumour and inguinal boost in phase II.

A 42-year old lady with locally advanced SCC of the vulva was referred to AMDI Oncology Centre for radiotherapy opinion. She was earlier seen by Gynae-Oncological surgeon and her tumour was deemed not resectable without significant morbidity. The patient written consent was obtained for publication of her case in this paper.

A month earlier patient presented with 2 weeks' history of labial swelling and a punch biopsy of the lesion revealed Moderately Differentiated SCC (see Figure 1). CT of the thorax/abdomen/pelvis then showed a heterogeneously enhancing lesion measuring 6.4 cm × 3.4 cm × 4.9 cm arising from left labia with suspicion of bladder, rectum and vaginal invasion. There were multiple inguinal nodes, largest measuring 2.7 cm. Otherwise no distant metastasis elsewhere (see Figure 2).

**Figure 1.** Tumour at presentation

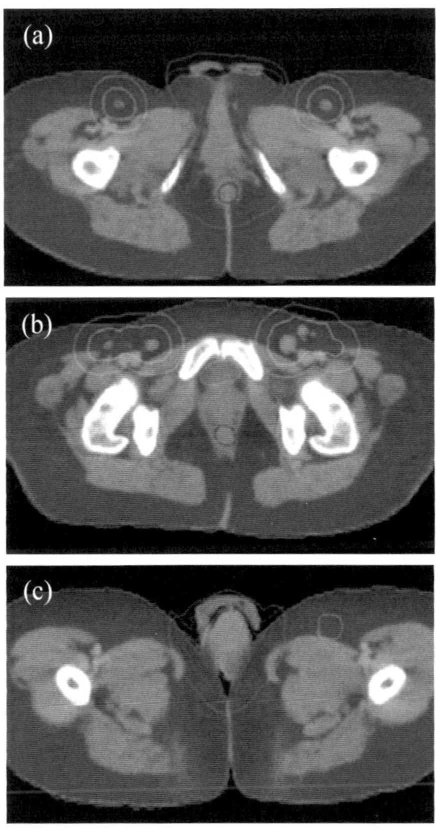


**Figure 2.** Primary marked red and lymph nodes purple in the CT images, (a) CT slice showing the primary tumour, (b) CT slice showing the lymph nodes and (c) CT slice showing the primary tumour on another slice

The disease extend and treatment expectation was discussed in length with the patient. Mutual decision to treat her condition with concurrent chemo-radiotherapy with definitive intent was made. The plan at that time was to treat in 3 phases: phase I (36 Gy in 20 fraction) treating the whole pelvis, phase II (9 Gy in 5 fraction) to the true pelvis and phase III (18 Gy in 10 fraction) boost to the GTV concurrent with IV Cisplatinum 40mg/m2 weekly (See Figure 3). Patient is to be re-assessed after the phase I for consideration of bilateral inguinal lymph node dissection.

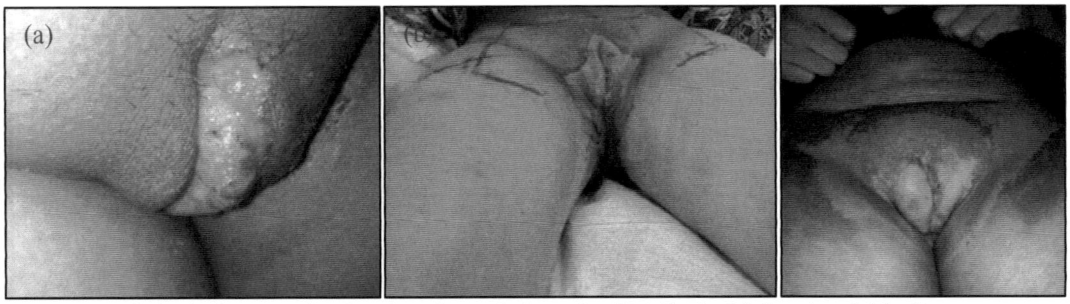


**Figure 3.** Clinical photograph after the planned phase I of 36 Gy in 20 fraction (a) before treatment, (b) 1 week and (c) 2 weeks after the full course of treatment.

CT simulation of the patient was repeated after the 18th fraction for the phase II planning since the tumour has shrunk significantly. The option of bilateral inguinal dissection was also discussed with the patient at that point. However, given the fact that inguinal nodes' PTV/CTV was extending into the phase II pelvic field and also the delays in the phase II of definitive radiotherapy that maybe caused by subjecting patient to inguinal dissection, it was decided to proceed with definitive chemo-radiotherapy albeit accepting the significantly higher risk of bilateral lympoedema in future. As mentioned earlier, since the inguinal nodes' PTV/CTV was extending into the true pelvic photon fields coupled with the fact that good bilateral femoral sparring was achieved with the MSBT technique it was decided to continue with the same phase I field arrangements to 46.8 Gy in 26 fractions. After completing phase I concurrent chemo-radiotherapy of 46.8 Gy the PTV-GTV is further boosted to 64.8Gy in 36 fractions.

## 2. Radiotherapy technique

### *2.1 CT-Simulation*
Patient was CT-simulated with full bladder, in supine position, both legs flexed and abducted (frog-leg position) to open up the inguinal folds and immobilised with vac bag. Primary tumour was exhautivey bolused with extra care to ensure there is no air gap especially the posterior part of the tumour abutting the anus. Thickness of the bolus is approximated to around 1 cm. 3 mm CT-images obtained with IV contrast from umbilicus to mid-thigh. The same process was repeated for the phase II planning (PTV-GTV boost)

### *2.2 Planning*
In phase I, patient was treated with large AP field covering the bilateral inguinal region anterior superior iliac spine (ASIS laterally), L5/S1 superiorly and 2 cm below the tumour inferiorly while shielding the lateral part of bilateral thighs (see Figure 4a). Posteriorly (PA), a smaller true pelvic field was used with both superior and inferior borders corresponding to the AP field however lateral borders were kept at 1.5 cm outside the pelvic brim (see figure 4b).

For the AP field, the ICRU prescription point was placed at the left inguinal region at 3 cm depth and 46.8 Gy was prescribed to that point. Dose was topped from the PA field to ensure uniform pelvic phase I dose coverage. This is achieved by iteratively changing the PA beam prescription point and also prescribing a dose that compensates the AP beam. The PA beam was also attenuated superiorly (treating full field before partially shielding the superior portion as to attenuate the dose from the superior portion of the PA beam). This optimisation process uses iterative trial of different beam energy, prescription points, shielding and etc. Finally, an optimal T-shaped plan was achieved, sparring the femoral heads as shown in Figure 5 and Figure 6.

The final plan was reviewed and we tried to conform to the ICRU 50 recommendations. The minimum PTV coverage of 95% for the volume of interest was aimed and at points where this is not achievable, the CTV coverage is ensured at least at 95% of prescribed dose. Though at a couple of points the maximum dose crossed the 107%, these spots were either in the bolus or inside the tumour itself.

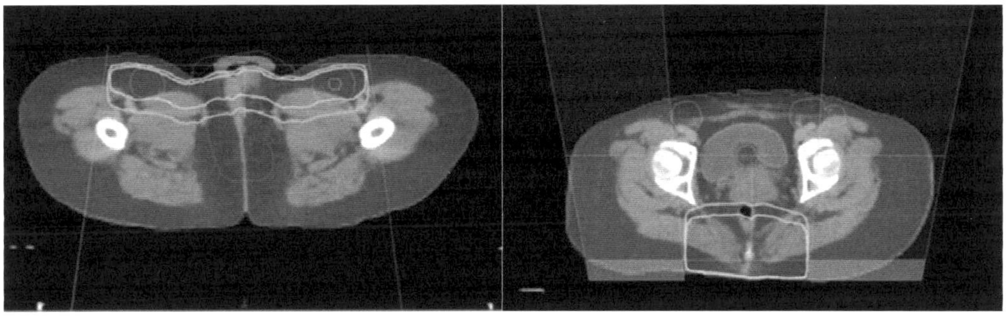

**Figure 4.** (a) large AP field with prescription point, (b) smaller PA field with prescription point

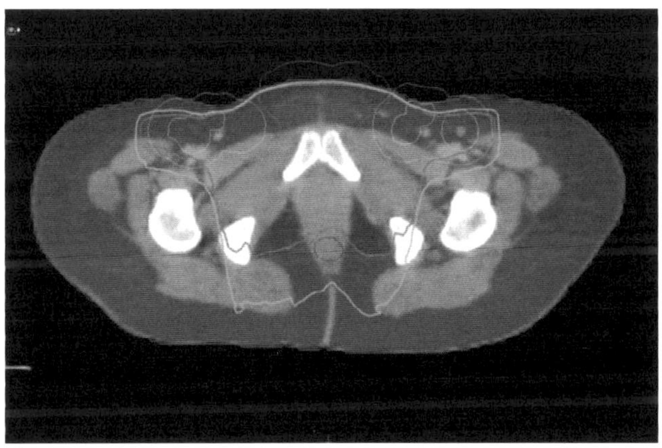

**Figure 5.** Phase I coverage – pelvic and inguinal coverage with sparring of the bilateral femorals (isodose lines: 95% in red, 90% in orange).

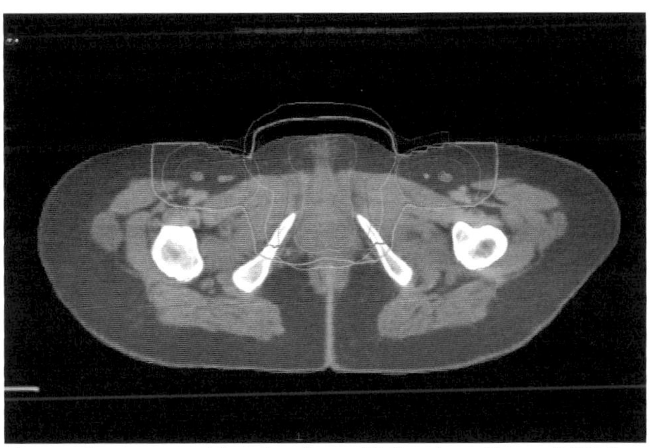

**Figure 6.** Electron-photon beam matching in phase II (isodose lines: 95% in red, 90% in orange).

In phase II, the primary tumour and the bilateral inguinal nodes were further boosted to 64.8 Gy in 38 fractions (a further 18 Gy in 10 fractions). The primary tumour was boosted with single direct anterior 10 MV photon beam matched bilaterally with electrons for the inguinal lymph nodes. The electron-photon beam was matched at the skin where the was overlap.

## 3. Treatment and verification

On the first day of the phase I treatment, extreme care was taken on bolusing the primary tumour to ensure there is no air gap and also correct bolus thickness. Cone beam CT was used to verify the positioning and also presence of any air gap.

Four TLDs were places at different corners of the primary tumour as to verify the tumour dose coverage. In phase II, TLDs were also placed on the skin surface where the electron beams covers the inguinal region. The TLD readings was compared to MONACO 5 planning system dose distribution as in Figure 7. Mostly there is an overdose rather than underdose to the tumour as recorded by the TLD readings. Cone beam CT was performed once a week to ensure patient positioning is the same as per simulation.

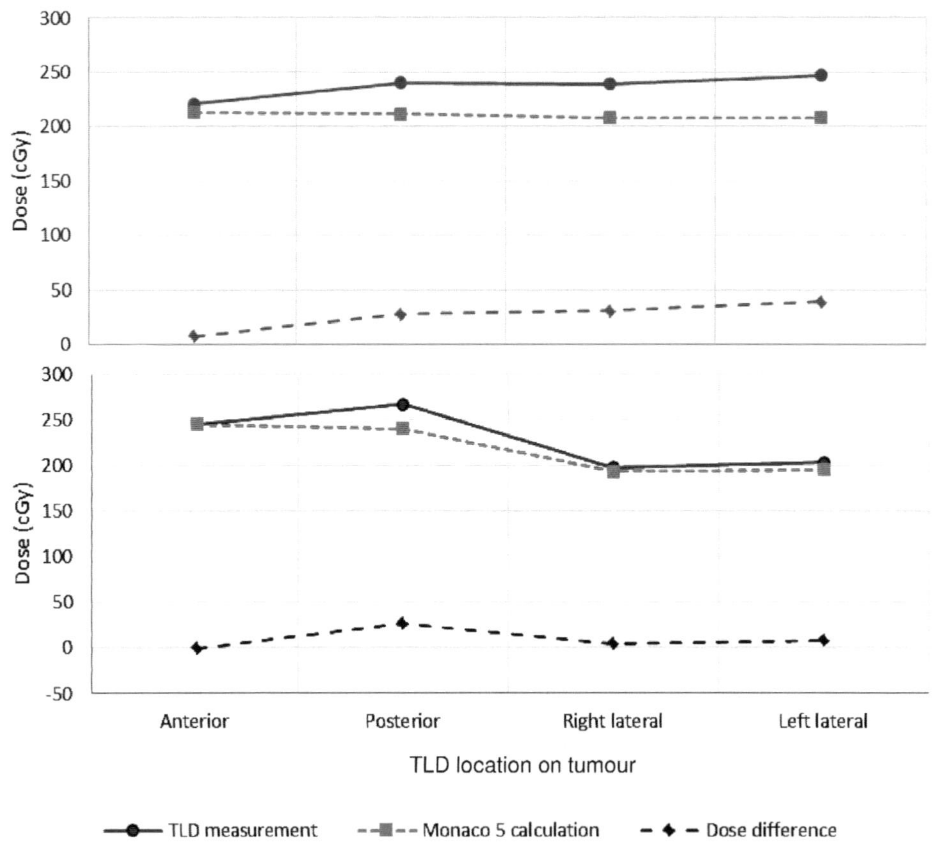


**Figure 7.** TLD readings at the start of phase I (a) and phase II (b)

## 4. Conclusions

Though HPV associated basaloid or condylomatous SCC is considered the more radiosensitive, keratinizing SCC of vulva also do respond well to radiotherapy as seen in this patient (complete clinical response, CR). The skills and training of the treating radiation oncologist, dosimetrists, physicist, radiation therapist and also nurses is of foremost importance is ensuring good clinical outcomes. This whole process had been very exhaustive, time consuming and resource intensive.

Though there were discrepancies in the actual TLD reading compared to MONACO planning system, our team had been happy that it is more of overdosing the tumour rather than underdose. This discrepancy can also be attributed to the fact that the precise point of posterior and lateral TLD placement could not be ascertained with high accuracy unlike the anterior TLD placement point which was easily visible and identifiable.

Though it is tempting to use IMRT in these type of difficult cases, there are many issues with IMRT such as confidence with not bolusing the tumour, or if bolused then ensuring the daily uniform thickness of the bolus, daily bladder variation, significant changes in the tumour dimensions during treatment and etc. As such, the use of 3D-CRT is the safest method in definitive radiotherapy of vulval carcinoma until a validated standard protocol on the use IMRT in vulval carcinomas is available.

## 5. References

[1] National Comprehensive Cancer Network (NCCN Guidelines 2016), Version 1.2017 – October 2016

[2] Improved treatment of pelvis and inguinal nodes using modified segmental boost technique: dosimetric evaluation. *Moran M, Lund MW, Ahmad M, Trumpore HS, Haffty B, Nath R. Int J Radiat Oncol Biol Phys. 2004 Aug 1;59(5):1523-30*

[3] IARC Publications : Pathology and Genetics of Tumours of the Breast and Female Genital Organs; Chapter 7: Tumors of the Vulva

## Acknowledgments

Authors would like to acknowledge the radiation therapists, nurses, medical officers & registrars and the physics team for their hard work in ensuring the best for this patient.